CONCEPTS
IN
THEORETICAL
ORGANIC
CHEMISTRY

CONCEPTS IN THEORETICAL ORGANIC CHEMISTRY

JERRY A. HIRSCH
Seton Hall University

Allyn and Bacon, Inc.
Boston

Copyright © 1974 by Allyn and Bacon, Inc.
470 Atlantic Avenue, Boston

All rights reserved. No part of the material
protected by this copyright notice may be
reproduced or utilized in any form or by
any means, electronic or mechanical,
including photocopying, recording, or
by any informational storage and retrieval
system, without written permission
from the copyright owner.

Library of Congress Catalog Card
Number: 73-88134

Printed in the United States of America

To My Wife

Contents

Preface   xi

PART I   STRUCTURE AND BONDING

Chapter 1   Quantum Chemistry   3
- 1-1   The Schrödinger Equation
- 1-2   The Hydrogen Molecule Ion—Molecular Orbital Theory
- 1-3   The Hydrogen Molecule Ion—Valence Bond Theory
- 1-4   Hybridization
- 1-5 π Systems—A Molecular Orbital Approach
- 1-6   Calculations and Experimental Observations

Chapter 2   Aromatic Character   44
- 2-1   Resonance Energies
- 2-2   Aromaticity and the "$4n+2$ Rule"
- 2-3   Heteroaromaticity, Homoaromaticity, and Bicycloaromaticity
- 2-4   Antiaromaticity

Chapter 3   The Woodward-Hoffmann Formalism   59
- 3-1   Electrocyclic Reactions
- 3-2   Sigmatropic Reactions
- 3-3   Cycloaddition Reactions
- 3-4   Secondary Effects
- 3-5   Summary

General References (Part I)   86

PART II   STRUCTURE AND REACTIVITY

Chapter 4   Substituent Effects   88
- 4-1   Polarity and Polarizability
- 4-2   Inductive and Field Effects

4-3 Resonance Effects
4-4 Alkyl-group Effects: Hyperconjugation
4-5 Hammett Equations for Equilibria

Chapter 5 The Rate of a Reaction   114
5-1 Kinetics
5-2 Partial Rate Factors
5-3 Transition-state Theory
5-4 Entropy of Activation and Solvent Effects
5-5 Salt Effects

Chapter 6 Substituent Effects on Rates   128
6-1 Hammett-type Equations for Rate Processes; Separation of Interaction Mechanisms
6-2 The Taft Equation; *ortho* and Steric Effects
6-3 The Swain-Lupton Approach
6-4 ρ Values and Mechanisms

Chapter 7 The Structure of the Transition State   147
7-1 Catalysis
7-2 Primary Kinetic Isotope Effects
7-3 Secondary Kinetic Isotope Effects; Steric Kinetic Isotope Effects
7-4 Solvent Isotope Effects
7-5 The Brønsted Catalysis Law
7-6 Acidity Functions
7-7 Acidity Functions and Transition-state Structure
7-8 Other Analyses of Transition-state Structure

Chapter 8 Nucleophilic Character   183
8-1 Nucleophilicity
8-2 The Edwards Equations
8-3 The Winstein Equations
8-4 Multiparameter Nucleophilicity Relationships
8-5 Solvation in Nucleophilic Processes
8-6 Ambident Anions; The Alpha Effect
8-7 Hard and Soft Acids and Bases
8-8 Charge-controlled and Frontier-controlled Reactions

Chapter 9 Medium Effects   207
9-1 General Treatment
9-2 Reactivity-based Solvent Scales
9-3 Solvatochromism Scales

    9-4 Multiple Correlations
    9-5 Solvent Effects on Electronic Transitions
    9-6 Solvent Effects on Nuclear Magnetic Resonance Transitions

General References (Part II)   **219**

PART III   STEREOCHEMISTRY AND CONFORMATIONAL ANALYSIS

Chapter 10   Stereochemistry   **222**
   10-1 Chirality
   10-2 Configuration—Relative and Absolute
   10-3 Diastereomers and *meso* Systems
   10-4 Prochirality
   10-5 Atropisomerism

Chapter 11   Conformational Analysis of Acyclic Systems   **234**
   11-1 Ethane Systems
   11-2 *n*-Butane Systems
   11-3 Unsaturated Systems
   11-4 Conformational Entropy
   11-5 Analysis of Configurational Equilibria
   11-6 F-strain and B-strain

Chapter 12   Conformational Analysis of Cyclic Systems   **249**
   12-1 Cyclohexane
   12-2 Substituted Cyclohexanes
   12-3 Conformation and Reactivity in Cyclohexanes
   12-4 Unsaturated and Heterocyclic Six-membered Rings
   12-5 Rings Other than Six-membered
   12-6 Methods of Conformational Analysis
   12-7 Chiroptical Methods

General References (Part III)   **276**

Author Index   **277**
Subject Index   **283**

Preface

This book has evolved as the result of a one-semester course given for several years at Seton Hall University. The course supplies an advanced senior-year elective for undergraduates and a required course in the area of organic chemistry for all graduate students regardless of area of specialization. For the graduate student interested in organic chemistry, the material presented in this book constitutes the first part of a three-course sequence, the latter two dealing with mechanism and synthesis.

It must be emphasized that this book is organized as a text. Coverage is broad and selective rather than detailed and comprehensive. The rigor and depth of treatment vary from section to section, since I consider it necessary to expose the reader to the ways in which selected concepts have developed, both in a historical sense and in the sense of an increase in understanding of the subtleties involved. Probing key concepts is a requirement if any book is to help fulfill the aims of its readers to further their chemical knowledge and personal scientific development.

It is my hope that the material contained here will bridge the gap between the average introductory undergraduate program (including one year each of organic chemistry and physical chemistry) and a more advanced study of chemistry through monographs and the original literature. The reader is encouraged to familiarize himself with the general references listed at the end of each of the three parts of this book and not to limit himself to the subject matter as presented herein. The footnotes are not intended to be comprehensive, and considerable emphasis is placed on monographs. To those researchers who may feel that their contributions to chemistry have been slighted, I offer assurance that this was not my intent. The decisions as to subject matter, depth of treatment, and references to be cited are necessarily subjective, were difficult to make, and are completely my responsibility.

The sections into which the material in this book is organized are as independent of one another as possible. Sections may be deleted or their order altered as desired. Many sections review material with which the reader will often be familiar (e.g., Chapters 1-1, 5-1, and 7-1, and much of Part III) and may be ignored as such in a classroom situation. However, little attempt has been made in Part III, Stereochemistry and Conformational Analysis, to separate introductory material from more advanced topics, since I feel an integrated approach is logical and more pedagogically useful in this area. I have encountered some sentiment that Part III should be omitted entirely in a book of this kind. My personal experience is that these topics are usually those which the student has grasped the least in his earlier training.

No undertaking of this kind is ever an individual effort. I owe a debt of deep gratitude to my parents, who constantly encouraged me to find a way of life which would be fulfilling, and to my teachers, especially Professors A. W. Ingersoll, R. H. Eastman, J. I. Brauman, and N. L. Allinger, who helped me develop to the point where I had the confidence and, hopefully, the ability to tackle this project. I am also indebted to my colleagues at Seton Hall University, Professors P. Ander, R. L. Augustine, Rev. A. V. Celiano, and D. P. Weeks, who read various sections and generously offered their opinions and encouragement. Special thanks go to Professors J. S. Swenton and J. B. Lambert, who read the entire manuscript and provided many valuable suggestions, and to the staff at Allyn and Bacon for their assistance.

However, my deepest thanks must be reserved for my wife, Sydney, for her assistance in correcting and enhancing my presentation, for her assistance with the mechanics of preparing the manuscript, and for the innumerable ways in which she helped and encouraged me. Lastly, I thank my daughter, Janine, who doesn't yet understand how she helped at all.

December 1972   *Jerry A. Hirsch*
South Orange, New Jersey

PART I

Structure and Bonding

Chemistry may be simply defined as the study of atoms and molecules and their various interactions. If we accept this definition, we accept, too, that ideas of atomic and molecular structure and of the various types of chemical bonding are essential to the structural theory of organic chemistry. In order to develop meaningful concepts of reactivity, it is imperative first to investigate various approaches to the interrelated concepts of structure and bonding.
General references for Part I will be found on page 86.

Chapter **1**

Quantum Chemistry

Any discussion of bonding after the mid-1930s must be done in the framework of quantum chemistry.[1] One of the fundamentals of this approach is the concept that electrons often may be described in terms of wavelike properties. Electrons may be diffracted just like ordinary light.

Consider a model of an electron in a one-dimensional box of length a and infinitely tall walls. If we assume that a wave starts and finishes at the ends of the box, we may draw diagrams to represent wave motion in the box (Fig. 1.1). Since for a wave of given amplitude the energy increases with increasing frequency, wave (a) must be of lower energy than wave (b) because it has a lower frequency. Each point where a wave crosses the axis other than at the ends of the box is called a *node*. Wave (a) has no nodes, wave (b) has one, wave (c) has two, and so forth for higher-frequency possibilities. The number of nodes may be used to represent relative energies. The energy of an electron relative to the energy of another electron therefore may be approximated in terms of the number of nodes encountered in the motion of the electron represented as a wave.

---

1. (a) For somewhat more detailed treatments of quantum chemistry, see W. G. Laidlow, *Introduction to Quantum Concepts in Spectroscopy*, McGraw-Hill, New York, 1970; I. N. Levine, *Quantum Chemistry*, vol. 1 and 2, Allyn and Bacon, Boston, 1970; M. W. Hanna, *Quantum Mechanics in Chemistry*, W. A. Benjamin, New York, 1969; J. C. Davis, Jr., *Advanced Physical Chemistry*, Ronald Press, New York, 1965; and W. Kauzmann, *Quantum Chemistry*, Academic Press, New York, 1957. (b) For treatments equivalent to that presented herein, see Reference A, Chapters 5–8; and References F, G, H, and I.

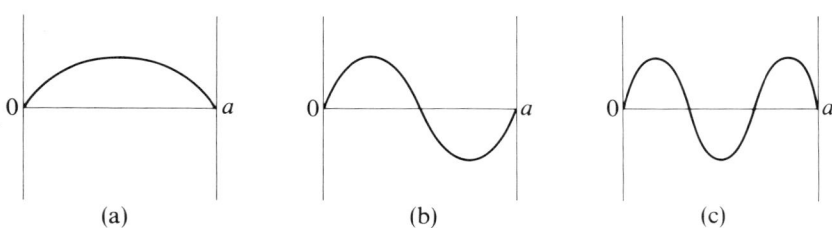

FIGURE 1.1 *An electron in a one-dimensional box.*

1-1 The Schrödinger Equation

The fundamental equation of wave mechanics is the wave (operator) equation, the most important example of which is the Schrödinger equation (Eq. 1.1):

$$\mathbf{H}\psi = E\psi \tag{1.1}$$

The wave nature of the particle under consideration, such as an electron, is described by ψ, a *wave function*. If ψ satisfies Eq. 1.1, it is called an *eigenfunction* of this equation. The wave function ψ has no physical reality; the term ψ^2 provides the connection with physical reality. A triple integral $\int_{-\infty}^{\infty} \int_{-\infty}^{\infty} \int_{-\infty}^{\infty} \psi^2 \, dx \, dy \, dz$ defines the probability that the electron (or particle in general) will be found in a three-dimensional volume defined by the integral limits. Using Cartesian coordinates and infinity limits for all the integrals (as above), the triple integral defines probability in all real space, and may be rewritten as:

$$\int_{-\infty}^{\infty} \psi^2 \, d\tau$$

The quantity $\mathbf{H}$ is the Hamiltonian operator (or total energy operator). An operator is a quantity that does something to a function on which it operates. For example, d/dx is an operator in

$$\frac{d}{dx}(x^2) = 2x \tag{1.2}$$

Quantum mechanical operators possess the special quality that they do not commute:

$$\mathbf{H}\psi \neq \psi\mathbf{H} \tag{1.3}$$

In classical mechanics, the Hamiltonian is composed of the kinetic energy T and the potential energy V; and the same holds true for the corresponding operators in quantum mechanics:

$$\mathbf{H} = \mathbf{T} + \mathbf{V} \tag{1.4}$$

This is an example of a general postulate that for every dynamical quantity in classical physics, there is an operator in quantum mechanics.

The symbol E in Eq. 1.1 is the energy of the system. If the Schrödinger equation is satisfied, E is said to be an *eigenvalue* of this equation. The value of E depends on the nature of the operator and wave function being used in the equation. The energy E is not an operator, obeys the usual algebraic rules, and, therefore, does commute:

$$E\psi = \psi E$$

For every energy state of the particle, there is a corresponding wave function. Whenever any dynamical variable is measured, the only possible results of this measurement are the eigenvalues obtained from the appropriate operator equation. This means that the only possible energies an electron may possess will be the eigenvalues E resulting from the Schrödinger equation for each ψ representing a state of the electron. The wave functions ψ for an electron may be obtained assuming wavelike motion in an area of definite volume.

Two problems arise in such an approach. The first is quantum mechanical. It involves the Heisenberg uncertainty principle:

$$\Delta x \cdot \Delta p \geq \hbar = \frac{h}{2\pi} \cong 10^{-27} \text{ erg-sec}$$

$$\Delta E \cdot \Delta t \geq \hbar \tag{1.5}$$

The Heisenberg uncertainty principle states that any particle as small as an electron must have a zero-point energy because it is permitted to move. Physically, this means that any determination of position x brings about an uncertainty in momentum p related to Planck's constant h, and vice versa. Alternatively, any determination of energy E introduces an uncertainty in the time t at which the measurement was performed. The act of measurement of one of these variables is inseparably related to an uncertainty in the complementary variable. We therefore must speak not of the location of an electron at a particular position, but of the probability of finding an electron at that position.

The second problem in adopting the wave-function representation of the electron is mathematical. While the Schrödinger equation can be completely solved for a few systems, such as the hydrogen atom, for more complicated systems one must be satisfied with exact solutions to approximate Hamiltonians, or approximate solutions to exact Hamiltonians, or, most often, approximate solutions to approximate Hamiltonians. Fortunately, reasonable results may be obtained from crude approximations.

For the hydrogen atom, consisting of one proton in the nucleus and one electron in an orbital, the potential energy is Coulombic:

$$V = -\frac{e^2}{r} \tag{1.6}$$

where e is the charge on the electron, and r is the distance separating the charges. The kinetic energy may be written as in Eq. 1.7, where m is the mass of the electron:

$$T = -\frac{h^2}{8\pi^2 m}\left(\frac{\partial^2}{\partial x^2} + \frac{\partial^2}{\partial y^2} + \frac{\partial^2}{\partial z^2}\right) \tag{1.7}$$

The sum of the second partial derivatives in Eq. 1.7 may be represented as ∇^2 (pronounced "del square"). Since E is equal to $V + T$, the Schrödinger equation for the hydrogen atom therefore becomes:

$$-\left(\frac{h^2}{8\pi^2 m}\nabla^2 + \frac{e^2}{r}\right)\psi = E\psi \tag{1.8}$$

The constants of integration resulting from the solution of this equation are the quantum numbers. The most precise solutions of this equation are called *hydrogenic orbitals*. We will use as solutions *Slater orbitals*, which only approximate the true hydrogenic orbitals but are very similar in electron density at the distances involved in chemical bonding. By using Slater orbitals, we sacrifice some accuracy for a more convenient mathematical form.

For the 1s orbital,

$$\psi_{1s} = N_{1s} e^{-cr}$$

N_{1s} is a *normalizing constant* used to satisfy the requirement that the electron is physically within the defined atom. This can be expressed mathematically by the *normalizing condition*:

$$\int \psi^2 \, d\tau = 1 \tag{1.9}$$

The exponent c is the *orbital exponent*, or *effective nuclear charge*, and is defined as

$$c = z - s \tag{1.10}$$

where z is the nuclear charge and s is a *screening constant*, or *shielding constant*, which evaluates the reduction in nuclear charge experienced by an electron because of the presence of other electrons closer to the nucleus. For the hydrogen atom, c is equal to 1 since there are no other electrons to screen the electron from the full nuclear charge. For carbon, on the

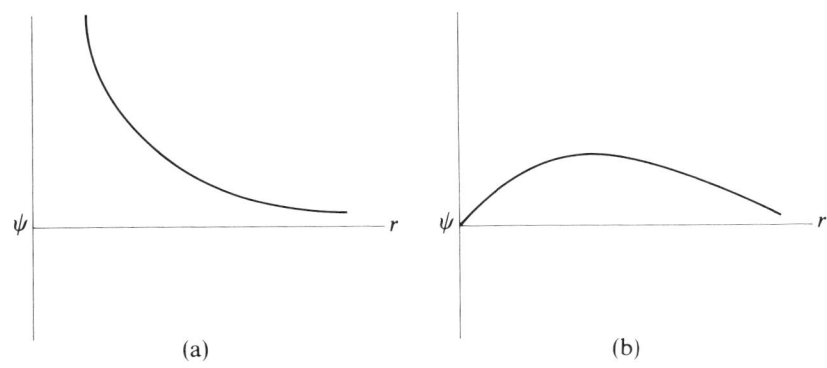

FIGURE 1.2 (a) *The 1s orbital;* (b) *the 2s orbital.*

other hand, c is 3.18. The 1s orbital has spherical symmetry (Fig. 1.2a), as do all s orbitals. The 2s orbital is drawn in Fig. 1.2(b) with

$$\psi_{2s} = N_{2s} r e^{-cr/2}$$

The p orbitals differ in that they possess directional character and are a function of one of the Cartesian coordinates x, y, or z, instead of the polar coordinate r. Since

$$\psi_{2p_x} = N_{2p} x e^{-cr/2}$$

the $2p_x$ orbital may be represented as in Fig. 1.3, with the corresponding representations of the $2p_y$ and $2p_z$ orbitals shown in Fig. 1.4. All three $2p$ orbitals are equivalent except for directional character and are linearly independent.

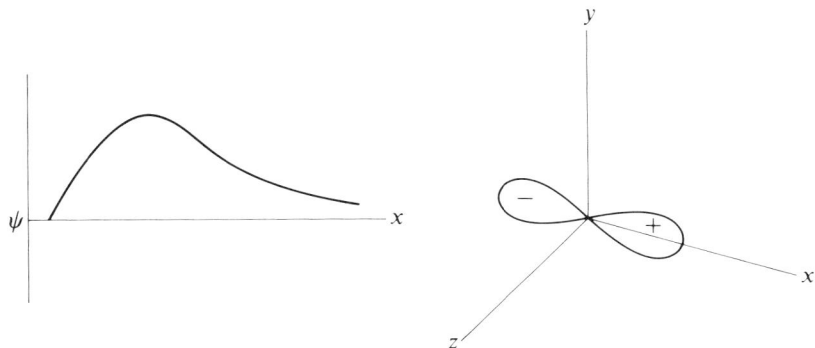

FIGURE 1.3 *The $2p_x$ orbital.*

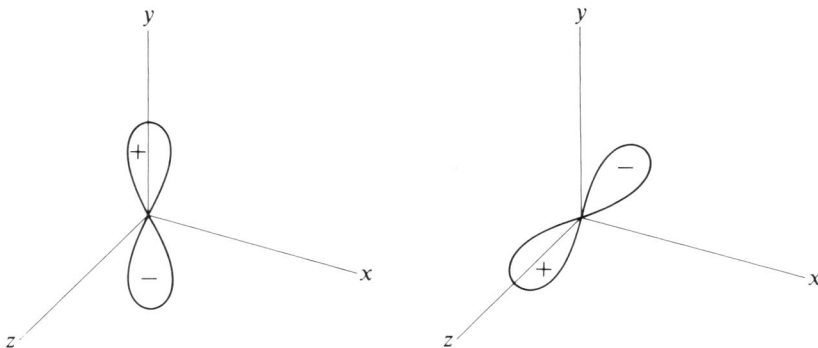

FIGURE 1.4 *The $2p_y$ and $2p_z$ orbitals.*

The Schrödinger equation for other species containing only one electron must include nuclear charge z in the Coulombic term, and may be generalized as

$$-\left(\frac{h^2}{8\pi^2 m}\nabla^2 + \frac{ze^2}{r}\right)\psi = E\psi \qquad (1.11)$$

For the He^+ ion, $z = 2$, while $z = 3$ for the Li^{2+} ion.

For the helium atom, complications result from the presence of two electrons. The Schrödinger equation must include a ∇^2 for each electron, a Coulombic attraction between nucleus and electron for each electron, and a Coulombic repulsion between the two electrons:

$$-\left[\frac{h^2}{8\pi^2 m}(\nabla_1^2 + \nabla_2^2) + \frac{ze^2}{r_1} + \frac{ze^2}{r_2} - \frac{e^2}{r_{12}}\right]\psi = E\psi \qquad (1.12)$$

An equation such as 1.12 can no longer be solved exactly, so approximation methods must be introduced.

Regardless of these mathematical problems, the chemist is still able to use the concept of electrons in orbitals to build up the periodic table. Because of the Pauli exclusion principle, which states that the total wave function for two electrons must be antisymmetrical, no two electrons can be assigned the same four quantum numbers. Therefore, two electrons in the same orbital must have opposite spins, and only two electrons can be in a given orbital. This fact, coupled with Hund's rule, which states that electrons fill equivalent orbitals one at a time with parallel spins before pairing in one of the orbitals, allows us to assemble the periodic table in a straightforward manner.

1-2 The Hydrogen Molecule Ion—Molecular Orbital Theory

The simplest system with two nuclei is the hydrogen molecule ion, H_2^+, which has two hydrogen nuclei and one electron.[1] If the two nuclei are far apart, we can let χ_1 be the wave function of the electron when it is under the influence of nucleus 1 and χ_2 be the wave function when the electron is under the influence of nucleus 2. When the nuclei are closer together, such as at bonding distances, the electron is under the influence of both nuclei simultaneously, and its motion must be considered in terms of molecular orbitals with wave functions ψ which involve the entire molecule. The simplest way to picture this molecular orbital is to form it by a linear combination of the atomic orbitals for the constituent atoms when they do not interact (LCAO method). The mathematical expression,

$$\psi = c_1\chi_1 + c_2\chi_2 \tag{1.13}$$

is mathematically legitimate, since the sum of solutions of a differential equation is also a solution of the equation. Additionally, it is a good approximation to physical reality in terms of the atoms-and-molecules model of chemical bonding taken for granted by most chemists, provided three conditions are satisfied:

1. The atomic orbitals must have the same symmetry with respect to the bond axis.
2. The atomic orbitals must have roughly comparable energies.
3. The overlap must be as large as possible.

The coefficients c_1 and c_2 represent these three conditions.

Molecular orbitals are similar to atomic orbitals in their primary characteristics. Each electron is described by a wave function which corresponds to a molecular orbital. Each molecular orbital is assigned a set of quantum numbers similar to those used to represent atomic orbitals. The energy associated with each molecular orbital is an ionization potential. Just as for atomic orbitals, electrons are placed in molecular orbitals in order of increasing energies based on the Pauli exclusion principle and Hund's rule.

Assuming stationary nuclei (the Born-Oppenheimer principle), let us use this LCAO method to solve the Schrödinger equation for the hydrogen molecule ion:

$$\mathbf{H}\psi = E\psi \tag{1.1}$$

Multiplying by ψ from the left,

$$\psi\mathbf{H}\psi = \psi E\psi$$

Rearranging and integrating over all space gives

$$\int \psi \mathbf{H} \psi \, d\tau = E \int \psi^2 \, d\tau$$

which may be rearranged to give

$$E = \frac{\int \psi \mathbf{H} \psi \, d\tau}{\int \psi^2 \, d\tau}$$

Using the LCAO method, we can write

$$E = \frac{\int (c_1 \chi_1 + c_2 \chi_2) \mathbf{H}(c_1 \chi_1 + c_2 \chi_2) \, d\tau}{\int (c_1 \chi_1 + c_2 \chi_2)^2 \, d\tau}$$

$$= \frac{\int [(c_1 \chi_1 \mathbf{H} c_1 \chi_1) + (c_1 \chi_1 \mathbf{H} c_2 \chi_2) + (c_2 \chi_2 \mathbf{H} c_1 \chi_1) + (c_2 \chi_2 \mathbf{H} c_2 \chi_2)] \, d\tau}{\int [c_1^2 \chi_1^2 + 2 c_1 c_2 \chi_1 \chi_2 + c_2^2 \chi_2^2] \, d\tau}$$

Since the integral of a sum is the sum of the integrals, and since the coefficients may be removed from under the integrals,

$$E = \frac{c_1^2 \int \chi_1 \mathbf{H} \chi_1 \, d\tau + c_1 c_2 \int \chi_1 \mathbf{H} \chi_2 \, d\tau + c_1 c_2 \int \chi_2 \mathbf{H} \chi_1 \, d\tau + c_2^2 \int \chi_2 \mathbf{H} \chi_2 \, d\tau}{c_1^2 \int \chi_1^2 \, d\tau + 2 c_1 c_2 \int \chi_1 \chi_2 \, d\tau + c_2^2 \int \chi_2^2 \, d\tau} \quad (1.14)$$

Let us now define some of these integrals. The *Coulomb integral*, H_{ii}, is defined by Eq. 1.15:

$$H_{ii} = \int \chi_i \mathbf{H} \chi_i \, d\tau \quad (1.15)$$

The simplest interpretation of the Coulomb integral is from the electrostatic point of view. The interaction between two particles viewed as unit point charges is:

$$V = e^2 \left(\frac{1}{r}\right) = e^2 \left[(1) \frac{1}{r} (1)\right]$$

If, bearing in mind the wave nature of the electron, we consider the interaction between a unit point charge and an electron, we arrive at the *nuclear attraction integral*:

$$V = \int e^2 \left[\chi_1 \left(\frac{1}{r}\right) \chi_1\right] d\tau \quad (1.16)$$

This integral represents the potential energy of an electron in the field of a nucleus. The integration is required because of the wave-function representation of the electron and the Heisenberg uncertainty principle.

The *resonance integral*, or *exchange integral*, H_{ij}, is defined by Eq. 1.17. This is the energy of an electron in the field of two nuclei, with H_{ij} equal to H_{ji} for all real systems.

$$H_{ij} = \int \chi_i \mathbf{H} \chi_j \, d\tau \qquad (i \neq j) \tag{1.17}$$

The various integrals in the denominator of Eq. 1.14 are called *overlap integrals*. These overlap integrals, S_{ij} (Eq. 1.18), measure the physical congruence of the two wave functions χ_i and χ_j. If $i = j$, the identical

$$S_{ij} = \int \chi_i \chi_j \, d\tau \tag{1.18}$$

wave functions are being compared, so $S_{ii} = 1$ since $\int \chi_i^2 \, d\tau$, the probability, must equal 1. If χ_i and χ_j do not overlap in any way, $S_{ij} = 0$. The general result, therefore, is:

$$0 \leq S_{ij} \leq 1$$

If $S_{ij} = 0$, the orbitals are called *orthogonal orbitals*. Any two normalized orbitals on the same atom must be orthogonal if the wave functions used are meaningful.

What would happen to S_{ij} as two $2p$ orbitals on different atoms are brought together head-on along a bond axis (Fig. 1.5)? At large distances, (a), S has a value of 0. As the orbitals come closer together, overlap begins to be appreciable, (b), and S increases. Maximum overlap is attained when similar lobes completely occupy the same space, (c). If each of the two orbitals continues to move in its original direction, overlap starts to decrease, (d), until the point is reached when the two orbitals overlap lobe for lobe in a cancelling sense (plus lobe of one occupying the same space as minus lobe of the other), (e). The overlap integral then begins to increase again, producing a situation identical to views (a) through (d) in reverse order with opposite signs for the lobes from those in (a) through (d); compare (f) and (b). The importance of the overlap integral may be stated quite simply: The greater the overlap integral between two orbitals, the stronger the bond formed from these orbitals. Hence, the third item in our list of conditions above.

Returning to H_2^+, we can now rewrite Eq. 1.14:

$$E = \frac{c_1^2 H_{11} + 2c_1 c_2 H_{12} + c_2^2 H_{22}}{c_1^2 S_{11} + 2c_1 c_2 S_{12} + c_2^2 S_{22}} \tag{1.19}$$

If we could evaluate H_{11}, H_{12}, H_{22}, S_{11}, S_{12}, and S_{22}, we could solve this equation for the energy E and the coefficients c_1 and c_2. Several methods are available to help us, the easiest of which is the *variation method*.[2] The

---

2. An alternate, slightly more sophisticated, technique is the *perturbation method*.

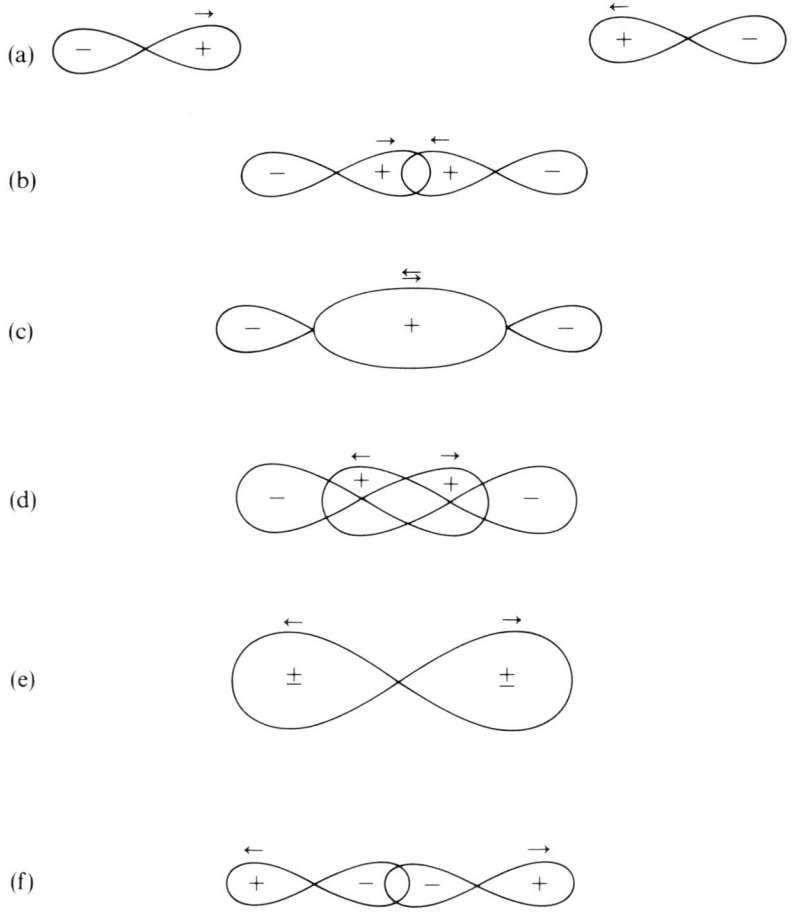

FIGURE 1.5 *The interaction between two $2p_x$ orbitals approaching along the x axis.*

true, "physically real" orbital will have the lowest possible energy; any approximation to this situation must have a higher energy. For a given set of wave functions, the best possible approach to physical reality (within the limitations of this set of wave functions) would be that approach that offers the lowest energy. The approach, therefore, is to minimize the energy by setting both $\partial E/\partial c_1$ and $\partial E/\partial c_2$ equal to 0. The resulting equations from Eq. 1.19 are the *secular equations*:

$$c_1(H_{11} - S_{11}E) + c_2(H_{12} - S_{12}E) = 0 \\ c_1(H_{12} - S_{12}E) + c_2(H_{22} - S_{22}E) = 0 \quad (1.20)$$

It is relatively easy to solve such simultaneous equations using determinants:[3]

$$\begin{vmatrix} (H_{11} - S_{11}E) & (H_{12} - S_{12}E) \\ (H_{12} - S_{12}E) & (H_{22} - S_{22}E) \end{vmatrix} = 0 \quad (1.21)$$

In order to simplify Eq. 1.21, we will use the simple Hückel MO theory. First, assume *neglect of overlap*:

$$S_{ii} = 1, S_{ij} = 0 \quad \text{(for } i \neq j\text{)} \quad (1.22)$$

Next, let the Coulomb integral H_{ii} be expressed in units of α, where $H_{ii} = \alpha$ for all hydrogen (or carbon) nuclei. The symbol α is a negative number except at infinite distances, when it is equal to zero. Similarly, the resonance integral H_{ij} can be expressed in units of β, a negative number whose size depends on the distance between nuclei i and j and which is also zero at infinite distances. *Neglect of nonbonded interactions* is also assumed, giving $H_{ij} = \beta$ if i is directly bonded to j, and $H_{ij} = 0$ if i and j are not directly bonded. Determinant 1.21 then becomes determinant 1.23:

$$\begin{vmatrix} \alpha - E & \beta \\ \beta & \alpha - E \end{vmatrix} = 0 \quad (1.23)$$

Solving this determinant leads to Eq. 1.24:

$$\begin{aligned} (\alpha - E)^2 - \beta^2 &= 0 \\ \alpha^2 - 2\alpha E + E^2 - \beta^2 &= 0 \\ E^2 - 2\alpha E + (\alpha^2 - \beta^2) &= 0 \end{aligned} \quad (1.24)$$

which is a quadratic equation in E with solutions

$$E = \alpha + \beta \quad \text{and} \quad E = \alpha - \beta \quad (1.25)$$

Since both α and β are negative numbers, the term $E = \alpha + \beta$ has lower energy than $E = \alpha - \beta$.

Having now solved for the energy, we may use these energies in the secular equations within the Hückel MO framework to solve for c_1 and c_2 for each energy level:

$$\begin{aligned} c_1(\alpha - E) + c_2 \beta &= 0 \\ c_1 \beta + c_2(\alpha - E) &= 0 \end{aligned} \quad (1.26)$$

Substituting $E = \alpha + \beta$ into Eq. 1.26, we find that $c_1 = c_2$ (or $c_1/c_2 = 1$) for the lowest energy level, the *ground state*. Using $E = \alpha - \beta$, c_1 is found to equal $-c_2$ (or $c_1/c_2 = -1$) for the higher-energy system, the *excited state*.

For the ground state,

$$\psi_1 = \chi_1 + \chi_2$$

---
3. Reference B, Appendix 2, and Reference G, Chapters 2 and 3.

However, this molecular wave function must still be normalized (Eq. 1.9):

$$\int \psi_1{}^2 \, d\tau = \int (\chi_1 + \chi_2)^2 \, d\tau$$

$$= \int \chi_1{}^2 \, d\tau + 2 \int \chi_1 \chi_2 \, d\tau + \int \chi_2{}^2 \, d\tau$$

Using the neglect-of-overlap approximation,

$$\int \psi_1{}^2 \, d\tau = 1 + 0 + 1 = 2$$

The normalized ground state wave function should therefore be written:

$$\psi_1 = \frac{1}{\sqrt{2}} (\chi_1 + \chi_2)$$

A similar normalization for the excited state leads to:

$$\psi_2 = \frac{1}{\sqrt{2}} (\chi_1 - \chi_2)$$

In general, the normalization constant N has a value of $1/C$ where $C = \sqrt{c_1{}^2 + c_2{}^2 + \cdots}$.

We now have found the Hückel LCAO–MO solutions for $H_2{}^+$ (Fig. 1.6). The single electron is placed in the lowest energy level for the ground

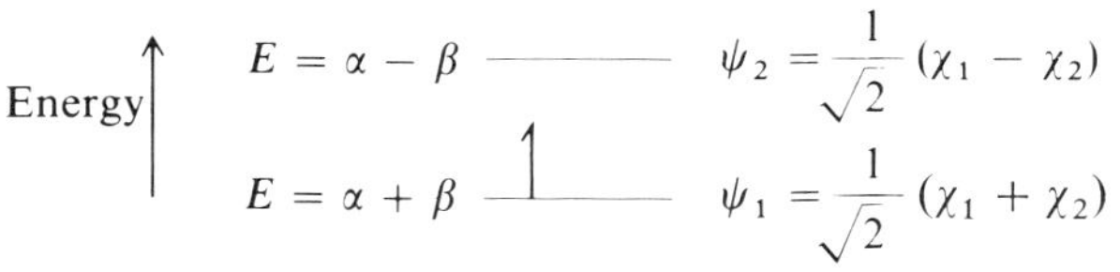

FIGURE 1.6 *The energy-level diagram for the hydrogen molecule ion.*

state situation. The total energy in this approximation is equal to $\alpha + \beta$, a magnitude dependent on the numerical evaluation of these symbols.

What do these wave functions and energies mean? Let us look first at the lower-energy wave function ψ_1. The total energy is the sum of the electronic energy and the internuclear Coulomb repulsion energy, and all are functions of the internuclear separation r (Fig. 1.7). The total energy curve exhibits an energy minimum at about 1.06 Å, suggesting the

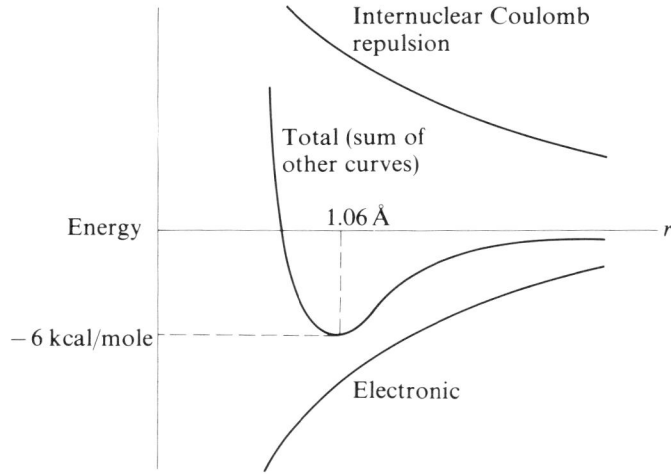

FIGURE 1.7 *The energy curves for the bonding orbital in H_2^+.*

existence of a stable form of the H_2^+ species at this internuclear separation. We may use ψ_1^2 to evaluate electron density (Fig. 1.8). Since

$$\psi_1^2 = \tfrac{1}{2}(\chi_1^2 + 2\chi_1\chi_2 + \chi_2^2)$$

there is more electron density between the nuclei ($\chi_1\chi_2$) than would be obtained from just the two atomic orbital wave functions ($\chi_1^2 + \chi_2^2$). This increased electron density between the two nuclei suggests increased stability and the formation of a chemical bond, so ψ_1 is called a *bonding state*. Qualitatively speaking, a bonding state has the charge cloud concentrated to a greater extent between the nuclei and possesses enough (negative) potential energy as the result of this electronic interaction to overcome the (positive) nuclear repulsion energy and the (positive) kinetic energy and thereby produce an energy minimum at a given internuclear distance.

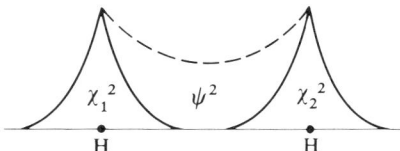

FIGURE 1.8 *A plot of ψ^2 for the bonding* MO *and the component* AO's.

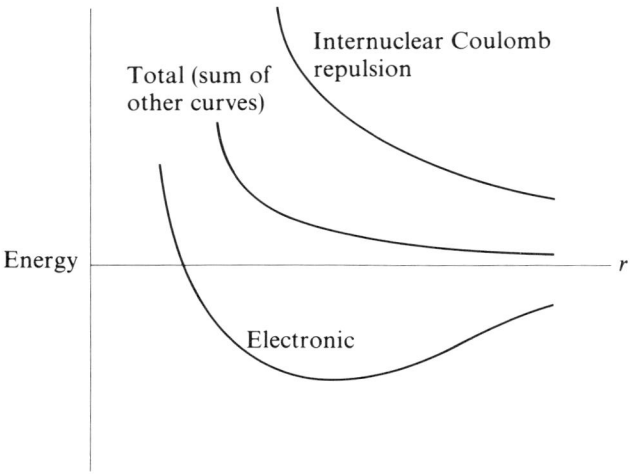

FIGURE 1.9 *The energy curves for the antibonding orbital in H_2^+.*

For the higher-energy wave function ψ_2, the total energy picture (Fig. 1.9) shows no energy minimum, suggesting instability relative to isolated atoms. In terms of electron density,

$$\psi_2^2 = \tfrac{1}{2}(\chi_1^2 - 2\chi_1\chi_2 + \chi_2^2)$$

so there is *less* electron density between the nuclei $(-\chi_1\chi_2)$ than for the isolated atoms at any given internuclear distance, resulting in the existence of a node (Fig. 1.10). This unstable state is called an *antibonding state*, and is designated by a superscript *.[4]

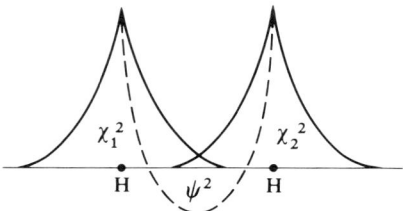

FIGURE 1.10 *A plot of ψ^2 for the antibonding* MO *and the component* AO's.

---

4. Should the electron density in the molecular orbital be identical with that in the component atomic orbitals, a *nonbonding state* exists.

These general concepts, both the calculations according to Hückel LCAO–MO and variation methods and the bonding–antibonding ideas, can be extended to all molecules. However, serious problems arise because of the interactions between the electrons, paralleling the situation in calculations for atoms. If we consider only valence electrons and their interactions, it is often possible to obtain results that are qualitatively satisfactory. The concept of a bonding state or an antibonding state may be expanded to ideas of net bonding or antibonding states (or configurations). For example, an orbital involving four nuclei with two bonding interactions and one antibonding interaction exhibits a net bonding interaction and will be termed a bonding state.

Orbitals that are cylindrically symmetrical about the bond axis are called σ orbitals (Fig. 1.11). They will have a zero component of angular

FIGURE 1.11 *Possible combination of AO's to give σ orbitals. Not shown are $s + d$, $p + d$, or $d + d$ in the same endwise manner.*

momentum about the bond axis and will be nondegenerate. Orbitals that result from a parallel combination, have a node in the plane of the bond axis, and have axial symmetry are called π orbitals (Fig. 1.12a). The π molecular orbital has different signs for the two streamer regions (Fig. 1.12b), but the two regions are an inseparable part of the wave function. A π orbital does have angular momentum about the bond axis and is doubly degenerate.

In the H_2^+ system, $|c_1|$ is equal to $|c_2|$, because this system is *homopolar*, or comprised of like atomic nuclei. For any *heteropolar*

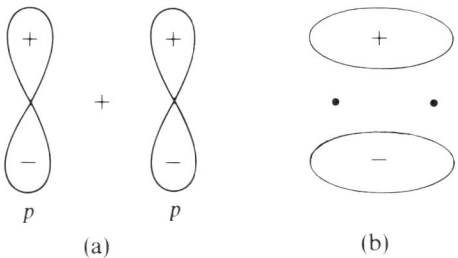

FIGURE 1.12 *The π orbital:* (a) *the AO's to be combined;* (b) *the resulting MO.*

diatomic molecule, $|c_1|$ is not equal to $|c_2|$. Therefore, the electron density is greater about one of the nuclei of the heteropolar diatomic molecule, leading to the existence of a dipole moment (see Chapter 4-1).

1-3 The Hydrogen Molecule Ion—Valence Bond Theory

An alternate and historically older approach to the problem of bonding is the Heitler-London valence bond (VB) approach. In this approach, the component atoms are treated as complete entities which are allowed to interact to form a molecule, a depiction comparable to the usual Lewis picture of bonding. Two initially distant atomic orbitals, each containing one electron, are brought together to form a covalent bond as overlap becomes appreciable. Ionic structures may be included as needed by considering additional structures formed by the combination of an empty atomic orbital on one atom with an orbital containing two electrons on the other atom.

Mathematically one assumes that the wave function of a diatomic system at any distance of separation may be represented as the product of the atomic wave functions that will interact:

$$\psi = \chi_1(1)\chi_2(2) \tag{1.27}$$

The symbolism used is to let χ_i be the wave function of an atomic orbital associated with nucleus i occupied by the electron j indicated in parentheses immediately following each χ. Equation 1.27 therefore indicates electron 1 near nucleus 1 and electron 2 near nucleus 2. Since electrons cannot be labeled as such, the molecular wave function must reflect this "exchange phenomenon" at distances approximating the distances involved in chemical bonding. Equation 1.27 must therefore be rewritten as:

$$\psi = c_1\chi_1(1)\chi_2(2) + c_2\chi_2(1)\chi_1(2) \tag{1.28}$$

For any homopolar diatomic system, $c_1 = \pm c_2$ in Eq. 1.28, with the lower-energy state utilizing the positive sign:

$$\psi_1 = \chi_1(1)\chi_2(2) + \chi_2(1)\chi_1(2) \tag{1.29}$$

Solution for the energy levels may be accomplished by the variation method as shown previously for the MO method. The symbolism used for the Coulomb integral and the exchange integral is not consistent from worker to worker, but the net result for H_2^+, assuming neglect of overlap, is still of the form

$$E = \alpha \pm \beta$$

Molecular Orbital Approach vs. Valence Bond Approach. Any comparison of MO and VB is very dependent on the level of sophistication used in either approach.[5] At the simplest levels, the exchange of electrons between nuclei is perfectly correlated in VB unless ionic structures are added. That is to say, the electrons involved in the bond under consideration are always situated in such a way that there is an electron in close association with each of the two nuclei. Ionic structures involving both electrons near one of the nuclei must be added to remove this constraint. No such correlation of electrons occurs in MO, the electrons being dissociated from the individual nuclei and associated only with the molecular orbitals encompassing both of the nuclei. Ionic structures are therefore overemphasized in MO.

This difference disappears at the next level of sophistication. Inclusion of ionic structures in VB with weighting coefficients λ gives

$$\psi = \chi_1(1)\chi_2(2) + \chi_2(1)\chi_1(2) + \lambda[\chi_1(1)\chi_1(2) + \chi_2(1)\chi_2(2)] \quad (1.30)$$

for a homopolar diatomic molecule in its lowest-energy state. Inclusion of *configuration interaction* in the MO wave function results in the addition of a weighted amount μ of higher-energy wave functions. For example, some of the antibonding wave function for the hydrogen molecule is added to the bonding wave function to give Eq. 1.31 (which is written in VB style for easier comparison with Eq. 1.30):

$$\psi = (\chi_1 + \chi_2)(1)(\chi_1 + \chi_2)(2) + \mu[(\chi_1 - \chi_2)(1)(\chi_1 - \chi_2)(2)] \quad (1.31)$$

Polarity appears as the relative weight of the coefficients and as the relative weight of the ionic structures in VB, while in the MO approach polarity depends only on the coefficients. For questions of electron density and dipole moment, MO is simpler but VB is closer to the Lewis pictures to which most chemists are accustomed.

In general, MO is more adaptable than VB to one-electron properties, such as electronic excitation (see the discussion in Chapter 1-6). However, MO without configuration interaction suffers from lack of electron correlation even for many one-electron properties. For example, an electron in a given orbital would be predicted under MO to behave in the same manner whether there is one electron or two in the given orbital, which is a physically unreasonable result. The MO approach automatically handles one-electron and three-electron bonds (as in the boron hydrides), while VB requires a new approach with a new bond type.

---

5. See Reference F, Chapter VI.

1-4 Hybridization

Theories of electronic bonding must not only be consistent with electronic properties and energy considerations, but must also be in consonance with structural features. Any consideration of bonding in methane must give results in agreement with the known tetrahedral nature of the carbon atom and the known equivalence of the hydrogen ligands in the CH_4 system. A carbon atom with the $(2s)^2(2p)^2$ configuration can use only the unpaired p electrons to form C—H bonds, resulting in a CH_2 molecule with an H—C—H angle of 90° instead of tetrahedral methane. The mathematical manipulation of atomic orbitals to produce new orbitals with different electronic configurations and with electronic concentrations in different spatial regions is called *hybridization*. Such a procedure is permissible since the individual orbitals in a determinant are not physically significant. The pragmatic approach of the hybridization concept is to combine a set of normalized orthogonal orbitals in a Slater determinant, using that orbital set which best satisfies the symmetry and geometry requirements of the molecule under consideration.[6] Hybridization involves making linear combinations of valence-shell atomic

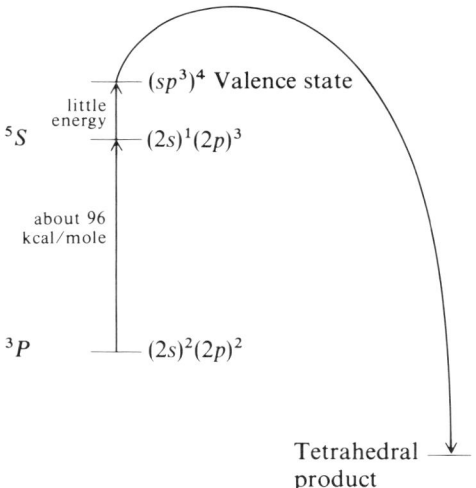

FIGURE 1.13 *The energy diagram for transformation of atomic carbon into sp^3 compounds.*

---

6. F. A. Cotton, *Chemical Applications of Group Theory*, Interscience, New York, 1963, Chapter 6; I. Cohen and J. Del Bene, *J. Chem. Educ.* **46**, 487 (1969); Reference B, Section 1-7.

orbitals to create new "hybridized" wave functions that may be used in MO or VB treatments. These new hybrid orbitals usually possess greater directional character than the original atomic orbitals, thereby creating a situation with potentially greater internuclear orbital overlap and stronger bonds.

For atomic carbon[7] (^3P in spectroscopic notation), approximately 96 kcal/mole is required to excite a $2s$ electron into a $2p$ orbital, thereby producing a $(2s)^1(2p_x)^1(2p_y)^1(2p_z)^1$ configuration (^5S), which is an sp^3 state with spherical symmetry. A slight amount of additional energy produces the sp^3 *valence state*, which exhibits tetrahedrally directed orbitals and is not a spectroscopic state. The energy of excitation required to transform ground state atomic carbon to this sp^3 valence state is recovered in the formation of new bonds—in our example, four C—H bonds to give CH_4 (Fig. 1.13). Mathematically, the sp^3 orbitals may be written as:

$$\psi = \tfrac{1}{2}(s + \sqrt{3}p) \tag{1.32}$$

Approximately seven eighths of the electron density in the valence state is concentrated in the direction in which the bond will be formed. The tetrahedral character of the valence state demonstrates an important

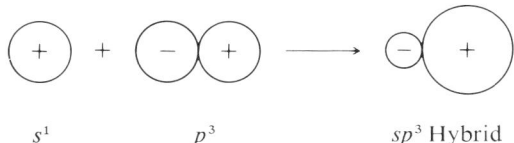

s^1 p^3 sp^3 Hybrid

characteristic of most chemical structures and the hybrid orbitals used to rationalize such structures: Stable structures have orbital arrangements which will minimize interorbital electronic repulsions and interligand interactions. The most stable arrangement of four groups equidistant from a central point is the tetrahedron.

The most stable arrangement of three groups equidistant from a given position is an equilateral triangle. Trigonal sp^2-hybridized carbon,

$$\psi = \frac{1}{\sqrt{3}}(s + \sqrt{2}p)$$

possesses such 120° interligand angles. Such sp^2 hybrid orbitals do not use one p orbital and electron in the σ systems. This unused electron may combine with an adjacent similar p electron to form a π bond, such as in ethylene (Fig. 1.14). The ethylene molecule demonstrates the typical

---

7. Reference F, Chapter VIII.

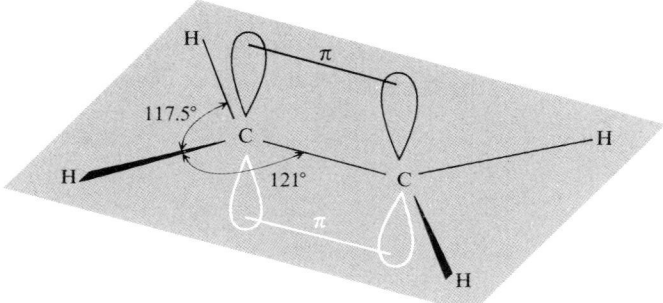

FIGURE 1.14 *The ethylene molecule: sp^2 hybridization.*

coplanarity of sp^2 systems, with the double bond lending rigidity to the C—C unit.

Acetylenic systems require digonal sp hybrids at each carbon with two unhybridized p electrons from each carbon combined to form two π bonds. The resulting compounds must be linear with axial symmetry at the C—C bond axis.

Whenever the atoms attached to the central atom whose hybridization is under consideration are dissimilar, nonequivalent hybrid orbitals are required.[8] Because different atoms have different electronegativities, as ligands they will require different electronic responses on the part of the central atom. The hybrid orbitals of the central atom must therefore be a function of the ligands attached to the central atom. Since the only limiting conditions for the hybrid orbitals are orthogonality, normalization, and summation to the total contributions of the component atomic orbitals, a large number of sets of hybrid orbitals are possible. Further, any given set of hybrid orbitals need not be an integral combination of atomic orbitals; an $sp^{2.7}$ hybrid orbital is just as acceptable as an sp^3 one. Such differences in hybridization within a given set of hybrid orbitals will require differences in molecular geometry. As shown in Fig. 1.14, the H—C—H angle in ethylene is smaller than the H—C—C angle and neither is the 120° expected for pure sp^2 hybrid orbitals. Chloromethane should be composed of three orbitals with identical hybridization and a unique fourth orbital.

These ideas about hybridization can be generalized for a central carbon atom.[8] If the ith bonding orbital is represented by

$$s + \lambda_i p$$

---

8. Reference F, Chapter VIII, and Reference M, Chapter 1.

where λ_i is the *mixing coefficient*, or *hybridization parameter*, the hybridization of this orbital may be written as

$$sp^{\lambda_i^2}$$

where λ_i^2 is called the *hybridization index*. For carbon, the sum of the fractional s-orbital contributions must equal 1:

$$\sum_i \frac{1}{1+\lambda_i^2} = 1 \tag{1.33}$$

while that of the p orbitals must equal the number of p orbitals mixed into the hybridization process:

$$\sum_i \frac{\lambda_i^2}{1+\lambda_i^2} = 1, 2, \text{ or } 3 \tag{1.34}$$

Assuming that the bonding orbitals are directed along the internuclear axis,

$$1 + \lambda_i \lambda_j \cos \theta_{ij} = 0 \tag{1.35}$$

where θ_{ij} is the angle between the two bonding orbitals i and j.[9]

Using the generalizations developed above, it is possible to use experimental bond angles to calculate hybridizations and then use these calculated hybridizations to predict and/or explain other physical or chemical properties. The measured H—C—H angle of 117.5° in ethylene (Fig. 1.14) may be substituted into Eq. 1.35:

$$1 + \lambda_i^2 \cos 117.5° = 0$$
$$\lambda_i^2 = 2.1$$

If the two C—H hybrid orbitals are $sp^{2.1}$ and if a p orbital, in order that it may form the π bond, is not involved in the hybridization, either Eq. 1.33,

$$\frac{1}{1+2.1} + \frac{1}{1+2.1} + \frac{1}{1+\lambda_j^2} = 1$$

or Eq. 1.34,

$$\frac{2.1}{1+2.1} + \frac{2.1}{1+2.1} + \frac{\lambda_j^2}{1+\lambda_j^2} = 2$$

leads to

$$\lambda_j^2 = 1.82$$

for the C—C hybrid σ orbital. Substitution of this result into Eq. 1.35 suggests a 120°43′ C—C—H angle, in excellent agreement with the 121° angle obtained experimentally.

---

9. The internuclear angle and interorbital angle are different quantities, as we shall see in the discussion that follows.

It must be emphasized that hybridizations calculated from bond angles in this manner have no physical significance. The entire scheme of bonding—MO, VB, hybridization, or whatever—is merely a model that provides a way of thinking which suggests results close to those observed from the actual physical situation. Any discrepancy between theory and experiment may be rationalized in any way desired, provided the basic assumptions inherent in the various models are recognized.

One example of the difficulties inherent in the hybridization model is the question of the hybridization of cyclopropane (Fig. 1.15). Using Eq. 1.33–1.35, an H—C—H angle of 114° suggests $sp^{2.5}$ orbitals for the C—H bonds, $sp^{3.7}$ orbitals for the C—C bonds, and C—C—C angles of 105.5°. Since the three carbon atoms must be identical and therefore occupy the corners of an equilateral triangle, the internuclear C—C—C angles must be 60°. The $sp^{3.7}$–$sp^{3.7}$ bonds therefore must be bent around the outside of the equilateral triangle to maintain the 105.5° interorbital angles. The assumption that the internuclear angles and interorbital angles coincide is not valid for cyclopropane. This fact may be recognized and the $sp^{3.7}$–$sp^{3.7}$ bonds considered a special type of bonding situation (called *bent bonds*, *banana bonds*, or *τ bonds*); or a new model must be developed to replace this type of hybridization picture and remove such discrepancies.

Another problem arises in the question of the hybridization of the nitrogen atom in ammonia. The experimental H—N—H angles are 107°. Atomic nitrogen exhibits a $(2s)^2(2p)^3$ valence electron configuration, so one might predict 90° angles for ammonia, with a pair of electrons in a 2s orbital. Since 90° angles are not optimum for three ligands attached to a central atom, interligand and interorbital repulsions would tend to increase these angles, conceivably producing a 107° compromise between the electronic preference for 90° and the repulsive forces.

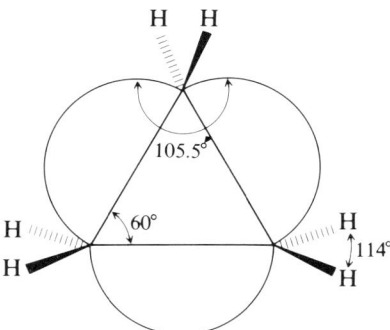

FIGURE 1.15 *The bonding in cyclopropane: "bent bonds".*

A second approach to the bonding in ammonia is to consider the nitrogen to be sp^3 hybridized, giving an $(sp^3)^5$ configuration. Such hybridization should cause tetrahedral (109°28′) angles, with a pair of electrons at one of the apices of the tetrahedron. Lone pair–hydrogen repulsions could lead to a decrease in the H—N—H angles to the 107° value.

A third approach to the ammonia problem would be one involving Eq. 1.33–1.35. The known bond angle could be used to calculate the hybridizations in the orbitals involved in the N—H bonds and in the unique orbital occupied by the lone pair of electrons. The orbital containing the lone pair of electrons must be somewhat similar to the bonding orbitals; yet it must be slightly different, precisely because it is not a bonding orbital in ammonia. The resulting nonintegral hybridizations should include all repulsive interactions which might be present in such a simple system but, as mentioned previously, have no physical significance.

The problems inherent in the ammonia molecule result from the fact that promotion of an s electron into a partly filled p orbital is not the same as promotion into an empty p orbital, as is readily apparent from atomic ionization potentials and the electronic configurations of many elements in the periodic table.

Accepting the idea of nonintegral hybridizations, one may use the fraction of s character in a given hybrid orbital to predict many physical properties (Table 1.1). A plot of the overlap integral S between two similar hybrid orbitals of the same atomic element against fraction s character (Fig. 1.16) exhibits a maximum at approximately sp hybridization. Since maximum overlap leads to the strongest (and shortest) bond, the energy required to homolytically cleave a C—H bond involving an sp-hybridized carbon should be greater than the energy required to cleave a C—H bond involving a carbon atom with less s character in the C—H orbital. This prediction agrees with the experimentally observed trends (Table 1.1).

TABLE 1.1

Hybrid	Molecule	Energy of C—H Bond Dissociation (kcal/mole)	C—H Bond Length (Å)	Force Constants	
				C—H Stretch (dynes/cm $\times 10^5$)	C—H Bend (ergs/radian $\times 10^{11}$)
sp	Acetylene	121	1.060	5.88	0.24
sp^2	Ethylene	106	1.085	5.05	0.60
sp^3	Ethane	104	1.094	4.88	0.66
p	C—H radical	81	1.120	4.09	—

NOTE: See Chapter VIII in Reference F and Section 1-7 in Reference B for similar tabulations.

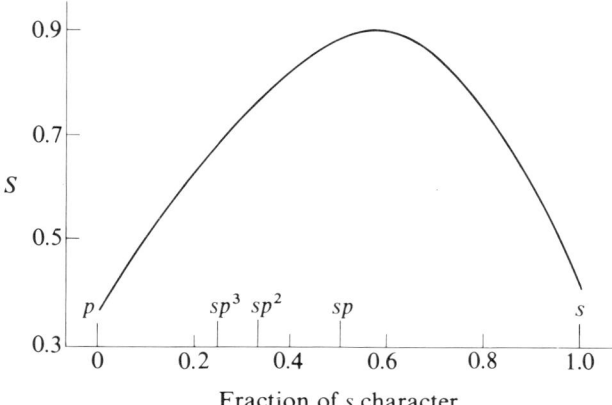

FIGURE 1.16 *The overlap integral as a function of the fraction of s character.*[10]

Similarly, a stronger bond should be more difficult to stretch. The C—H bond in acetylene does exhibit a greater force constant for such a stretching vibrational mode than the C—H bond in ethylene or ethane. However, since a hybrid orbital with greater *s* character has less directional commitment, the C—H bond in acetylene should be more susceptible to bending motions than such a bond involving carbon hybrid orbitals with less *s* character, again in agreement with experiment (Table 1.1).

Somewhat more difficult to visualize is the fact that a hybrid orbital with greater *s* character is more electronegative than one with less. Since an electron in the 2*s* atomic orbital of carbon is more stable than an electron in a 2*p* orbital of carbon, increased *s* character lends stability to any system which might have increased electron density on carbon. Just as an *sp*-hybridized C—H bond is stronger than an *sp*²-hybridized bond, an anion containing two electrons in an *sp*-hybrid orbital should be more stable than an anion with two electrons in an *sp*² hybrid. The implications of this analysis are that acetylene should be more acidic than ethylene, which should be more acidic than ethane. This order of acidity is in complete agreement with the experimental result.

Several physical measurements reportedly indicate the amount of *s* character in a hybrid orbital. Several groups[11] have investigated a relationship of the form

$$J_{^{13}C-H} = \frac{500}{1 + \lambda_i^2} \quad (J \text{ value in Hz}) \qquad (1.36)$$

---

10. A. Maccoll, *Trans. Faraday Soc.* **46**, 369 (1950)—discussed in Reference F, Chapter VIII, and Reference B, Section 1-7.
11. Reference C, p. 193; P. Laszlo and P. Stang, *Organic Spectroscopy*, Harper and Row, New York, 1971, Chapter 3; and references in these sources.

between the ^{13}C—H nuclear magnetic resonance (nmr) spin-spin coupling constant $J_{^{13}C-H}$, and the hybridization parameter λ_i^2. Nuclear quadrupole resonance also offers such hybridization information;[12] however, it is often difficult to evaluate how much of the measured electron distribution inhomogeneity is caused by hybridization and how much by partial ionic character in many of the covalent bonds.

1-5 π Systems—A Molecular Orbital Approach

Within the framework of the simplest MO theory,[13] systems containing π bonds are treated as if the σ and π bonding electrons did not interact in any way and could therefore be considered independent entities. The energy, electron densities, and reactivity of such a system are calculated as a function of the π electrons independent of the σ framework.

Using this *π-electron approximation*, calculations for ethylene are identical with those calculations performed for the hydrogen molecule, both compounds containing two electrons and two nuclei. Neglecting electron correlations and repulsive forces between the two electrons, the secular equations may be written as Eq. 1.20 and solved to yield:

$$E = \alpha \pm \beta$$

$$\psi = \frac{1}{\sqrt{2}}(\chi_1 \pm \chi_2)$$

Since both electrons would occupy the bonding orbital in the ground state of ethylene, an energy-level diagram may be drawn as shown in Fig. 1.17. The total energy of ethylene, E_π, would be

$$E_\pi = 2(\alpha + \beta)$$

ignoring electron correlation.

FIGURE 1.17 *The energy-level diagram for π energy for ethylene.*

---

12. For a more detailed discussion, see Reference C, Chapter 6, and references therein.
13. For analogous treatments, see Reference A, Chapter 6; Reference F, Chapter IX; Reference G, Chapter 2; and Reference H, Chapter 2.

Butadiene presents additional problems, since the component double bonds may be twisted in an infinite number of angles with respect to each other. If the π orbitals were all coplanar, the π clouds of the two double bonds would overlap to the greatest extent, thereby creating what is presumably the most stable arrangement of the carbon atoms. Even assuming coplanarity, two possible conformations are still possible: the

FIGURE 1.18 *Conformational isomers of butadiene.*

s-trans and the *s-cis* (Fig. 1.18). Within the Hückel LCAO–MO approximations (including zero overlap),[14] the secular equations yield the determinant

$$\begin{vmatrix} H_{11} - E & H_{12} & H_{13} & H_{14} \\ H_{12} & H_{22} - E & H_{23} & H_{24} \\ H_{13} & H_{23} & H_{33} - E & H_{34} \\ H_{14} & H_{24} & H_{34} & H_{44} - E \end{vmatrix} = 0 \qquad (1.37)$$

for either butadiene conformer. As defined for Eq. 1.23, $H_{ii} = \alpha$, $H_{ij} = \beta$ for i adjacent to j, and $H_{ij} = 0$ for i not adjacent to j (*neglect of non-neighbor interactions*). Since all non-neighbor interactions are thereby ignored, *s-cis-* and *s-trans*-butadiene become mathematically indistinguishable in such a treatment. Equation 1.37 becomes:

$$\begin{vmatrix} \alpha - E & \beta & 0 & 0 \\ \beta & \alpha - E & \beta & 0 \\ 0 & \beta & \alpha - E & \beta \\ 0 & 0 & \beta & \alpha - E \end{vmatrix} = 0 \qquad (1.38)$$

Dividing by β and letting $(\alpha - E)/\beta = x$,

$$\begin{vmatrix} x & 1 & 0 & 0 \\ 1 & x & 1 & 0 \\ 0 & 1 & x & 1 \\ 0 & 0 & 1 & x \end{vmatrix} = 0$$

Solving,[3]
$$x^4 - 3x^2 + 1 = 0$$
$$x = \pm 1.618, \pm 0.618$$

---

14. For simple treatments using definite values for the overlap integral, see Reference B, Sections 1-10 and 1-11; Reference J; and Reference K.

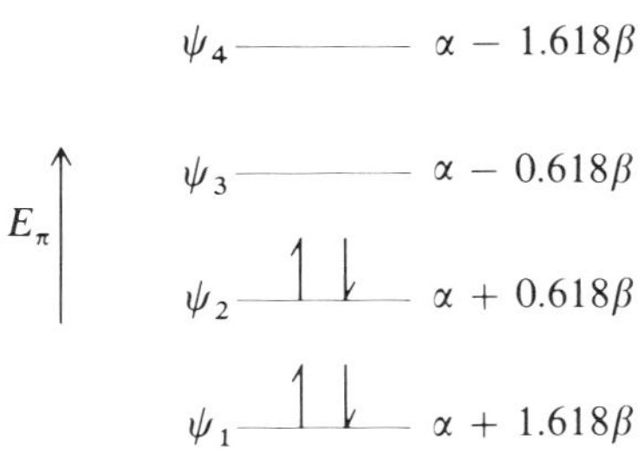

$$\psi_4 \underline{\qquad} \alpha - 1.618\beta$$

$$\psi_3 \underline{\qquad} \alpha - 0.618\beta$$

$$E_\pi$$

$$\psi_2 \underline{\uparrow\downarrow} \alpha + 0.618\beta$$

$$\psi_1 \underline{\uparrow\downarrow} \alpha + 1.618\beta$$

FIGURE 1.19 *The energy-level diagram for π energy for butadiene.*

Therefore, the energies are

$$E = \alpha \pm 1.618\beta \quad \text{and} \quad E = \alpha \pm 0.618\beta$$

and the energy-level diagram for π energy may be represented as in Fig. 1.19. Were butadiene merely two noninteracting ethylene-like double bonds, the π energy of butadiene would have been

$$4(\alpha + \beta) = 4\alpha + 4\beta$$

The π energy actually calculated for butadiene is

$$2(\alpha + 0.618\beta) + 2(\alpha + 1.618\beta) = 4\alpha + 4.472\beta$$

The difference between these two π energies, the calculated one and the hypothetical one, for the noninteracting double bonds is called the *delocalization energy* (DE), or the *resonance energy* (to be discussed in Chapter 2). This energy,

$$4\alpha + 4.472\beta - (4\alpha + 4\beta) = 0.472\beta$$

represents the additional stabilization present in the butadiene system because of the "delocalization" of the π electrons when the σ framework defines a plane.

Solution of the secular equations of butadiene for the coefficients leads to the following wave functions in decreasing order of stability:

$$\psi_1 = 0.371\chi_1 + 0.600\chi_2 + 0.600\chi_3 + 0.371\chi_4 \qquad (1.39)$$

$$\psi_2 = 0.600\chi_1 + 0.371\chi_2 - 0.371\chi_3 - 0.600\chi_4 \qquad (1.40)$$

$$\psi_3 = 0.600\chi_1 - 0.371\chi_2 - 0.371\chi_3 + 0.600\chi_4 \qquad (1.41)$$

$$\psi_4 = 0.371\chi_1 - 0.600\chi_2 + 0.600\chi_3 - 0.371\chi_4 \qquad (1.42)$$

These wave functions may be represented schematically as in Fig. 1.20. Notice the presence of an antibonding node whenever the wave function

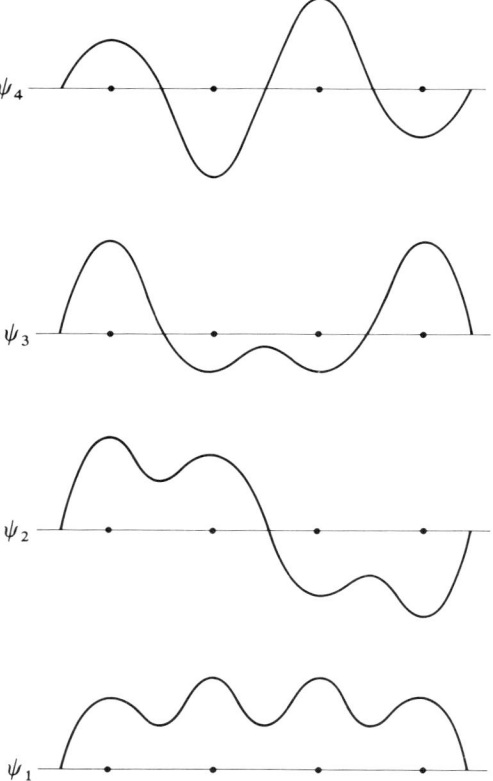

FIGURE 1.20 *Schematic representation of the* MO's *of butadiene.*

changes sign, as well as the increase in the number of nodes as the energy increases. Smoothed-out versions of the orbitals represented in Fig. 1.20 may be drawn, based on the one-dimensional particle-in-the-box idea (Fig. 1.1). Rough schematic orbital representations like those in Fig. 1.20 are embodied in the *free-electron model*,[15,16] and can be produced quickly without mathematical computation merely by considering the number of nodes in each possible representation. As we shall see, degenerate states may require some caution.

The Hückel secular determinant, such as Eq. 1.38, may be created without any mathematical manipulation. If $x = (\alpha - E)/\beta$, set every ii

---

15. J. R. Platt, et al., *Free Electron Theory of Conjugated Molecules*, Wiley, New York, 1964.
16. W. C. Herndon and E. Silber, *J. Chem. Educ.* **48**, 502 (1971).

determinant element equal to x and set every ij determinant element connecting two adjacent atoms participating in the π system equal to 1; all other ij elements are set equal to 0. For benzene, the secular determinant for the delocalized π system may be written as:

$$\begin{vmatrix} x & 1 & 0 & 0 & 0 & 1 \\ 1 & x & 1 & 0 & 0 & 0 \\ 0 & 1 & x & 1 & 0 & 0 \\ 0 & 0 & 1 & x & 1 & 0 \\ 0 & 0 & 0 & 1 & x & 1 \\ 1 & 0 & 0 & 0 & 1 & x \end{vmatrix} = 0 \qquad (1.43)$$

Solving,[3]

$$x^6 - 6x^4 + 9x^2 - 4 = 0$$
$$x = 2, 1, 1, -1, -1, -2$$
$$E = \alpha + 2\beta, \alpha + \beta, \alpha + \beta, \alpha - \beta, \alpha - \beta, \alpha - 2\beta$$

and the energy-level diagram for π energy may be represented as in Fig. 1.21. The total π energy of benzene is found to be

$$2(\alpha + 2\beta) + 4(\alpha + \beta) = 6\alpha + 8\beta$$

giving a delocalization energy of 2β. The orbitals may be drawn schematically by a cyclic variation of the free-electron model developed by Platt[15] called the *perimeter free-electron model*. As shown in Fig. 1.22, the electron is treated as a particle constrained to follow a continuous circular path. The most stable energy level, ψ_1, is characterized by a wave function with positive overlap (no nodes) at each atom in the system. The presence of degenerate states (ψ_2 and ψ_3, and ψ_4 and ψ_5) requires an orbital representation for each member of a degenerate pair which are equal in energy and therefore have the same number of nodes. The π energy of an orbital in a perimeter free-electron model approach may be evaluated qualitatively by the amount of net bonding or antibonding

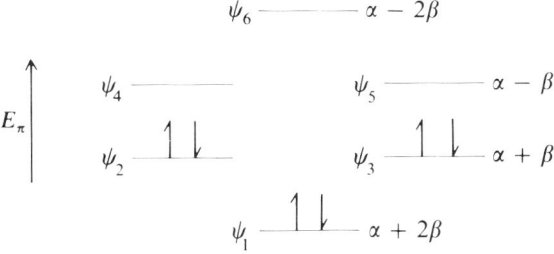

FIGURE 1.21 *The energy-level diagram for π energy for benzene.*

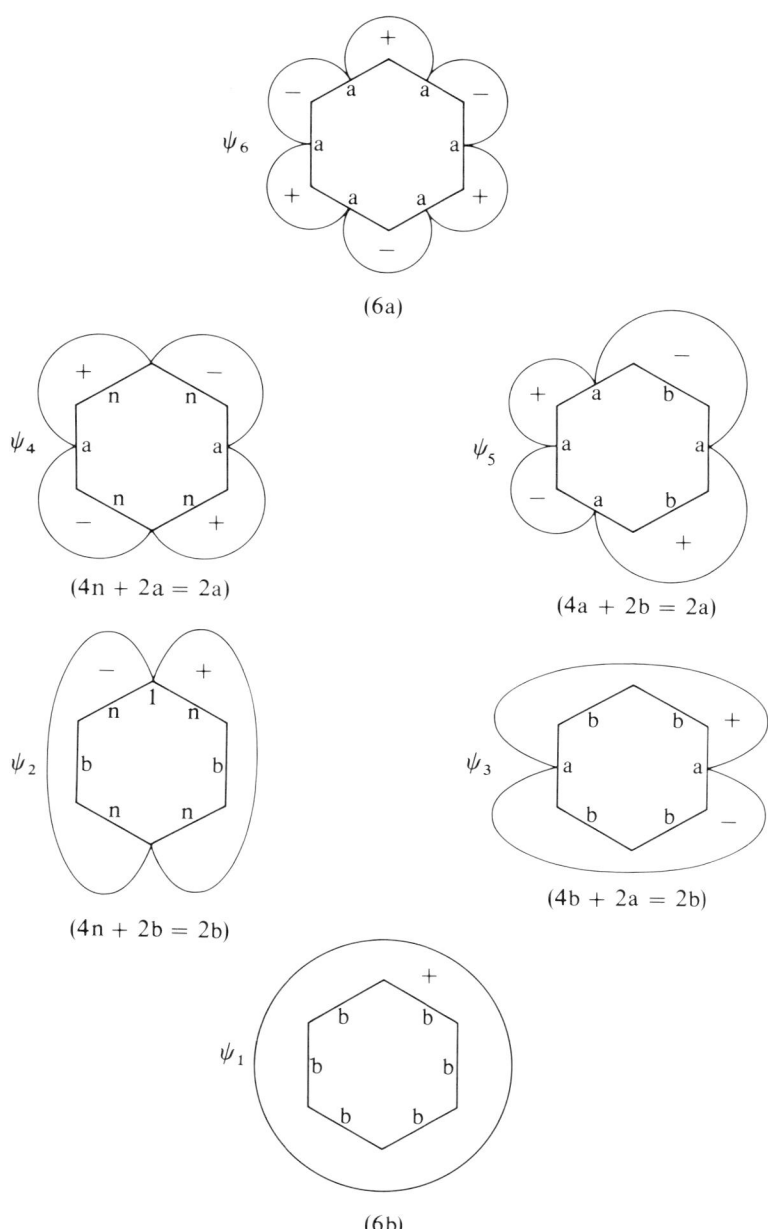

FIGURE 1.22 *Perimeter free-electron model representations of the benzene* MO's (b = *bonding,* n = *nonbonding,* a = *antibonding*).

character through simple addition. This assumes that an antibonding portion and a bonding portion of an orbital are equal in energy but opposite in sign, and that a nonbonding situation is energetically equivalent to an isolated atomic orbital ($E_\pi = \alpha$):

$$\text{Bonding} + \text{antibonding} = \text{nonbonding}$$

Because of the symmetry of the benzene molecule, individual carbon atoms as such in a given orbital are not uniquely bonding, nonbonding, or antibonding. For example, the carbon atom denoted as number 1 in ψ_2 of Fig. 1.22 is both nonbonding and bonding in the various symmetry-related equivalent representations of this orbital.

If this same Hückel technique is applied to cyclobutadiene, the secular determinant may be written as

$$\begin{vmatrix} x & 1 & 0 & 1 \\ 1 & x & 1 & 0 \\ 0 & 1 & x & 1 \\ 1 & 0 & 1 & x \end{vmatrix} = 0$$

and solved to give

$$x^4 - 4x^2 = 0$$
$$x = 0, 0, +2, -2$$
$$E = \alpha + 2\beta, \alpha, \alpha, \alpha - 2\beta$$

The energy-level diagram is shown in Fig. 1.23. Notice that use of the Pauli exclusion principle and Hund's rule leads to a prediction that cyclobutadiene is a square planar triplet with no delocalization energy (see Chapter 2-4).

Alternant Hydrocarbons. The symmetry of the energy levels about $E_\pi = \alpha$ in compounds such as ethylene (Fig. 1.17), butadiene (Fig. 1.19), benzene (Fig. 1.21), and cyclobutadiene (Fig. 1.23) suggests the possibility of a general phenomenon for certain types of π systems. Coulson and Lon-

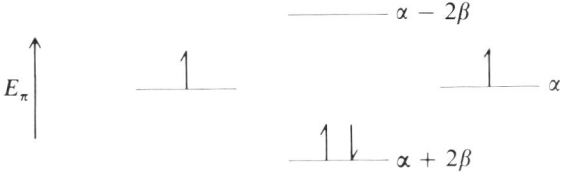

FIGURE 1.23 *The energy-level diagram for π energy for cyclobutadiene.*

guet-Higgins,[16,17] in one such generalization, have defined an *alternant hydrocarbon* (AH) as a planar conjugated system having no odd-membered rings, in which the carbon atoms may be divided into two sets such that each set has only members of the other set adjacent to it. Using stars (*) to represent one set, examples of alternant hydrocarbons are shown in Fig. 1.24. These hydrocarbons and radicals have the same number of π electrons on each carbon atom: one. All other planar conjugated systems are called nonalternant hydrocarbons.

Alternant hydrocarbons may be *even*, where the two sets either contain the same number of atoms (a, b, and c in Fig. 1.24) or differ by an even number of components; or they may be *odd*, where the two sets differ by an odd number of atoms (d and e). *For all alternant systems, the energy levels are symmetrically disposed about the zero energy level* ($E_\pi = \alpha$). The coefficients of the pair of molecular orbitals with energies $\alpha \pm x\beta$ are numerically the same, but differ in sign for one set of atomic orbitals (see Eq. 1.39 and 1.42). Even alternant hydrocarbons[18] possess no nonbonding energy levels (that is, no E_π's having a value of α) if the two sets contain the same number of carbon atoms,[19] and an even number of nonbonding

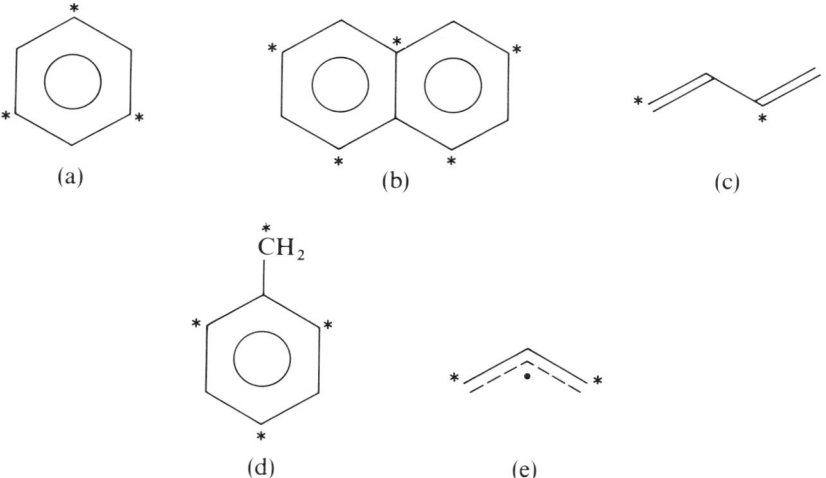

FIGURE 1.24   *Alternant hydrocarbons.*

---

17. Reference G, Chapter 2.8; Reference J, Chapters 4.3 and 5; Reference K, Chapter 6.
18. Simple Hückel theory works best for alternant hydrocarbons. All alternant hydrocarbons lead to an expression for the expansion of the secular determinant having only even or only odd powers of x; see references listed in note 17.
19. Cyclobutadiene is an exception. See the discussion of antiaromaticity in Chapter 2-4.

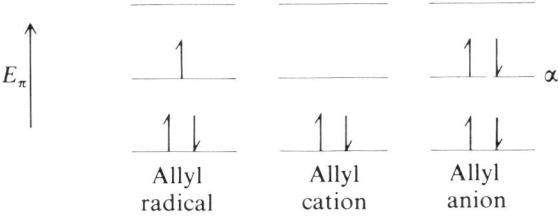

FIGURE 1.25 *The energy-level diagrams for π energy for the allyl radical, cation, and anion.*

energy levels if the two sets differ in number of components.[20] Odd alternant hydrocarbons possess an odd number of nonbonding energy levels.

If an odd alternant hydrocarbon is a radical, such as Fig. 1.24(d) or (e), the delocalization energies of this radical and of the corresponding cation and anion will be equal. As shown in Fig. 1.25 for the allyl radical (e), the electron removed from the radical to form the cation or added to the radical to form the anion is added or removed at the nonbonding energy level, $E_\pi = \alpha$. Since α is the energy of a localized electron, no change occurs in the delocalization energy. This approach to the delocalization energies of such odd alternant systems involves the critical approximation inherent in any approach using one-electron orbitals: that repulsions between electrons are not considered.

Odd alternant systems also exhibit the unique behavior that the coefficients of the atomic orbitals combined to form the nonbonding molecular orbital (NBMO) may be determined without solving the secular equation. If the odd alternant system is organized in such a way that the starred set has the greater number of components, the starred atoms may be assigned values in such a way that the sum of the starred atoms about any unstarred atom is zero, the *zero-sum rule*. As shown in Fig. 1.26 for the allyl system (Fig. 1.24e) and in Fig. 1.27 for the benzyl system (Fig. 1.24d), the starred numbers are squared, their summation is obtained, and the normalization requirement is satisfied. The values of the starred numbers

---

20. No classical structures can be written for this type of even alternant hydrocarbon, one example of which is

The converse of this statement is not true; see Reference K.

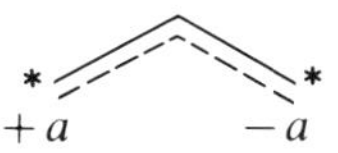

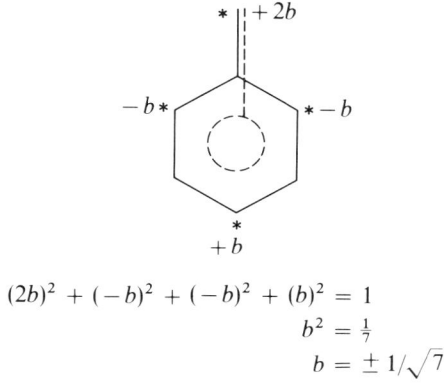

FIGURE 1.26 *The NBMO coefficients in the allyl system.*

$$(2b)^2 + (-b)^2 + (-b)^2 + (b)^2 = 1$$
$$b^2 = \tfrac{1}{7}$$
$$b = \pm 1/\sqrt{7}$$

FIGURE 1.27 *The NBMO coefficients in the benzyl system.*

obtained from this procedure are the corresponding NBMO coefficients. The squares of these coefficients with the appropriate signs correspond to the charge densities at each atom in the odd alternant hydrocarbon ions (see Eq. 1.47 and 1.48).[21]

The energy levels for cyclic hydrocarbons containing continuous π systems may also be calculated without solving the secular determinants.[22] The polygon is inscribed in a circle of radius 2β with a vertex at the lowermost point of the circle (see Fig. 1.28). The energy levels may be evaluated trigonometrically, one energy level per vertex.

A quick comparison of the molecular orbital approach to π systems with the valence bond approach to such molecules is informative. A VB

---

21. Other applications of these concepts of alternant character, such as in the simplification of secular determinants and in the evaluation of aromatic character, are discussed in Herndon and Silber, note 16, and by Riggs and Dewar in their monographs, References J and K.
22. A. Frost and B. Musulin, *J. Chem. Phys.* **21**, 572 (1953); see also H. Zimmerman, *Accounts Chem. Res.* **4**, 272 (1971).

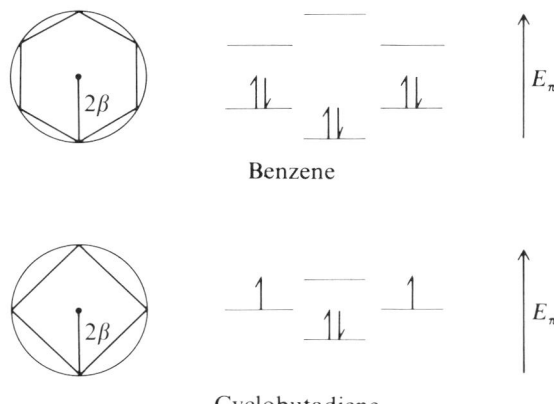

FIGURE 1.28 *The energy-level diagrams for π energy for benzene and cyclobutadiene, determined by the polygon-in-a-circle method.*[22]

analysis of benzene requires that relative importance be assigned to the five major localized structures (**I–V**). This problem of relative importance,

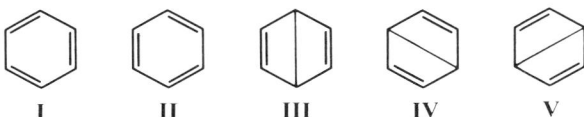

as well as the complexity of the calculations to be performed, quickly becomes prohibitive as the systems increase in size.

1-6 Calculations and Experimental Observations

The usefulness of any set of approximations can be evaluated only in terms of the correlation of the results of such calculations with the pertinent experimental observations. As we shall see, this correlation is often complicated by the interpretations inherent in the conversion of the actual physical observations into a chemically useful form.

The results from MO calculations hitherto presented require the evaluation of terms such as β in order to be placed into a numerical framework. More sophisticated MO techniques can lead to meaningful heats of formation,[23] but the simple Hückel approach provides only the delocalization

---

23. See Reference K.

energies of the π systems, which are defined relative to localized models. As will be discussed in Chapter 2, the experimental evaluation of these delocalization energies is full of problems.

Many properties should be related to the E_π energy-level diagram structure, particularly to the energies of the highest occupied molecular orbital (HOMO) and the lowest unoccupied molecular orbital (LUMO). The electron affinity (EA) of a molecule, for example, is defined as the energy liberated when an electron joins with the molecule in the gas phase to give a radical anion. The most stable radical anion would result when the added electron occupies the LUMO. While few electron affinities have been determined, the polarographic half-wave reduction potential provides a reasonable experimental alternative which may be obtained much more readily. Correlation of the reduction potential with the LUMO energy is very good for hydrocarbon systems.[24]

Similarly, the ionization potential (IP)—the energy required to remove an electron from a molecule in the gas phase—should correlate with the energy of the HOMO. The various methods used to measure the ionization potential (spectroscopy, electron impact, and photoionization) disagree with one another to some extent, primarily because each method involves some ambiguity as to the exact purely electronic energy levels required to evaluate the ionization potential accurately. Correlation of the ionization potential with the HOMO energy is only fair;[24] correlation of the polarographic half-wave oxidation potentials with the HOMO energies is somewhat better but still not entirely satisfying.[24] Apparently, the neglect of electron repulsion effects in Hückel MO calculations is the source of these difficulties. While Hückel MO predicts that all radicals of odd alternant hydrocarbons should have one electron in an NBMO and, therefore, the same ionization potential, different values are obtained experimentally for different radicals of this type.

Excitation of an electron from a filled molecular orbital to an unfilled one requires visible or ultraviolet light, whose energy may be determined by the readily available electronic spectra.[25] The lowest-energy excitation should be that of an electron from the HOMO to the LUMO and should correlate with the difference between these energy levels. However, excitation is not such a straightforward process. The σ–π separation approximation often breaks down; the geometry of the excited state may differ significantly from that of the ground state, introducing significant solvent effects; and the excitation may involve a change in vibrational level, thereby encompassing an additional energy term. Hückel MO

---

24. Reference G, Chapter 7.
25. Reference A, Chapter 10; Reference B, Section 1-18; Reference G, Chapter 8; H. H. Jaffé and M. Orchin, *Theory and Applications of Ultraviolet Spectroscopy*, Wiley, New York, 1962.

calculations ignore spin states and assume "center of gravity" energies somewhere between the values associated with the actual, physical singlet and triplet energy states.

Several *selection rules* govern the intensity and probability of electronic excitations:[25]

1. Many-electron excitations are forbidden.
2. Excitations between states differing in both electronic configuration and spin are forbidden; i.e., transitions from singlet to triplet do not occur.
3. Certain transitions between states of different symmetry (as analyzed by group theory)[25,26] are forbidden.

To say that an excitation is "forbidden" in character means that the excitation either is not observed or is of weak intensity. The well-known $n \to \pi^*$ band of carbonyl compounds, which has an ε of less than 200 in saturated systems, is an example of a forbidden transition.

The effects which substituents exert on an electronic excitation may often be predicted using empirical rules rather than by quantum chemical calculations. Woodward has formulated rules for predicting the position of maximum absorption for the longest-wavelength $\pi \to \pi^*$ transition of dienes in ethanol, while Woodward and the Fiesers have developed a set of similar rules for the same transition in conjugated unsaturated carbonyl compounds[27] (see Table 1.2). Cross-conjugated systems of various types (**VI** and **VII**) and enediones[28] (**VIII**) cannot be readily handled by any set of empirical rules. However, resolution into the various possible

$$\begin{array}{cc} \text{C} & \text{O} \\ \| & \| \\ \text{C}=\text{C}-\text{C}-\text{C}=\text{C}-\text{C}=\text{C} \quad & \text{C}=\text{C}-\text{C}-\text{C}=\text{C}-\text{C}=\text{C} \\ \textbf{VI} & \textbf{VII} \end{array}$$

$$\begin{array}{cc} \text{O} & \text{O} \\ \| & \| \\ -\text{C}-\text{C}=\text{C}-\text{C}- \\ \textbf{VIII} \end{array}$$

---

26. F. A. Cotton, *Chemical Applications of Group Theory*, Wiley, New York, 1963; H. H. Jaffé and M. Orchin, *Symmetry in Chemistry*, Wiley, New York, 1965.

27. H. H. Jaffé and M. Orchin, *Theory and Applications of Ultraviolet Spectroscopy*, Wiley, New York, 1962, Chapter 10; D. J. Pasto and C. R. Johnson, *Organic Structure Determination*, Prentice-Hall, Englewood Cliffs, N.J., 1969, Sections 3.5.1 and 3.5.2; J. B. Hendrickson, D. J. Cram, and G. S. Hammond, *Organic Chemistry*, 3rd ed., McGraw-Hill, New York, 1970, pp. 248–252.

28. J. A. Hirsch and A. J. Szur, *J. Heterocycl. Chem.* **9**, 532 (1972), and references therein.

TABLE 1.2 *Empirical Rules for* $\pi \to \pi^*$ *Absorption Maxima*[27]

System	Structure[a]	λ_{max}[b]	$\alpha = R$[c]	Increment[b] $\beta = R$[c]	$\gamma, \delta = R$[c]
Diene	$\overset{\alpha}{\underset{\alpha}{>}}C=C-C=C\overset{\alpha}{\underset{\alpha}{<}}$ with α, α on middle C's	217[d,e,f]	5	—	—
α,β-Unsaturated ketone	$\overset{\beta}{\underset{\beta}{>}}C=C-C-R$ with α, O	215[d,e]	10	12	—
Dienone	$\overset{\delta}{\underset{\delta}{>}}C=C-C=C-C-R$ with γ, β, α, O	245[d,e]	10	12	18

[a] *s-trans* stereochemistry implied.
[b] In nanometers in ethanol.
[c] R = alkyl group or ring residue; see sources listed in footnote 27 for examples of other substituents.
[d] If the conjugation is extended by C=C, add 30 nm; if a C=C is exocyclic to a 5- or 6-membered ring or endocyclic in a 5- or 7-membered ring, add 5 nm.
[e] If the diene is homoannular, add 39 nm.
[f] 214 nm for steroids.

linearly conjugated fragments and consideration of that fragment which would be expected to absorb at the longest wavelength are often useful, particularly for systems resembling **VI**. Whenever the double bonds are not more or less planar so that delocalization is maximal, all of the empirical rules will fail. Solvent effects (Chapter 9-5) must be included whenever the solvent is not ethanol.

Wherever energy correlations between quantum chemical calculations and experimental observations are noted, problems inherent in each energy criterion in each system must be proportionately related. The correlation is usually more a fortunate accident than an unambiguous relationship.

If an LCAO–MO is written as

$$\psi_j = c_{j1}\chi_1 + c_{j2}\chi_2 + \cdots + c_{jr}\chi_r \tag{1.44}$$

we may define the *electron density* at atom r in molecular orbital j as

$$n_j c_{jr}^2$$

where n is the number of electrons in this molecular orbital j. The *total electron density* at atom r, q_r, may be obtained by summing these orbital electron densities over all the occupied MO's:

$$q_r = \sum_{j=1}^{\text{occupied}} n_j c_{jr}^2 \tag{1.45}$$

For all neutral alternant hydrocarbons (including radicals), the electron density at each atom will be unity. Since each atom in such systems has the same electron density, there can be no dipole moment since there is no charge separation (see Chapter 1-2). For nonalternant hydrocarbons, the unequal distribution of the electron densities leads to the existence of a dipole moment.

Since total electron densities are positive numbers, it is often more convenient to discuss *charge densities*,

$$\xi_r = 1 - q_r \tag{1.46}$$

Charge densities more accurately demonstrate electron-richness or electron-poorness as they correspond in sign to the electrical charge on the atoms being considered. For alternant hydrocarbons with an NBMO, relative charge densities are determined entirely by the coefficients of the NBMO. For an odd alternant hydrocarbon anion,

$$\xi_r = -(c_{\text{NBMO}})_r^2 \tag{1.47}$$

while for the corresponding cation

$$\xi_r = +(c_{\text{NBMO}})_r^2 \tag{1.48}$$

Using the allyl system (Fig. 1.26), Eq. 1.47 suggests that the allyl anion be written with half of a negative charge at each of the terminal carbon atoms, and Eq. 1.48 suggests that the allyl cation be written with half of a positive charge on the same terminal atoms. Note that this corresponds to the usual picture of a resonance hybrid midway in character between the two equally important resonance contributions:

$$\overset{x}{C}H_2-CH=CH_2 \longleftrightarrow CH_2=CH-\overset{x}{C}H_2$$

$$\overset{\delta\ x}{C}H_2-CH-\overset{\delta\ x}{C}H_2$$

For the benzyl system (Fig. 1.27), in either the cation or the anion less than half of the charge is delocalized into the *ortho* and *para* positions of the benzene ring:

$$\begin{array}{c} 4/7\ x \\ CH_2 \\ | \\ 1/7\ x \bigcirc 1/7\ x \\ 1/7\ x \end{array}$$

In a radical species, the distribution of the unpaired electron, the *spin density*, parallels that of the charge in the analogous charged species.

For example, the unpaired electron in the benzyl radical would be expected to be three-sevenths delocalized into the *ortho* and *para* positions of the benzene ring.[29] Molecular orbital theory permits prediction not only of these spin densities[30] but also of electron spin resonance hyperfine spin couplings in aromatic radical ions. McConnell[30] has proposed that the hyperfine splitting induced in aromatic radical ions by the interaction of a benzenoid hydrogen with an unpaired π-electron spin density is proportional to the spin density on the carbon atom to which the given hydrogen is bonded. The spin density calculated by quantum chemical techniques can, in fact, be directly related to the experimentally observable hyperfine coupling strength with reasonable success. Whenever Hückel MO theory is employed, the symmetry inherent in the highest bonding orbital and the lowest antibonding orbital of a given alternant species leads to a prediction that the spin densities and hyperfine splittings in a positive aromatic radical ion and the corresponding negative species will be identical. Deviations from this prediction are often observed,[30] since the Hückel method neglects interactions between π electrons and does not permit negative spin densities, both of which are physically logical phenomena. More sophisticated calculations[30] often produce excellent correlations between the experimental and calculated hyperfine splittings, and may therefore be used with some confidence in evaluating spin densities.

For nonalternant hydrocarbons, the unequal distribution of electron densities permits calculation of their dipole moments. However, Hückel MO theory invariably predicts too large a dipole moment. Presumably, the σ electrons create a dipole moment opposing that resulting from the π electrons, an instance where the assumption of σ–π separation becomes unsatisfactory.

Electron density or charge density may be used to define various other quantities[31] related to rates of reactivity and/or positions of attack by a given type of reagent. While such indices are often qualitatively useful for internal comparisons, such as in the prediction of the favored position of electrophilic attack in an aromatic system, they are of limited utility for either intermolecular comparisons or quantitative analyses. Too many chemists have adopted various criteria to justify or explain their results without sufficiently questioning the limitations of these criteria. Proper criteria properly used are of considerable assistance in the analysis of many types of phenomena.

---

29. Reference J, pp. 160–162.
30. J. D. Memory, *Quantum Theory of Magnetic Resonance Parameters*, McGraw-Hill, New York, 1968.
31. Reference B, Section 1-12 and 1-14; Reference G, Chapters 6 and 11–14; Reference J, Chapters 4–6; Reference K, Chapter 8.

Another important quantity is the *mobile π-bond order*, p_{rs}. For the LCAO–MO of Eq. 1.44 and for adjacent atoms r and s,

$$p_{rs} = \sum_{j=1}^{\text{occupied}} n_j c_{jr} c_{js} \qquad (1.49)$$

These π-bond orders range from 0 to 1.00 for olefinic and aromatic compounds in bonding situations. A negative bond order denotes an antibonding situation. The total bond order may be defined as the π-bond order plus 1, since there must be one and only one σ bond between two atoms. For butadiene, the total C—C bond orders calculated from Eq. 1.39 and 1.40 are 1.89 for the end bonds and 1.45 for the central bond. In VB terminology, the central bond has partial double-bond character and the end bonds have a double-bond character less than that associated with an isolated double bond.

Since bond order should be related to bond strength, π-bond orders should correlate with bond lengths, a higher bond order corresponding to a shorter bond length. This type of correlation has been particularly effective with aromatic hydrocarbons.[32] Bond orders can indicate a significant amount of "partial bond fixation" in aromatic systems (in agreement with an analysis based on resonance contributors), and therefore correlate with relative reactivities toward double-bond reagents[33] such as ozone or dienes.

Nonaromatic systems provide more ambiguity in bond order–bond length correlations.[32] Is the central bond in butadiene 1.46 Å because it has partial double-bond character or because it is a single bond connecting two sp^2 carbon atoms? An sp^2–sp^2 single bond is stronger and shorter than an sp^3–sp^3 single bond (see discussion of Fig. 1.16), but how much stronger and shorter is open to question.

---

32. M. J. S. Dewar and H. N. Schmeising, *Tetrahedron* **11**, 96 (1960); Reference B, Section 1-12; Reference F, Chapter IX; Reference G, Chapter 6.7; Reference H, Chapter 3; Reference J, Chapter 4.2; and Reference K, Chapters 3.9 and 5.3–5.9.
33. Reference G, Chapter 14.

Chapter 2

Aromatic Character

One of the quantities obtained from quantum mechanical calculations is the *delocalization energy* (DE), the π-energy difference between the system treated as if the π electrons were delocalized and the system considered as if the double bonds were isolated and noninteracting with each double bond contributing $2\alpha + 2\beta$ to the total energy (see Chapter 1-5). Strictly speaking, the delocalization energy is a calculated stabilization energy relative to an analogous system with identical geometry but without electron delocalization, while the *resonance energy* is an energy postulated because an experimental result suggests that the compound cannot be adequately represented by a single Lewis structure.[1,2] The experimental result may be thermodynamic, kinetic, or spectroscopic, as we shall see, but only thermodynamic measurements can provide quantitative "energies".

To review some fundamental thermochemical definitions, the *bond energy* is defined as the enthalpy required to split a molecule into its component atoms in the gas phase at 1 atmosphere and 298°K. It may be

---

1. For simple treatments of resonance see R. T. Morrison and R. N. Boyd, *Organic Chemistry*, 3rd ed., Allyn and Bacon, Boston, 1973, Chapters 6, 8, and 10; J. B. Hendrickson, D. J. Cram, and G. S. Hammond, *Organic Chemistry*, 3rd ed., McGraw-Hill, New York, 1970, Chapter 5; T. A. Geissman, *Principles of Organic Chemistry*, 3rd ed., W. H. Freeman, San Francisco, 1968, Chapters 10 and 19.
2. (a) For treatments equivalent to that herein see Reference A, Chapter 7; Reference G, Chapters 9 and 10; and Reference K, Section 5.6. (b) For a general approach to this chapter not discussed herein, see M. J. Goldstein and R. Hoffmann, *J. Amer. Chem. Soc.* **93**, 6193 (1971).

determined spectroscopically or thermochemically.[3] The "bond energy of methane" implies the reaction

$$CH_4(g) \longrightarrow 4H(g) + C(g)$$

and is the average value of the four successive bond dissociation steps shown in Fig. 2.1, each of which has its own characteristic *bond strength*, or *bond dissociation energy* (BDE). Although bond dissociation energies have the greatest practical value for organic chemists, they are far more difficult to obtain than bond energies.[3] Therefore, bond energies are used as admittedly poor approximations when bond dissociation energies are desired. However, since alternative reaction paths often differ by only a few kcal/mole, bond energies are often unsatisfactory approximations.

The *heat of formation*[4] of a compound is regarded as the sum of the bond energies, or, more specifically, as the enthalpy difference between the compound and its component atoms in the standard gaseous state at 298°K. Isomers, such as *n*-butane and isobutane, do not have experimentally identical heats of formation, suggesting that prediction of heats of formation by addition of bond energies is unsatisfactory unless environmental dependencies are included in the evaluation of the bond energies.

Reaction	Bond Dissociation Energy (kcal/mole)
$CH_4 \rightarrow CH_3\cdot + H\cdot$	104
$CH_3\cdot \rightarrow [CH_2]: + H\cdot$	104
$[CH_2]: \rightarrow [CH]: + H\cdot$	108
$[CH]: \rightarrow C + H\cdot$	81
Sum: $CH_4 \rightarrow C + 4H$	397
Bond energy $= \frac{397}{4} =$ 99 kcal/mole	

FIGURE 2.1 *The relationship between the bond dissociation energies and the bond energy in methane.* [*Values suggested by J.A. Kerr, Chem. Rev.* **66**, 465 (1966).]

---

3. T. L. Cottrell, *The Strengths of Chemical Bonds*, Butterworth and Co., London, 1958; M. Szwarc, *Quart. Rev.* **5**, 22 (1951); S. W. Benson, *J. Chem. Educ.* **42**, 502 (1965); J. A. Kerr, *Chem. Rev.* **66**, 465 (1966).
4. Reference B, Section 2-4.

The *heat of combustion* is the energy involved in converting the compound into completely oxidized components upon treatment with oxygen; e.g., complete oxidation of a hydrocarbon in the gas phase and standard state to carbon dioxide in the gas phase and water in the liquid phase at 1 atmosphere and 298°K. Similarly, the *heat of hydrogenation* is the enthalpy liberated in the conversion of an unsaturated compound in the standard state to the corresponding saturated system by the action of molecular hydrogen over a catalyst, again requiring the gas phase and standard state for all reactants and products.

2-1 Resonance Energies

Any of the three types of heat quantities defined above—heat of formation, heat of combustion, or heat of hydrogenation—may be used to evaluate the resonance energy of an organic system,[2] provided some method exists for the evaluation of the same type of heat quantity for the hypothetical system with the same structure but without electron delocalization. The major problem in such an approach to resonance energy determination is the evaluation of the energy of this hypothetical reference system. Tables of group contributions[2,3] to heats of combustion or heats of formation may be used to assemble the energy of the hypothetical localized structure, or heats of hydrogenation of simpler molecules may be combined to create the required model (also see Chapter 2-4).

Using benzene as an example, "cyclohexatriene" made up of three cyclohexene-like double bonds may be used as the localized model. The hydrogenation of cyclohexene to cyclohexane liberates 28.6 kcal/mole, so this "cyclohexatriene" should liberate 3×28.6 (or 85.8) kcal/mole. Benzene itself liberates only 49.8 kcal/mole on complete hydrogenation to cyclohexane. The smaller observed heat of hydrogenation shows benzene to be 36.0 kcal/mole more stable than the hypothetical localized "cyclohexatriene" structure.[5]

A statement that benzene contains 36 kcal/mole of resonance energy is only as good as the hypothetical localized structure used in the comparison. The "cyclohexatriene" used as the reference would have alternating single and double bonds, and would not really correspond to a Kekulé structure. The difference in enthalpy between benzene and "cyclohexatriene" must therefore include not only the energy gained as the result of electron de-

---

5. A lower heat of hydrogenation indicates a more stable compound only if the hydrogenation product is identical for those compounds being compared. *trans*-2-Butene has a lower heat of hydrogenation than isobutylene, but no conclusion may be drawn as to the relative stability of these alkenes because *n*-butane and isobutane, the respective hydrogenation products, are dissimilar.

localization but also the energy required to stretch the double bonds and compress the single bonds in the "cyclohexatriene" to obtain the regular hexagon of benzene with its equal bond lengths. The 36 kcal/mole value is therefore called an *empirical resonance energy* (Fig. 2.2). Coulson[6] estimates the distortion energy at 27 kcal/mole, leaving a *vertical resonance energy* of 63 kcal/mole, which should correspond to the actual delocalization energy. However, Dewar[2,7] has pointed out that this analysis is still incomplete. When cyclohexene is hydrogenated, not only is an sp^2–sp^2 C=C bond converted to an sp^3–sp^3 C—C, but two sp^3–sp^2 C—C bonds become sp^3–sp^3 C—C bonds. In the hydrogenation of benzene, every σ bond is converted from sp^2–sp^2 to sp^3–sp^3. Since there is no reason to assume similar energies for sp^2–sp^2, sp^2–sp^3, and sp^3–sp^3 single bonds, and since the hybridization changes in the hydrogenation of benzene are different from those postulated for "cyclohexatriene" or a single Kekulé

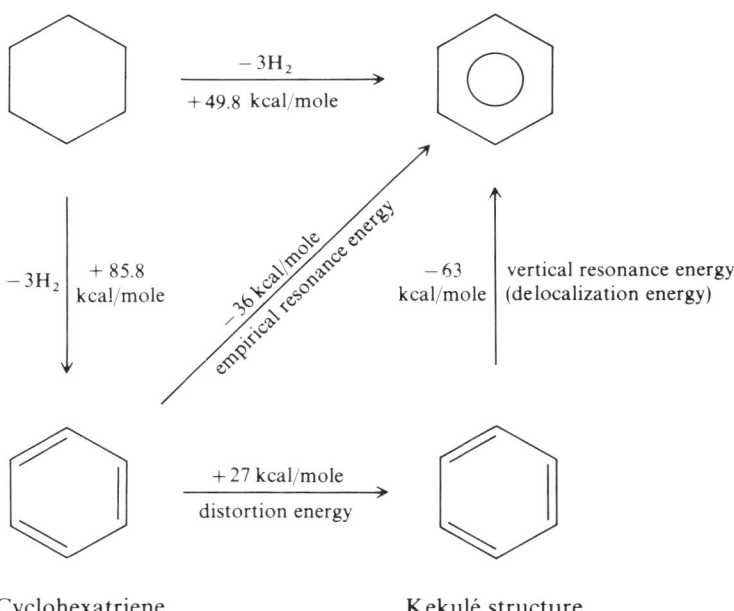

FIGURE 2.2 *Resonance energy in benzene.*

---

6. C. A. Coulson and S. L. Altmann, *Trans. Faraday Soc.* **48**, 293 (1952); discussed in Reference F, Chapter IX, or Reference G, Chapter 9.
7. M. J. S. Dewar and H. M. Schmeising, *Tetrahedron* **11**, 96 (1960); and Reference K, Section 5.6.

structure of benzene, various hybridization energies must be introduced as adjustments in the vertical resonance energy. Dewar finally arrives at a value of 20 kcal/mole for the "resonance energy of benzene" (see also the discussion of Dewar resonance energy in Chapter 2-4).

Having now considered the problems inherent in quantitative evaluation of the resonance energy, it should be apparent that resonance energies cannot be equated to Hückel calculated delocalization energies and used to assign a numerical value to β—just as we predicted they could not in Chapter 1-6. Literature values of β range from 10 to 60 kcal/mole.[2,7,8] None of these values has real physical significance, nor can it even be argued that β must be a constant for all chemical systems.

However large the resonance energy assigned to benzene, thermochemical stabilization as the result of electron delocalization is much greater for benzenoid hydrocarbons than for conjugated systems such as butadiene. In the simplest evaluation using heats of hydrogenation, the resonance energy of butadiene is found to be 3.5 kcal/mole when that of benzene is 36 kcal/mole.

It is important to realize that resonance energy is not an energy as such. It merely explains a difference between an observed result and a hypothetical expectation.

2-2 Aromaticity and the "$4n + 2$ Rule"

An *aromatic* compound usually is defined (casually) as benzene or an organic compound resembling benzene.[1,2] The problem in such a definition lies in the choice of the criteria[2,9] on which to base this resemblance. Early definitions of aromaticity centered on kinetic stability toward reagents that attack double bonds, coupled with a preference for substitution reactions over addition reactions. Later definitions relied on thermochemical stability as measured by the size of the resonance energy, a procedure full of problems (see Chapter 2-1). More recently, spectroscopic and magnetic criteria have been advocated. Magnetic anisotropy can be induced in planar π-electron systems and then detected by the shift to lower fields in the proton magnetic resonance (pmr) spectrum[10] or by the exaltation in

---

8. F. A. Cotton, *Chemical Application of Group Theory*, Wiley, New York, 1963, pp. 129–131 and Appendix IV.
9. A. J. Jones, *Rev. Pure Appl. Chem.* **18**, 253 (1968); R. D. Breslow, *Chem. & Eng. News* 90 (June 28, 1965); G. M. Badger, *Aromatic Character and Aromaticity*, Cambridge University Press, New York, 1969; D. Lloyd and D. R. Marshall, *Angew. Chem. Intern. Ed. Engl.* **11**, 404 (1972).
10. L. M. Jackman, *Applications of Nuclear Magnetic Resonance Spectroscopy in Organic Chemistry*, Pergamon, New York, 1959; J. A. Pople, W. G. Schneider, and H. J. Bernstein, *High-Resolution Nuclear Magnetic Resonance*, McGraw-Hill, New York, 1959.

the diamagnetic susceptibility measurement.[11] The π-electron current also produces electronic spectra significantly different from the spectra exhibited by simple conjugated alkenes.[12] The ultimate physical criterion is the observation of equal bond lengths and coplanarity for the entire aromatic system, an analysis requiring x-ray crystallography, microwave spectroscopy, or electron diffraction techniques.

Molecular orbital calculations suggest another criterion for aromatic character.[2] The occupancy of the orbitals appears to have more fundamental significance than the magnitude of the delocalization energy. An energy-level diagram for the π energy of an aromatic compound will exhibit completely filled bonding orbitals and either completely filled or completely empty nonbonding orbitals, creating a *closed-shell* or *filled-shell* system. As shown in Fig. 1.21 and 1.23, benzene would be aromatic by this criterion and cyclobutadiene would not be.[13]

Ideally, the organic chemist would like to be able to evaluate aromaticity without performing either an experiment or a calculation. This is possible using the Hückel $4n + 2$ *rule*,[2] developed in the early 1930s. In its most general form, the $4n + 2$ rule states:

> Those monocyclic coplanar systems of trigonally hybridized atoms which contain a quantity $4n + 2$ of π electrons, where n is an integer, will exhibit aromatic character.

In reality, this number $4n + 2$ of π electrons corresponds to a closed-shell system in a Hückel MO calculation on an alternant hydrocarbon. Table 2.1 lists some examples of systems which obey the $4n + 2$ rule.[2,9,14]

Those compounds in Table 2.1 with an n-value of 0 or 1 other than the cyclobutenyl dication have been prepared and extensively studied. However, problems arise when n has a value of 2. Both the cyclooctatetraenyl dianion and the cyclononatetraenyl anion have been prepared and appear to be aromatic by several criteria. Cyclodecapentaene[15] can exist as various

---

11. H. J. Dauben, Jr., J. D. Wilson, and J. L. Laity, *J. Amer. Chem. Soc.* **91**, 1991 (1969), and **90**, 811 (1968); and *Nonbenzenoid Aromatics*, vol. 2, ed. by J. P. Snyder, Academic Press, New York, 1971, p. 167.
12. H. H. Jaffé and M. Orchin, *Theory and Applications of Ultraviolet Spectroscopy*, Wiley, New York, 1962.
13. Cyclooctatetraene is an outstanding example of a system which, if planar, would have half-filled NBMO's and a large delocalization energy. Experimentally (see Chapter 2-4), it exists in a nonplanar form and is not aromatic. For a discussion of such systems, see Reference J, Chapter 5.4, or Reference G, Chapter 10.7.
14. For a general discussion of compounds which do and do not obey this rule, see P. J. Garratt and M. V. Sargent, *Advan. Org. Chem.* **6**, 1 (1969), and *Nonbenzenoid Aromatics*, vol. 2, ed. by J. P. Snyder, Academic Press, New York, 1971, p. 208.
15. T. L. Burkoth and E. E. van Tamelen, "The Cyclodecapentaene Problem," in *Nonbenzenoid Aromatics*, vol. 1, ed. by J. P. Snyder, Academic Press, New York, 1969; E. E. van Tamelen, et al., *Chem. Commun.* 601 (1971), and *J. Amer. Chem. Soc.* **93**, 6111, 6120 (1971); S. Masamune, et al., *J. Amer. Chem. Soc.* **93**, 4966 (1971), and *Accounts Chem. Res.* **8**, 272 (1972).

TABLE 2.1 *Compounds Satisfying the $4n + 2$ Rule*

π Electrons	n	Structure	Name
2	0	△ (+)	Cyclopropenyl cation
		□ (+ +)	Cyclobutenyl dication
6	1	□ (−−)	Cyclobutadienyl dianion
		⬠ (−)	Cyclopentadienyl anion
		⬡	Benzene
		⬣ (+)	Cycloheptatrienyl cation or Tropylium ion
10	2	⯃ (− −)	Cyclooctatetraenyl dianion
		(9-gon) (−)	Cyclononatetraenyl anion
		(10-gon)	Cyclodecapentaene or [10]-Annulene
14	3	[14-annulene shape]	[14]-Annulene

geometrical isomers, any, all, or none of which might be aromatic. For reasons of this type, the term *annulene*[14,16] is used to describe macrocyclic hydrocarbons formally consisting of alternating single and double bonds. Unsubstituted [10]-annulenes have been prepared recently.[15] The instability of the parent [10]-annulenes is ascribed to ease of thermal cyclization to bicyclic systems and, in one isomer (**I**), to crowding between the internal hydrogens (circled in **I**). This crowding may be relieved by bridging

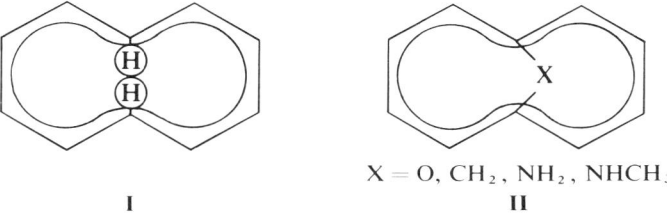

$X = O, CH_2, NH_2, NHCH_3$

I   II

the two positions in question, effectively replacing two hydrogens with one atom to give 1,6-bridged-[10]-annulenes (**II**). Several of these bridged compounds[14,15] have been prepared by Vogel and Sondheimer and their coworkers and found to be aromatic.

The [14]-annulene ($n = 3$)[14,16] has been prepared and, surprisingly at first glance, has been found *not* to be aromatic.[17] For the system to be planar, severe interactions would have to take place between two pairs of hydrogens (circled in **III**). As with [10]-annulene, these nonbonded interactions can be removed by replacing the interacting hydrogens with a bridge. The resulting dihydropyrenes have been prepared by Boekelheide and his

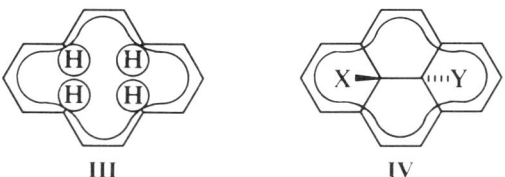

III   IV

group[14,17] with a variety of *trans*-15,16 substituents (**IV**), and all satisfy the various criteria for aromatic character.

The [18]-annulene ($n = 4$),[14,16] however, does appear to be aromatic. Crystallographic study shows that the ring is almost planar and that there is no alternation from single to double bonds in a successive manner around

---

16. F. Sondheimer, *Proc. Roy. Soc.* **A297**, 173 (1967); *Accounts Chem. Res.* **5**, 81 (1972); *Pure Appl. Chem.* **28**, 331 (1971); R. M. McQuilkin and F. Sondheimer, *J. Amer. Chem. Soc.* **92**, 6341 (1970); B. W. Metcalf and F. Sondheimer, ibid., **93**, 5271, 6675 (1971).
17. A low-temperature pmr study suggests some aromaticity—see references in note 16. See also E. Vogel, *Pure Appl. Chem.* **28**, 353 (1971), and C. C. Chiang and I. C. Paul, *J. Amer. Chem. Soc.* **94**, 4741 (1972).

the ring. However, all the bonds are not exactly equal in length and tend to group into two categories, neither of which corresponds to pure single or double bonds. The heat of combustion and the pmr behavior both suggest aromatic character. The only known [22]-annulene ($n = 5$) also exhibits aromatic characteristics.[16]

Various [26]- and [30]-annulenes ($n = 6$ and 7, respectively)[14,16] have been synthesized. Some of the [26]-annulenes are planar, while others and the [30]-annulenes are nonplanar with alternating single and double bonds and little or no π-electron current, suggesting that a large ring has difficulty achieving a sufficiently effective electron delocalization to be considered an aromatic compound.

The $4n + 2$ approach to various aromatic systems points out another weakness in VB theory. One of the tenets of the simplest VB approach is that the more equivalent resonance contributors that can be drawn, the more stable the resonance hybrid will be. Comparing the equilibria in Eq. 2.1 and 2.2, we find that there are seven equivalent resonance contributors for

$$\text{cycloheptatriene} \rightleftharpoons \text{V} + H^+ \quad (2.1)$$

$$\text{cyclopentadiene} \rightleftharpoons \text{VI} + H^+ \quad (2.2)$$

anion **V** and only five for anion **VI**. The VB approach would suggest greater stability for **V** than for **VI** relative to the hydrocarbons, leading to a prediction that cycloheptatriene is more acidic than cyclopentadiene. Experimentally, the pK_a of cycloheptatriene is greater than 37, and the pK_a of cyclopentadiene is about 15. The greater acidity of the cyclopentadiene may be quickly appreciated using the $4n + 2$ rule.

A statement of Hückel's $4n + 2$ rule includes only monocyclic systems. However, similar arguments may be applied with some success to polycyclic systems by counting only the peripheral (planar) electrons. For example, azulene may be treated as a compound containing ten π electrons, and it does, in fact, exhibit significant aromatic character.[18] On the other

---

18. Azulene may also be viewed as a cyclopentadienyl anion fused to a tropylium cation:

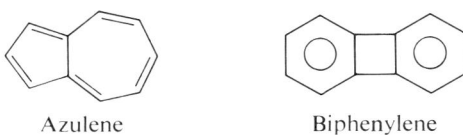

Azulene   Biphenylene

hand, biphenylene should be looked on as two independent six-π benzenoid systems joined by carbon–carbon bonds, with the possibility of unusual behavior resulting from a cyclobutadiene-like central ring being recognized as well. The third possible aromatic arrangement is that exhibited by the polynuclear aromatic hydrocarbons such as naphthalene or phenanthrene.[19a,b] In these systems the π systems overlap in such a way that the π cloud is not limited to just one ring. For example, the π cloud in naphthalene may be viewed as a large doughnut with two holes.

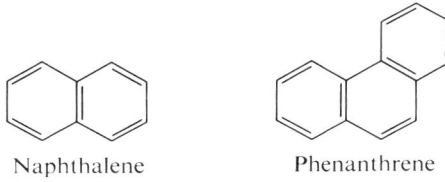

Naphthalene   Phenanthrene

2-3 Heteroaromaticity, Homoaromaticity, and Bicycloaromaticity

The basic concept of aromatic character can be extended in several different ways. The most common extension is *heteroaromaticity*,[20] involving similar systems but removing the restriction that the compounds be hydrocarbons. Heteroaromatic compounds may be divided into two general classes: those compounds which utilize a lone pair of electrons from the heteroatom in the aromatic π system, and those which do not. Examples of the first class are furan, pyrrole, and thiophene. In these compounds, a

---

19. (a) See R. T. Morrison and R. N. Boyd, *Organic Chemistry*, 3rd ed., Allyn and Bacon, Boston, 1973, Chapter 30; L. F. Fieser and M. Fieser, *Advanced Organic Chemistry*, Reinhold, New York, 1961, Chapter 27, and *Topics in Organic Chemistry*, Reinhold, New York, 1963, Chapter 1. (b) A circle in a ring will be used to represent an electron current. For this reason, a single Kekulé structure will be used whenever electron currents are not continuous in a linear manner.
20. R. T. Morrison and R. N. Boyd, *Organic Chemistry*, 3rd ed., Allyn and Bacon, Boston, 1973, Chapter 31; J. S. Hendrickson, D. J. Cram, and G. S. Hammond, *Organic Chemistry*, 3rd ed., McGraw-Hill, New York, 1970, Chapter 24; L. F. Fieser and M. Fieser, *Topics in Organic Chemistry*, Reinhold, New York, 1963, Chapter 2; and the various monographs on heterocyclic compounds such as those by Acheson, Badger, Katritzky and Lagowski, Palmer, and Paquette.

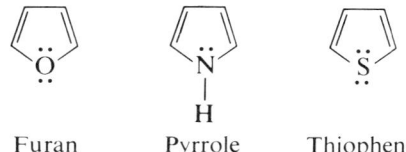

Furan   Pyrrole   Thiophene

lone pair from the oxygen, nitrogen, or sulfur and the four π electrons from the diene portion combine to give a six-π $4n + 2$ delocalized system. The amount of aromatic character exhibited by these compounds is directly related to the energies of the orbitals containing the lone pairs and the p orbitals constituting the π system of the diene. These are the orbitals that must be combined to give the delocalized π system. Experimentally, thiophene exhibits greater aromatic character than pyrrole, with furan exhibiting the least.

Some examples of the second class of heteroaromatics are pyridine and the pyrylium cation. In pyridine, the lone pair of electrons on the

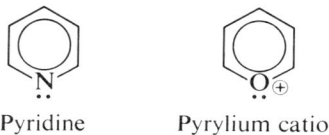

Pyridine   Pyrylium cation

nitrogen atom are not involved in the aromatic stabilization.

Molecular orbital calculations are difficult for heteroaromatic systems, because new values of α and β must be defined for the heteroatom.[21] Inductive effects[21] also complicate comparisons of calculations with experimental observations in heteroatomic systems.

A newer, somewhat more complex concept is *homoaromaticity*[2b,22]. A homoaromatic system obeys the Hückel $4n + 2$ rule but possesses additional atoms which break the physical continuity of the delocalized π system without affecting the delocalized system as such. The additional atoms neither participate in the delocalized π system nor prevent it from existing. One simple example is the monohomotropylium ion (**VII**), in which the intervening group is a methylene group outside the plane defined

VII   VIII   IX

---

21. Reference G, Chapter 5; Reference H, Chapter 6; Reference J, Chapters 4 and 6; Reference K, Chapter 9.
22. S. Winstein, *Quart. Rev.* **23**, 141 (1969).

by the delocalized six-π seven-carbon system. The designation of a compound such as **VII** as monohomoaromatic has no relation to the number of atoms inserted between any two atoms which are part of the delocalized system. It merely specifies the number of sides in which the σ backbone of the delocalized π system is broken by intervening atoms. For example, the simplest trishomocyclopropenyl cation would possess structure **VIII** and would be the trimethylene example of the more general trishomocyclopropenyl cation **IX**. Homoaromatic compounds exhibit all of the properties associated with aromaticity, such as additional stability and the ability to sustain a ring current.

The monohomotropylium ion may be generated from cyclooctatetraene in strong acid, while the trishomocyclopropenyl cation was first postulated from an analysis of the solvolytic behavior of *cis*-bicyclo[3.1.0]hex-3-yl tosylate (**X**). The perhomocyclopentadienide anion (**XI**) is not currently known, and present evidence does not support a homoaromatic character for perhomobenzene (**XII**), suggesting instead the localized form, **XIII**.

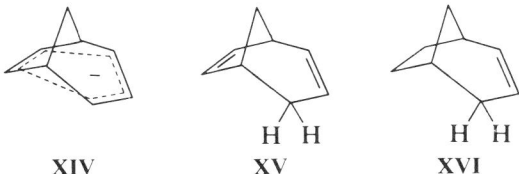

| X | XI | XII | XIII |

However, the bishomocyclopentadienide (**XIV**) has been prepared and behaves as expected for a homoaromatic system. This bicyclooctadienyl

anion is unusually stable and exhibits an appreciable ring current. The precursor diene (**XV**) exchanges the indicated hydrogens $10^{4.5}$ times faster than the monoalkene (**XVI**).

Goldstein[2b,23] has extended the concept of aromaticity into three dimensions. Consider a bishomoaromatic system such as the bishomocyclopentadienide anion (**XIV**). Inclusion of an additional π system from the remaining bridge would convert this bishomoaromatic ion into the bicyclo[3.2.2]nonatrienyl anion (**XVII**), which might exhibit enhanced

---

23. M. J. Goldstein, *J. Amer. Chem. Soc.* **89**, 6357 (1967); also see the discussion in *Ann. Repts.* **64B**, 138 (1967).

stability relative to the reference ion with the same number of trigonal carbons and π electrons. From simple MO calculations, Goldstein[23] pre-

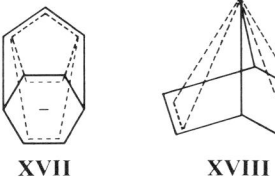

XVII   XVIII

dicted that ions with $4n$ π electrons such as **XVII** and **XVIII** will possess such enhanced stability, which he termed *bicycloaromaticity*, and that systems with $4n + 2$ π electrons will not show such aromatic character—which is the exact opposite of the Hückel $4n + 2$ rule for monocyclic compounds. Winstein[24] prepared the bicyclo[3.2.2]nonatrienyl anion and reported that it exhibits delocalized aromatic character. However, structure **XVIII**, in which both double bonds simultaneously overlap the positive charge, has been excluded except as the transition state interconverting the two homoaromatic ions in which only one double bond overlaps the positive charge. More experimental evidence is being generated and seems to support the generality of Goldstein's prediction,[2b] although Winstein[24] has concluded that monohomoaromatic character dominates and antiaromatic destabilization (see Chapter 2-4) is small in bicycloaromatic systems.

2-4 Antiaromaticity

In general, monocyclic conjugated systems with $4n$ π electrons are not aromatic, even though they have often been termed *pseudoaromatic*. For the simplest examples of $4n$ systems of this type—the cyclopropenyl anion (**XIX**), cyclobutadiene (**XX**), the cyclopentadienyl cation (**XXI**), and cyclooctatetraene (**XXII**)— MO calculations[25] (see p. 33 for cyclobutadiene) pre-

XIX   XX   XXI   XXII

dict triplet ground states for the coplanar symmetrical systems. Comparison of such a coplanar symmetrical system with the corresponding linear

---

24. J. B. Grutzner and S. Winstein, *J. Amer. Chem. Soc.* **90**, 6562 (1968), and **94**, 2200 (1972).
25. Reference G, Chapter 10, and Reference K, Section 5.4.

system (for example, cyclobutadiene with butadiene) leads to the prediction[2b,26] that the cyclic system is *destabilized* relative to the linear system by the π-electron delocalization (delocalization energies of 0 for cyclobutadiene and 0.472β for butadiene). Breslow[9,26] has proposed that the term *antiaromatic* be applied to such resonance-destabilized $4n$ planar systems.

The destabilization described by the term antiaromatic is manifested in several ways. In some systems, such as cyclobutadiene, the destabilization prevents the system from exhibiting maximum symmetry. Cyclobutadiene (**XX**),[27] a highly unstable system, appears to be a rectangular singlet rather than a square triplet in the ground state.[28] Cyclopentadienyl cations (**XXI**) exhibit singlet–triplet thermal equilibria,[26] with the singlet presumably distorted from perfect fivefold symmetry. In some other systems, such as cyclooctatetraene (**XXII**) the ground state is stable, but not as a planar form. Cyclooctatetraene exists in the "tub" conformation, **XXIII**.[13] Antiaromatic systems do not exhibit diamagnetic susceptibility

XXIII

exaltations[11] or pmr shifts to lower fields.[10] In fact, Untch[29] reports that a paramagnetic ring current is induced in antiaromatic compounds, leading to a pmr shift to *higher* fields for those protons outside the ring.

Quantitative evaluation of the destabilization in antiaromatic compounds is confronted with the same difficulties as quantitative evaluation of aromatic stabilization[2b] (Chapter 2-1). There exists one report[30] suggesting that, for aromatic and antiaromatic systems related to the triphenylmethyl species, the stabilization of the aromatic species is proportional to the destabilization of the corresponding antiaromatic species. More work is required to evaluate the generality of this idea.

The so-called Dewar resonance energy (DRE)[31] may provide a useful quantitative measure of the aromaticity, nonaromaticity, or antiaromaticity of a given compound. From various molecular orbital calculations on acyclic conjugated systems, Dewar noted that the bond energies assigned each type of bond remain constant regardless of the exact molecule in

---

26. R. Breslow, *Angew. Chem. Intern. Ed. Engl.* **7**, 565 (1968); *Chem. Brit.* 100 (1968); *J. Amer. Chem. Soc.* **89**, 4383 (1967); and *Pure Appl. Chem.* **28**, 111 (1971). See also Reference K, Section 6.15.
27. M. P. Cava and M. J. Mitchell, *Cyclobutadiene and Related Compounds*, Academic Press, New York, 1967.
28. R. Pettit, et al., *J. Amer. Chem. Soc.* **91**, 5888, 5890 (1969).
29. J. A. Pople and K. G. Untch, *J. Amer. Chem. Soc.* **88**, 4811 (1966); K. G. Untch and D. C. Wysocki, ibid., **89**, 6386 (1967).
30. M. Feldman and W. C. Flythe, *J. Amer. Chem. Soc.* **91**, 4577 (1969).
31. Reference K, Chapter 5; N. C. Baird, *J. Chem. Educ.* **48**, 509 (1971).

question, implying that conjugation has the same consequences in terms of energy in all nonaromatic unsaturated systems.[32] The DRE of a molecule therefore depends only on its calculated or experimental heat of formation and not on the exact values assigned the various bonds.[33] A positive value of DRE indicates an aromatic compound, a negative value an antiaromatic compound, and a value of zero a nonaromatic system; the magnitude of the value indicates the "amount" of such a character. Typical values using the method of Dewar and deLlano[31,34] are shown in Table 2.2.

TABLE 2.2 *Dewar Resonance Energies*[31]

Compound	DRE Value (kcal/mole)
Benzene	+21
Naphthalene	+33
Anthracene	+42
Phenanthrene	+49
Cyclobutadiene	−17
Cyclooctatetraene	−10
[10]-Annulene	+10
[14]-Annulene	+4
Azulene	+7
Biphenylene	+35
Styrene	+21
Stilbene	+42

Notice that the nonbenzenoid double bonds in styrene and stilbene do not contribute to the DRE's, and that the DRE's are not additive in the number of benzene rings in the polycyclic aromatic compounds such as naphthalene, anthracene, and phenanthrene. The DRE's may also be used to predict, for example, sites of additions to double bonds or across *para* positions in the polycyclic aromatics. It is highly probable that many other types of information will be correlated with the Dewar resonance energies in the ensuing years.

---

32. This does not mean that there is no conjugation in these systems.
33. The same conclusion has been reached using simple Hückel molecular orbital theory —see B. A. Hess, Jr., and L. J. Schaad, *J. Amer. Chem. Soc.* **93**, 305, 2413 (1971).
34. M. J. S. Dewar and C. J. deLlano, *J. Amer. Chem. Soc.* **91**, 789 (1969), and previous papers in series.

Chapter 3

The Woodward-Hoffmann Formalism[1,2]

Almost all of the previous discussion of molecular orbital theory and its applications has been concerned with the more-or-less static properties of molecules. It was not until 1965 that a truly general application of simple MO theory to reacting systems appeared, the sum of which is usually called the Woodward-Hoffmann rules.[1] These rules are based on the principle:

> Orbital symmetry is conserved in concerted reactions.

---

1. The basic reference for this chapter is Reference N, which has also been published as *Angew. Chem. Intern. Ed. Engl.* **8**, 781 (1969). The fundamental ideas were first presented by R. B. Woodward and R. Hoffmann in *J. Amer. Chem. Soc.* **87**, 395, 2046, 2511, 4388, 4389 (1965), in *Accounts Chem. Res.* **1**, 17 (1968), and in *Science* **167**, 825 (1970).

2. Treatments paralleling those presented herein may be found in (a) J. J. Vollmer and K. L. Servis, *J. Chem. Educ.* **45**, 214 (1968); (b) J. J. Vollmer and K. L. Servis, *J. Chem. Educ.* **47**, 491 (1970); (c) K. Fukui and H. Fujimoto, *Mechanisms of Molecular Migrations* **2**, 117 (1969); K. Fukui, *Accounts Chem. Res.* **4**, 57 (1971); and earlier papers; (d) Reference A, Section 20.5; (e) E. M. Kosower, *An Introduction to Physical Organic Chemistry*, Wiley, New York, 1968, Sections 1.8 and 1.9; (f) G. B. Gill, *Quart. Rev.* **22**, 338 (1968); (g) S. I. Miller, *Advan. Phys. Org. Chem.* **6**, 185 (1968); (h) L. Salem, *Chem. Brit.* **5**, 449 (1969); (i) J. B. Hendrickson, D. J. Cram, and G. S. Hammond, *Organic Chemistry*, 3rd ed., McGraw-Hill, New York, 1970, Chapter 21; (j) H. Katz, *J. Chem. Educ.* **48**, 84 (1971); (k) R. E. Lehr and A. P. Marchand, *Orbital Symmetry, A Problem-Solving Approach*, Academic Press, New York, 1972; (l) C. H. Depuy and O. L. Chapman, *Molecular Reactions and Photochemistry*, Prentice-Hall, Englewood Cliffs, N.J., 1972, Chapters 6 and 7; (m) T. L. Gilchrist and R. C. Storr, *Organic Reactions and Orbital Symmetry*, Cambridge University Press, 1972; (n) R. T. Morrison and R. N. Boyd, *Organic Chemistry*, 3rd ed., Allyn and Bacon, Boston, 1973, Chapter 29.

The essential notion of this approach is that a concerted reaction will be favored and therefore require a not unreasonable energy of activation if some bonding character is retained as the orbitals of the reactants are transformed into the orbitals of the products. Since both reactant orbitals and product orbitals are characterized by elements of symmetry, phase signs (+ or −), and nodes, retention of bonding character along the reaction coordinate (Chapter 5-3) requires that the symmetries and phase signs of the reactant orbitals be correlated in a 1:1 way with the symmetries and phase signs of the product orbitals, bonding orbitals with bonding orbitals and antibonding orbitals with antibonding orbitals. We therefore must consider the MO's of the reactants, the transition state (Chapter 5-3), and the products in order to decide whether a process is *symmetry-allowed* or *symmetry-forbidden*. Analysis thereby is restricted to (a) a reaction proceeding directly from reactants to products in a concerted manner without the intervention of any intermediates or (b) one step—reactant to intermediate, intermediate to succeeding intermediate, or intermediate to product—in a more complex mechanistic sequence.

A symmetry-allowed reaction involves a cyclic transition state in a concerted reaction which proceeds through neither ionic nor free-radical intermediates (unless both reactants and products have electronic character of either type) and is essentially independent of external effects such as solvent or catalysts. These *pericyclic* reactions exhibit high or complete stereospecificity because of the cyclic nature of their transition states. When the orbital symmetry is not conserved in a particular reaction, the process occurs only under higher-energy conditions (with a higher energy of activation) or takes place by a multistep mechanism involving discrete ionic or free-radical intermediates. A symmetry-forbidden reaction therefore may be well known in its own right and may also be concerted (p. 84).

Typical examples of those types of pericyclic reactions which will be discussed in this chapter are shown in Eq. 3.1–3.7.

$$\text{cis} \xrightarrow{176°} \text{cis,trans} \quad (3.1)$$

$$\text{trans} \xrightarrow{176°} \text{cis,cis} \quad (3.2)$$

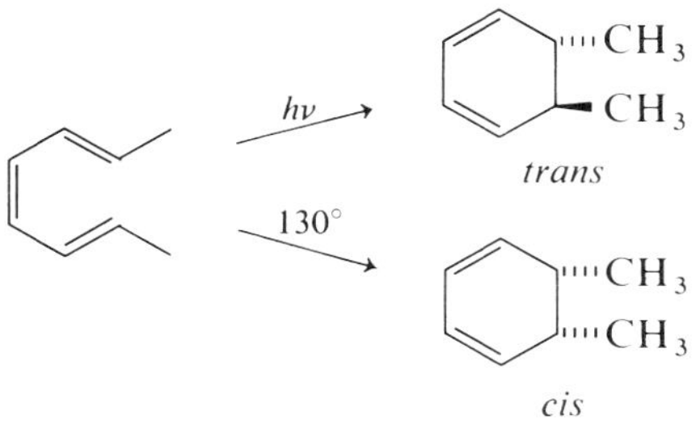

(3.3)

(3.4)

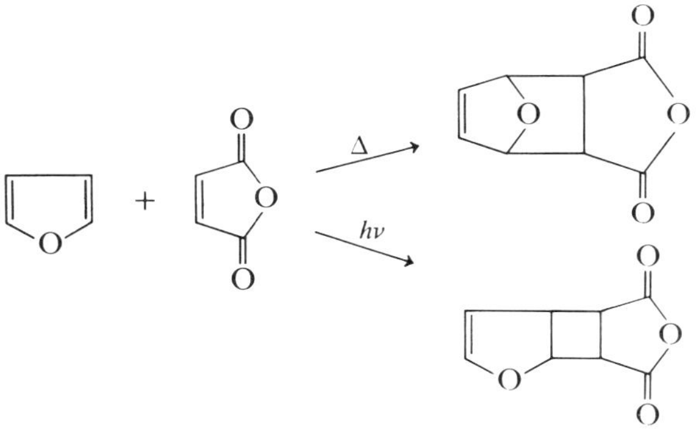

(3.5)

(3.6)

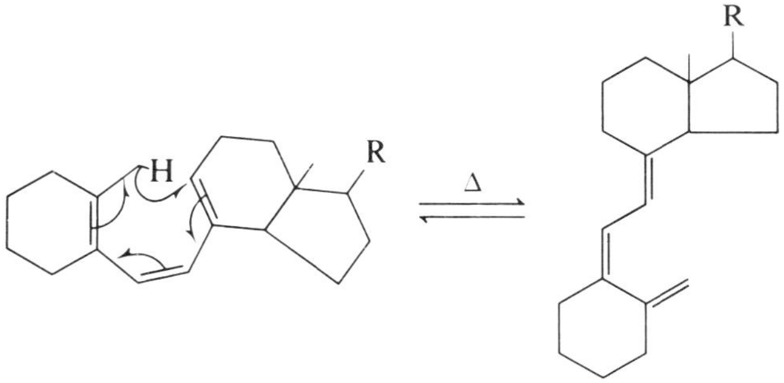

(3.7)

Precalciferol   Calciferol

3-1 Electrocyclic Reactions[1,2]

A reaction involving the interconversion of a linear polyene and its cyclic isomer formed by ring closure at the termini of the conjugated π-electron system of the polyene is called an *electrocyclic reaction*[1] (Eq. 3.1–3.4). An electrocyclic process effects the net conversion of a π bond to a σ bond, or the converse. Electrocyclic reactions are completely stereospecific, since the fixed geometrical isomerism of the polyene (see Chapter 10-3) is directly related to the isomerism imposed by the rigidity of the cyclic system if the reaction is concerted. This stereorelationship may be either *conrotatory*, in which *cis* ring substituents rotate in the same direction into the developing plane defined by the polyene system (Fig. 3.1a), or *disrotatory*, in which *cis* ring substituents rotate in opposite directions (Fig. 3.1b).[3] Using these terms, the thermal cyclization of the butadienes (Eq. 3.1 and 3.2) and the photochemical cyclization of the hexatriene (Eq. 3.3) are conrotatory processes, while the thermal cyclization of the hexatriene (Eq. 3.4) is disrotatory. If two alternate disrotatory processes or two alternate conrotatory processes are possible and physically differentiable, steric factors will usually determine which of each pair will predominate *after* orbital symmetry requirements control whether the process should be disrotatory or conrotatory.

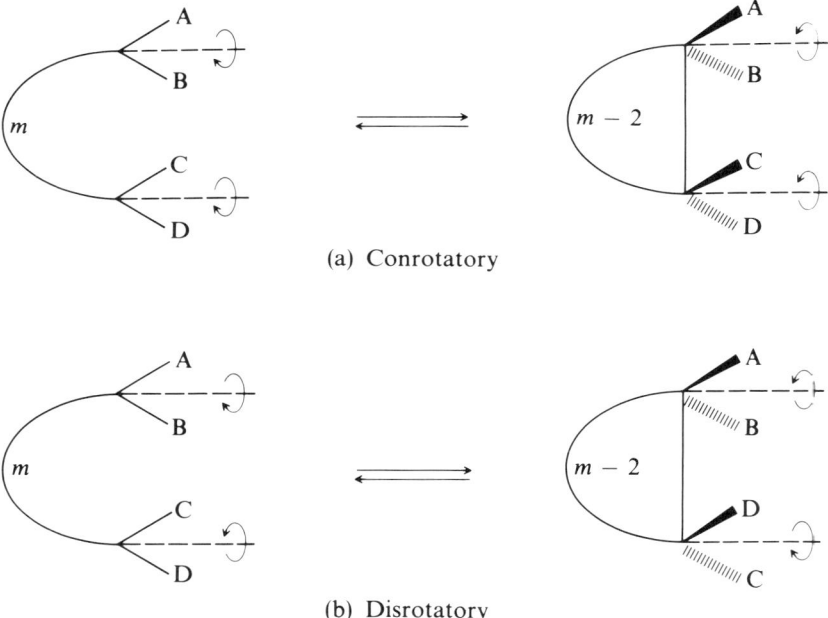

(a) Conrotatory

(b) Disrotatory

FIGURE 3.1 *Electrocyclic processes*[1,2] *(m = number of π electrons in the conjugated polyene, $m - 2$ = number of π electrons in the cyclic isomer): (a) conrotatory process; (b) disrotatory process.*

---

3. Because of the principle of microscopic reversibility, symmetry considerations in the concerted conversion of a polyene to a cyclic isomer will be the same as those in the concerted conversion of the cyclic isomer to the polyene. Any electrocyclic process therefore may be analyzed in either direction—as a cyclization or as a ring closure.

The simplest approach to an analysis of the preferred direction of rotation is to consider the effect of the two possible motions on the highest occupied molecular orbital (HOMO) of the polyene.[4] Using butadiene as an example, the HOMO in the ground state would be ψ_2 (Fig. 1.19 and 1.20), which may be represented as

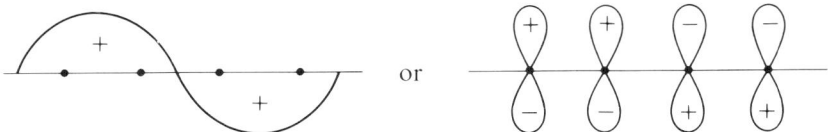

using the free-electron model. If we consider ψ_2 in an *s-cis* arrangement (Fig. 1.18), a disrotatory process would push a plus lobe toward a minus lobe (Fig. 3.2a), creating a destabilizing, repulsive interaction in the transition state which cannot lead to the formation of a σ bond. The conrotatory process, on the other hand, pushes a plus lobe toward a plus lobe (or the equivalent, a minus lobe toward a minus lobe) (Fig. 3.2b), which creates a stabilizing, attractive interaction which can culminate in the formation of a σ bond. The conclusion from this analysis would be that the thermal concerted cyclization of a butadiene to a cyclobutene should be a conrotatory process. The same type of HOMO approach may be applied to the thermal cyclization of a hexatriene (Eq. 3.4). A disrotatory process would be required in ψ_3 (Fig. 3.3) in order to bring about combination of similar orbital phases at the termini of the polyene system. Butadiene and hexatriene exemplify a recurring pattern in that the preferred mode of thermal cyclization will alternate from conrotatory

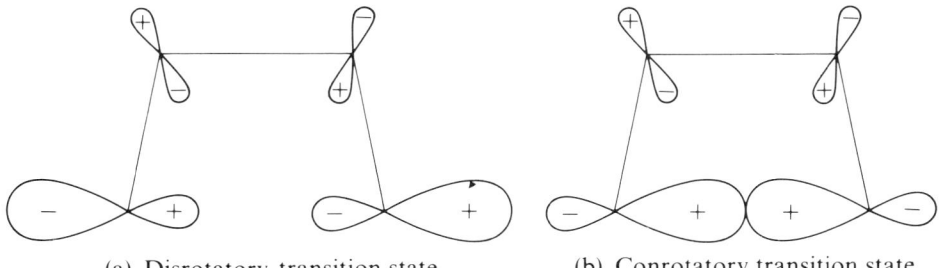

(a) Disrotatory transition state   (b) Conrotatory transition state

FIGURE 3.2 *The transition states for the different modes of cyclization of ψ_2 of butadiene:* (a) *disrotatory;* (b) *conrotatory.*

---

4. Such an approach often is called a "frontier orbital" approach, and was pioneered by Fukui (note 2c).

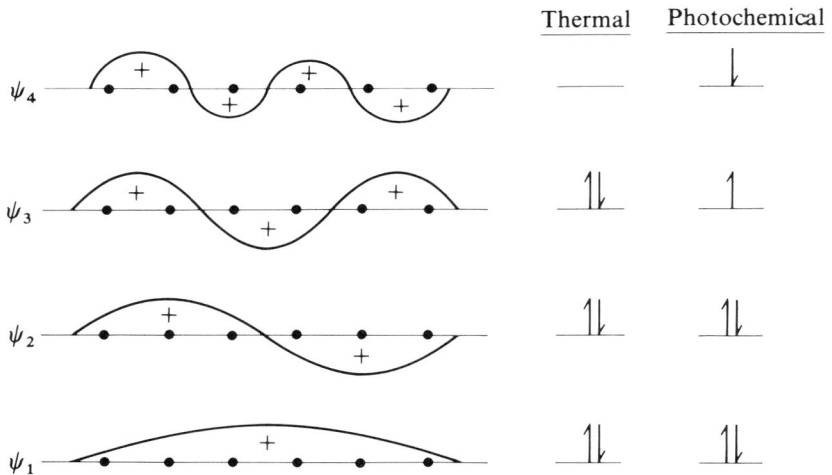

FIGURE 3.3 *Schematic hexatriene molecular orbitals.*

to disrotatory and back on the successive addition of conjugated double bonds to the polyene (Fig. 3.4).

If we view photochemical excitation as involving promotion of an electron from the HOMO to the lowest unoccupied molecular orbital (LUMO), photochemical cyclization should be analyzed in terms of the MO next higher in energy from the MO considered in the thermal reaction. Cyclization based on ψ_4 in hexatriene (Fig. 3.3) must be conrotatory to produce positive overlap (Eq. 3.3). The photochemical process and the thermal process in a given polyene system will always require opposite rotatory modes (Fig. 3.4). Odd-electron systems would follow the same course as the even-electron polyene with one additional electron, and

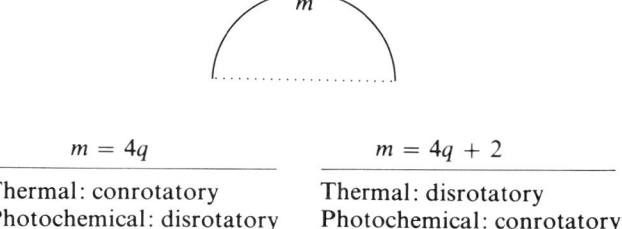

$m = 4q$	$m = 4q + 2$
Thermal: conrotatory	Thermal: disrotatory
Photochemical: disrotatory	Photochemical: conrotatory

FIGURE 3.4 *The Woodward-Hoffmann rules*[1,2] *for electrocyclic reactions* ($1\pi \rightleftarrows 1\sigma$) (*m = number of π electrons, q is an integer*).

neutral and charged species with the same number of electrons would behave similarly.

Longuet-Higgins[5] introduced a slightly more sophisticated approach emphasizing the symmetry of the MO's and introducing correlation diagrams, an idea adopted by Woodward and Hoffmann.[1,2] In a concerted conrotatory cyclization an axis of symmetry (C_2)[6] is maintained, while a plane of symmetry (σ) is present in a disrotatory process. The symmetry classification, S for symmetric or A for antisymmetric, of a particular orbital with respect to a particular symmetry element (C_2 or σ) cannot change in a concerted process if bonding interactions are to be present continuously. In order to predict which rotatory mode will be preferred, correlation diagrams relating the symmetry of each reactant orbital to the symmetry of each product orbital may be constructed for each of the rotatory modes.[7] Typical correlation diagrams for the interconversion of cyclobutenes and butadienes are shown in Fig. 3.5. The thermal reaction would be symmetry-allowed in the conrotatory mode, because every bonding orbital of the reactant may be related to a bonding orbital of the product, a correlation which does not exist for the disrotatory process in the

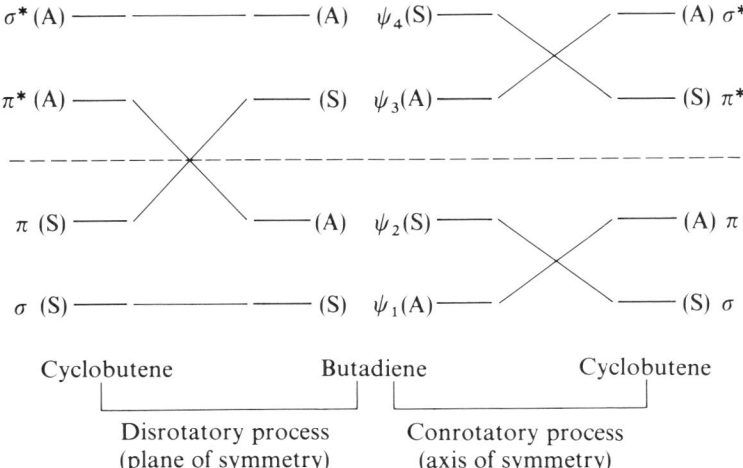

FIGURE 3.5 *Correlation diagrams for the disrotatory and conrotatory cyclobutene–butadiene processes.*

---

5. H. C. Longuet-Higgins and E. W. Abrahamson, *J. Amer. Chem. Soc.* **87**, 2045 (1965).
6. The symmetry element must bisect bonds being made or broken in the reaction.
7. For details, see sources listed in notes 1, 2a, 2i, and 5. Correlation diagrams will be discussed in more detail in Chapter 3-3.

ground state. On the other hand, the photochemical process would be disrotatory for these compounds, since only this mode maintains electrons in the lowest-energy orbitals in the excited states of reactant and product.

It is important to realize that a process that is symmetry-forbidden in a given energy situation can occur under more energetic reaction conditions. As discussed above, the thermal conrotatory ring cleavage of a *cis*-disubstituted cyclobutene produces a *cis,trans*-butadiene (Eq. 3.1), while a *trans*-disubstituted cyclobutene produces a *cis,cis*-butadiene (Eq. 3.2). Criegee[8] has utilized the fused systems (Fig. 3.6a and b) to demonstrate the importance of steric effects in such processes. The *anti* ring-fused isomers (Fig. 3.6a) undergo pyrolyses at quite reasonable temper-

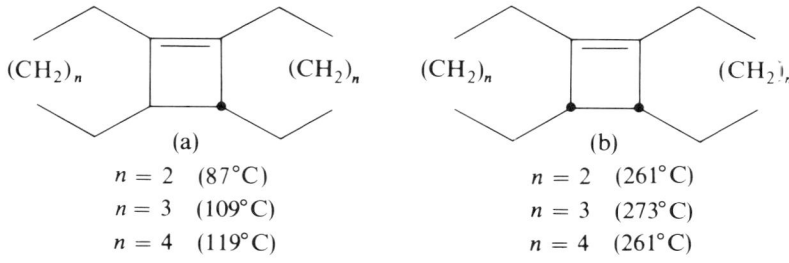

(a)
$n = 2$ (87°C)
$n = 3$ (109°C)
$n = 4$ (119°C)

(b)
$n = 2$ (261°C)
$n = 3$ (273°C)
$n = 4$ (261°C)

FIGURE 3.6 *Thermal cleavage of* (a) *anti and* (b) *syn fused cyclobutenes.*[8]

atures, producing products with *cis* double bonds in six-, seven-, or eight-membered rings. Those with *syn* backbones, (b), require much more extreme conditions for ring cleavage, the energies of activation being 12–18 kcal/mole greater than those for the *anti* compounds. Since *trans* double bonds are extremely unstable in six- and seven-membered rings, the *syn* compounds must be cleaving either by the symmetry-forbidden disrotatory process or by a nonconcerted mechanism. The decrease in the pyrolysis temperature in the eight-membered *syn* compound relative to its next smallest analog suggests the disrotatory process rather than a nonconcerted mechanism, since the eight-membered rings could better accommodate a *trans* double bond.

Various quantitative estimates of the difference in energy between a symmetry-allowed and a symmetry-forbidden thermal electrocyclic process have been reported.[9] These values range from a minimum difference in

---
8. R. Criegee, *Angew. Chem. Intern. Ed. Engl.* **7**, 559 (1968).
9. J. I. Brauman and D. M. Golden, *J. Amer. Chem. Soc.* **90**, 1920 (1968); E. C. Lupton, Jr., *Tetrahedron Lett.* 4209 (1968); G. A. Doorakian and H. H. Freedman, *J. Amer. Chem. Soc.* **90**, 5310, 6896 (1968); J. I. Brauman and W. C. Archie, Jr., ibid., **94**, 4262 (1972).

energy of activation of 7.3 kcal/mole to a maximum of about 15 kcal/mole. Experimentally,[9] a substituted cyclobutene molecule heated at 124°C for 51 days underwent 2.6×10^6 conrotatory openings without a disrotatory "error".

3-2 Sigmatropic Reactions[1,2]

A migration involving breaking a σ bond at an allylic position and forming a new σ bond at the other end of at least one conjugated π system (with the concomitant shift of the π electrons) is called a *sigmatropic rearrangement*.[1,2] A sigmatropic change of order $[i, j]$ is defined[1,2] as an uncatalyzed concerted intramolecular reaction involving the migration of a σ bond flanked by one or more π-electron systems to a new position whose termini are $(i - 1)$ and $(j - 1)$ atoms removed from the original bonded loci (Fig. 3.7). In this nomenclature, the thermal conversion of precalciferol to calciferol (Eq. 3.7) is a [1,7] sigmatropic reaction. When the migrating group is always associated with the same face of the π system, a *suprafacial* (s) migration occurs. If the migrating group forms the new σ bond on the opposite face of the π system from that of the σ bond which was broken, the process is termed *antarafacial* (a). These topologically distinct processes are shown for a [1,5] hydrogen migration in Fig. 3.8.

If a generalized $[1, j]$ hydrogen rearrangement in an all-*cis* polyolefin framework is considered (Eq. 3.8), the transition state may be treated as

$$\overset{1}{R_2CH}-(CH=CH)_k-CH=\overset{j}{CR_2'} \rightleftharpoons$$

$$\overset{1}{R_2C}=CH-(CH=CH)_k-\overset{j}{CHR_2'} \quad (3.8)$$

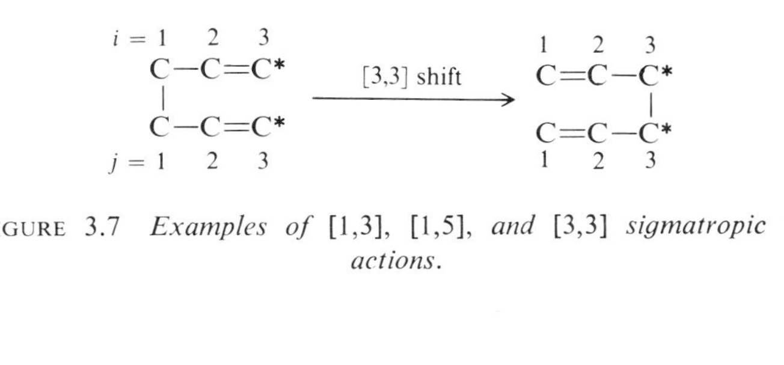

FIGURE 3.7 *Examples of* [1,3], [1,5], *and* [3,3] *sigmatropic reactions.*

FIGURE 3.8 *Suprafacial and antarafacial [1,5] hydrogen migrations.*[1,2]

a hydrogen atom combined with a radical containing $2k + 3$ π electrons. In the suprafacial process, the transition state would possess a plane of symmetry (Fig. 3.8), while in an antarafacial process, the transition state would be characterized by a twofold symmetry axis (Fig. 3.8). The highest occupied molecular orbital of the $2k + 3$ π-electron radical transition state would be a nonbonding allylic system (see pp. 35 and 41), a general representation of which would be

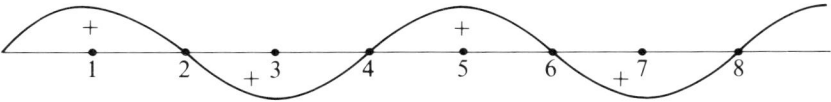

In order for positive overlap to be maintained between the migrating group and the π framework, symmetry considerations allow suprafacial thermal migrations for odd values of k and antarafacial thermal migrations for even values of k, and the reverse results for photochemical processes. These predictions, summarized in Fig. 3.9, assume retention of configuration in the migrating group.

Similar considerations may be applied to the general $[i,j]$ rearrangement when neither i nor j is equal to 1. Since the migrating group now also involves a π system, the topological descriptions must be applied to both π systems through which the σ bond is "moving". If a suprafacial cleavage of a σ bond involves either retention or inversion at both atoms and

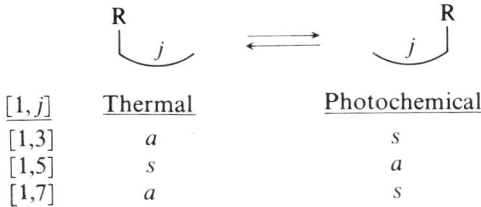

[1,j]	Thermal	Photochemical
[1,3]	a	s
[1,5]	s	a
[1,7]	a	s

FIGURE 3.9 *The Woodward-Hoffmann rules for symmetry-allowed [1,j] sigmatropic migrations.*[1,2]

an antarafacial cleavage involves inversion at one atom and retention at the other, the general rule is:[1,2i]

> A thermal reaction is symmetry-allowed if the total number of suprafacial regroupings of the participating bonds is an odd number.

Another way of stating this general rule is that *a thermal reaction is symmetry-allowed if the total number of* $(4q + 2)_s$ *and* $(4r)_a$ *components is odd* (where q and r are integers).[1] The reverse would be true for photochemical processes. For example, if a [1,3] migration occurs thermally with retention of configuration in the migrating group ($_\sigma 2_s$, or involvement of two σ electrons with retention), inversion must occur in the π system ($_\pi 2_a$, or involvement of two π electrons with inversion), producing an overall ($_\sigma 2_s + {}_\pi 2_a$) antarafacial rearrangement (Fig. 3.9). The resulting symmetry-allowed [i,j] reactions are summarized in Fig. 3.10.

Notice that a given symmetry-allowed process, such as a [1,3] migration, can occur (a) with retention in the migrating group and inversion in

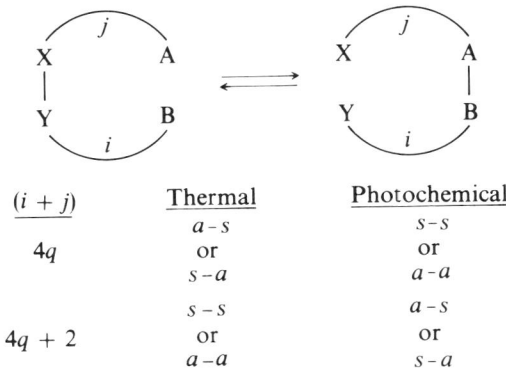

$(i + j)$	Thermal	Photochemical
$4q$	a-s or s-a	s-s or a-a
$4q + 2$	s-s or a-a	a-s or s-a

FIGURE 3.10 *The Woodward-Hoffmann rules for allowed [i,j] sigmatropic migrations.*[1,2]

the π system ($_\sigma 2_s + {}_\pi 2_a$), an antarafacial process with respect to the π system, or (b) with inversion in the migrating group and retention in the system (denoted $_\sigma 2_s + {}_\pi 2_a$, or $_\sigma 2_a + {}_\pi 2_s$), which would be a suprafacial process with respect to the π system. The important fact from the symmetry point of view is the net odd number of suprafacial components in this [1,3] system. Berson[1,2,10] has recently uncovered such a concerted symmetry-allowed π-suprafacial [1,3] migration with inversion of configuration at the migrating center (Eq. 3.9), thereby dramatically illustrating the im-

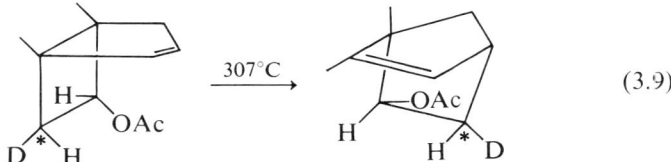

(3.9)

portance of the orbital symmetry in both components of the migratory process. This type of rearrangement with inversion in the migrating group is particularly important for [1,3] migrations, since antarafacial processes are impossible in small or medium rings and difficult at all times. If the migrating group possesses an available π orbital and the steric situation is not prohibitive, inversion at the migrating center therefore would occur quite readily.

Several additional considerations associated with the occurrence of a symmetry-allowed migration may be important in some situations. As for electrocyclic reactions, the rearrangement cannot involve a discrete intermediate. Any distortion of the carbon framework that causes serious π-electron localization or that prevents maximal delocalization may influence the course of the reaction. One implication of the usual drive for maximal linear conjugation is that, given a choice, the symmetry-allowed rearrangement with the largest possible value of $(i + j)$ will occur. Only the [1,7] hydrogen migration is observed in the calciferol series (Eq. 3.7), with none of the possible [1,3] or [1,5] products being encountered. A cyclopropane ring can replace a π bond in a framework, or a sigmatropic migration can occur within an ionic species, without changing the selection rules.

3-3 Cycloaddition Reactions[1,2]

The combination of two π systems to form a molecule containing at least one more ring, two more σ bonds, and two fewer π bonds than the original arrangement is termed a *cycloaddition reaction*.[1,2,11] Such $2\pi \to 2\sigma$

---

10. J. A. Berson, *Accounts Chem. Res.* **1**, 152 (1968).
11. The reverse process is called a *cycloreversion reaction*.

processes (Eq. 3.5 and 3.6) are classified by the number of conjugated π electrons m and n in each of the combining systems, and by the stereochemical features of the reaction, which may be $(s + s)$, $(a + a)$, $(a + s)$, or $(s + a)$ (Fig. 3.11). When the two reactants are identical, the $(a + s)$ and $(s + a)$ processes are indistinguishable. When m or n is greater than 2, the reactants may be able to approach in two different modes for each stereochemical possibility, leading to *exo* and *endo* processes in each case (see Chapter 3-4).

Let us consider the relative probabilities of 1,2- and 1,4- $(s + s)$ addition of ethylene to butadiene under thermal and photochemical reaction conditions. The 1,2-addition would be analogous to the dimerization of ethylene (Eq. 3.10), while the 1,4-addition to produce cyclohexene (Eq. 3.11)

$$H_2C=CH_2 + H_2C=CH_2 \rightleftharpoons \square \quad (3.10)$$

$$H_2C=CH_2 + H_2C=CH-CH=CH_2 \rightleftharpoons \hexagon \quad (3.11)$$

is formally the simplest example of the Diels-Alder reaction (Chapter 3-4). Since for the present this analysis is restricted to the $(s + s)$ addition modes, it is reasonable to assume that the reactants approach one another in such a parallel way that a plane of symmetry is maintained perpendicular to the planes of the approaching molecules and bisecting each molecular π system (assuming an *s-cis* conformation for the butadiene) (Fig. 3.12). All of the molecular orbitals therefore may be classified as symmetric (S) or antisymmetric (A) with respect to the bisecting plane.

Looking first at the ethylene dimerization, a $[2_s + 2_s]$ process, four orbitals from the two ethylene molecules must be considered: the π and π^* orbitals of each ethylene. Two π orbitals may approach one another in either of two ways (Fig. 3.13a and b), both of which are S with respect to the bisecting plane and bonding, while two π^* orbitals may approach in two ways which are A and antibonding (Fig. 3.13c and d). For reasons of symmetry, a π orbital of one ethylene molecule may not combine with a π^* orbital of a second ethylene in a concerted cycloaddition. The four molecular orbitals of the cyclobutane product may be written as shown in Fig. 3.14. Two of the orbitals—σ_1 and σ_2—are bonding orbitals, and two —σ_3 and σ_4—are antibonding. Two—σ_1 and σ_3—are S with respect to the bisecting plane, while two—σ_2 and σ_4—are A. If a correlation diagram is constructed (Fig. 3.15), we see that conservation of orbital symmetry for the thermal reaction requires that a bonding orbital of the ethylene reactants, π_2, be transformed into an antibonding orbital of the cyclobutane, σ_3, and vice versa, antibonding π_3 into bonding σ_2. Such a thermal $[2_s +$

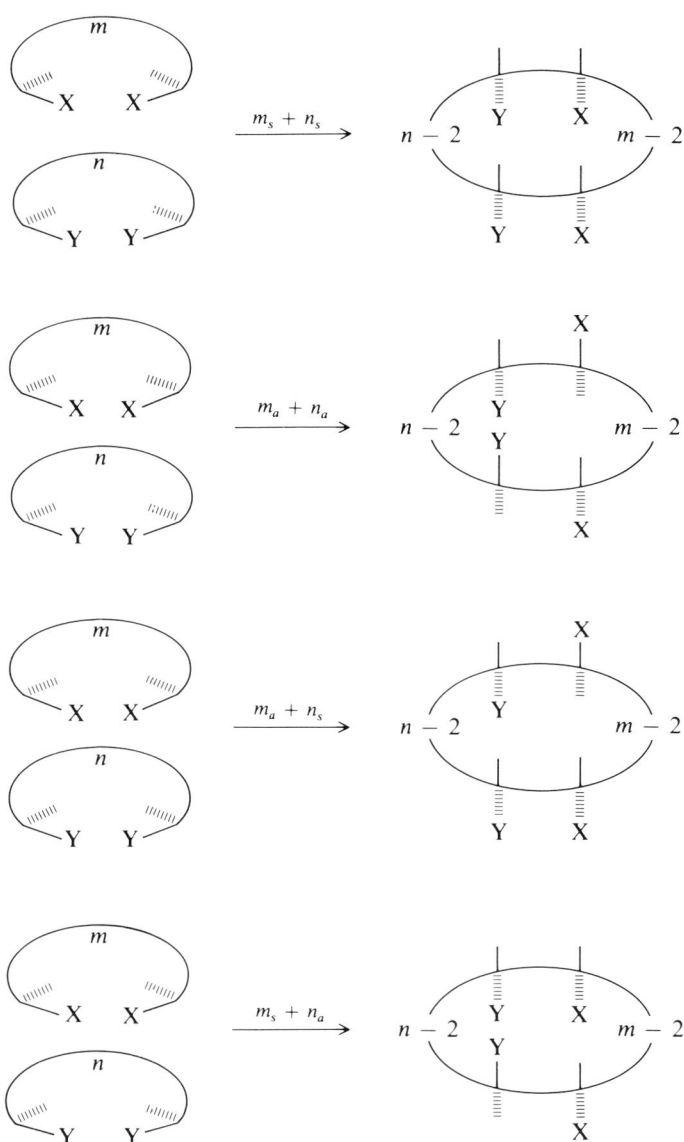

FIGURE 3.11 *The stereochemical possibilities of two-component cycloadditions.*[1,2]

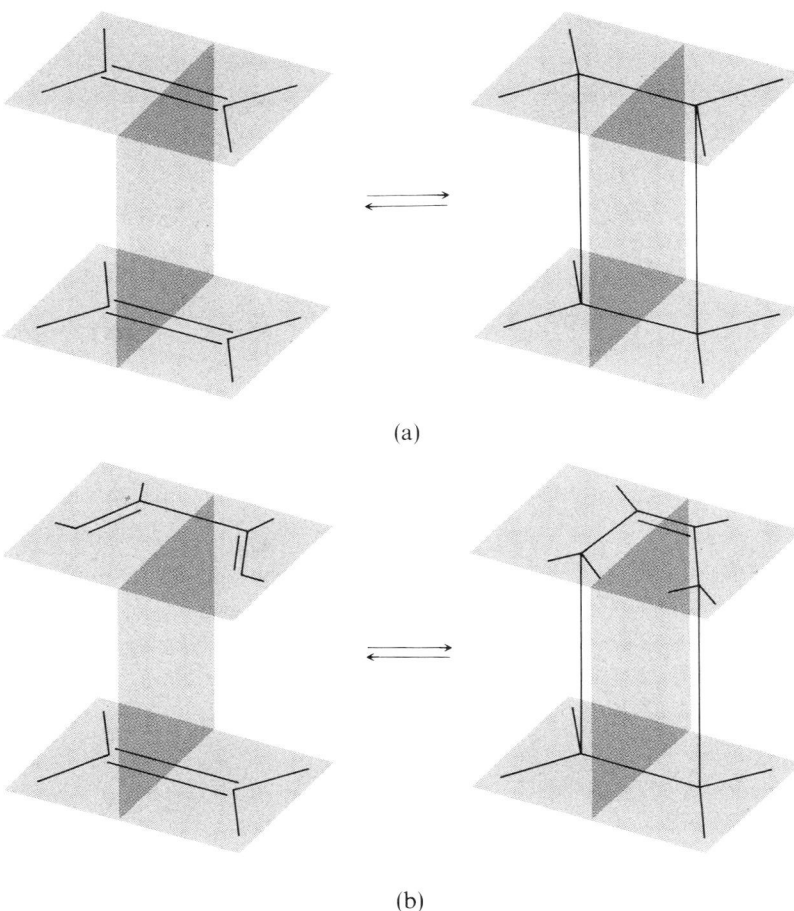

FIGURE 3.12 *The reaction paths[1,2b] for* (a) *the concerted dimerization of ethylene and* (b) *the concerted cycloaddition of ethylene and butadiene.*

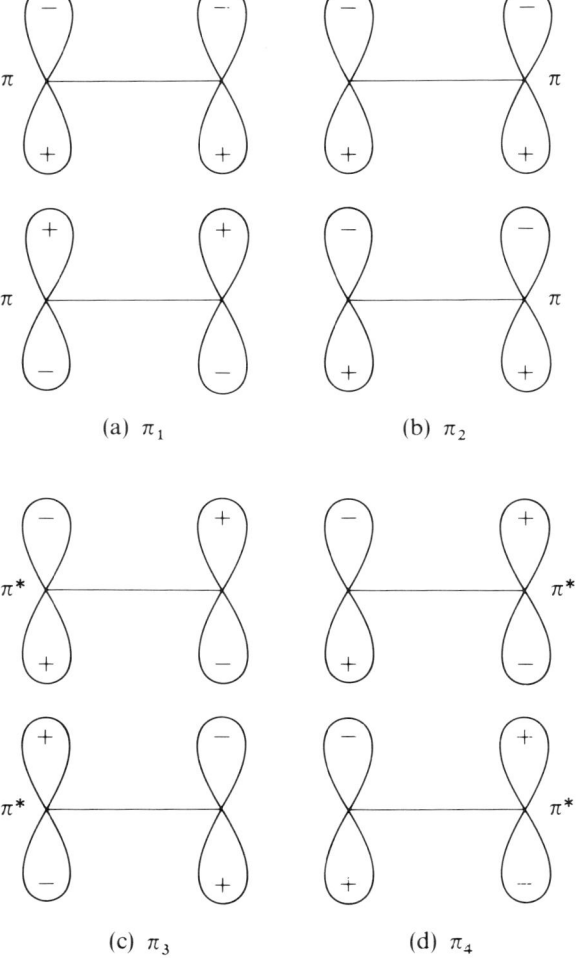

FIGURE 3.13 *The molecular orbitals of the two ethylenes involved in the $[2_s + 2_s]$ cycloaddition.*[1,2]

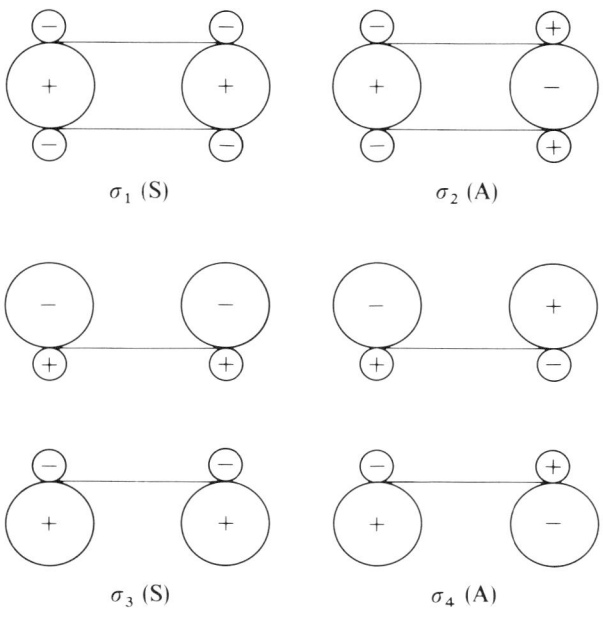

FIGURE 3.14 *The molecular orbitals of cyclobutane formed in the $[2_s + 2_s]$ cycloaddition.*[1,2]

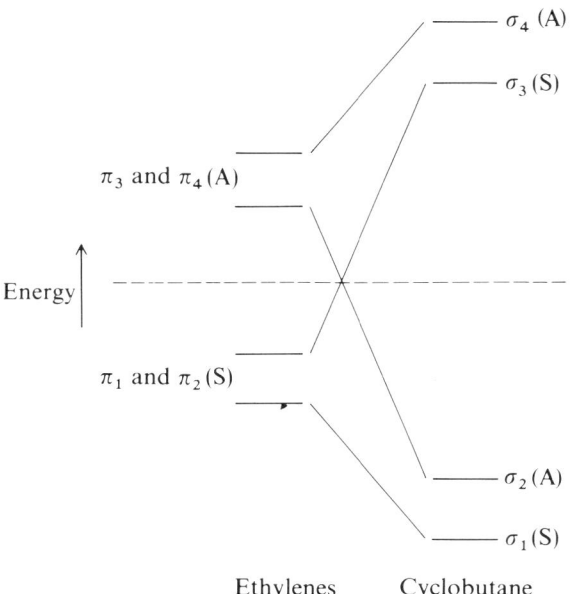

FIGURE 3.15 *Correlation diagram for $[2_s + 2_s]$ cycloadditions.*[1,2]

2_s] process therefore is symmetry-forbidden. As discussed in Chapter 3-2, the photochemical process would be symmetry-allowed for this $[2_s + 2_s]$ addition.

For the 1,4-addition process, a $[4_s + 2_s]$ reaction, the relevant orbitals will be the π and π^* of ethylene, which are S and A, respectively; the four molecular orbitals of butadiene (Fig. 1.20), of which ψ_1 is S, ψ_2 is A, ψ_3 is S, and ψ_4 is A; and the six molecular orbitals of cyclohexene shown in Fig. 3.16. The correlation diagram, Fig. 3.17, indicates that the thermal $[4_s + 2_s]$ reaction would be symmetry-allowed, while the corresponding photochemical process would be symmetry-forbidden.

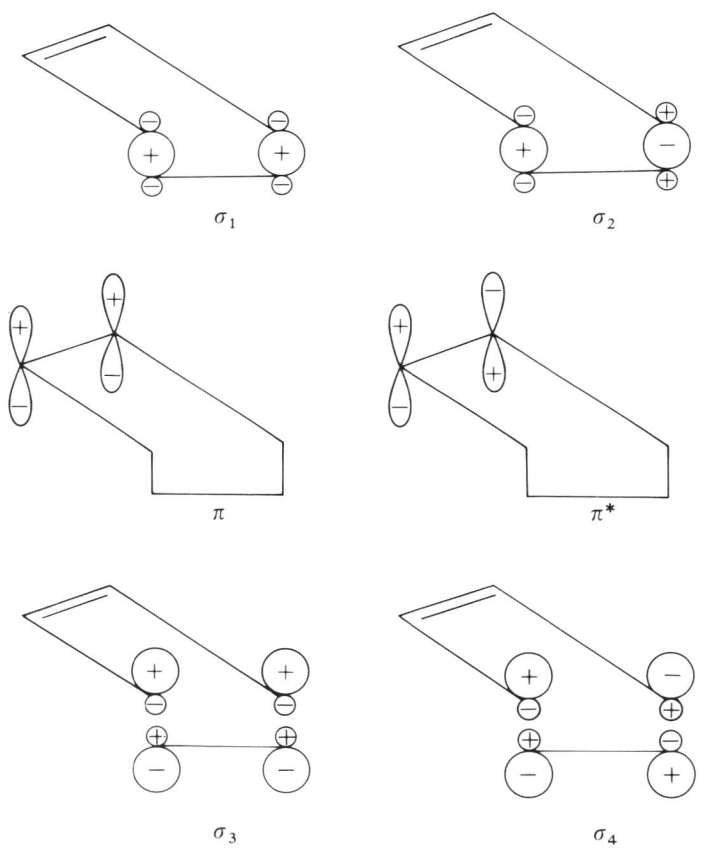

FIGURE 3.16 *The molecular orbitals of cyclohexene formed in the* $[2_s + 4_s]$ *cycloaddition.*[1,2]

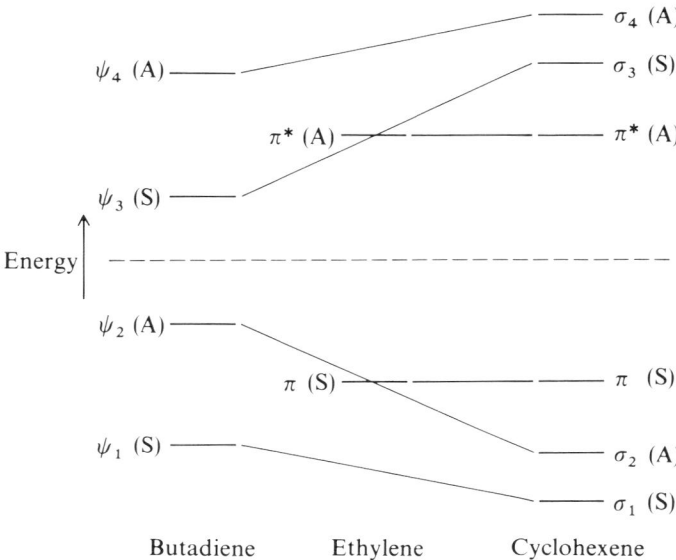

FIGURE 3.17 *Correlation diagram for* [$2_s + 4_s$] *cycloadditions.*[1,2]

Comparison of the correlation diagrams for ($s + s$) 1,2-addition (Fig. 3.15) and ($s + s$) 1,4-addition (Fig. 3.17) leads to the conclusion that the 1,4-product, cyclohexene, would be favored thermally and the 1,2-product, a cyclobutane, would be favored photochemically, which agrees with the experimental results exemplified in Eq. 3.5 and 3.6. Notice that these results could have been predicted using the general rule presented for sigmatropic reactions (p. 69), which is really the general rule for all pericyclic reactions.[1,2] The resulting symmetry-allowed $2\pi \to 2\sigma$ cycloadditions are summarized in Fig. 3.18.

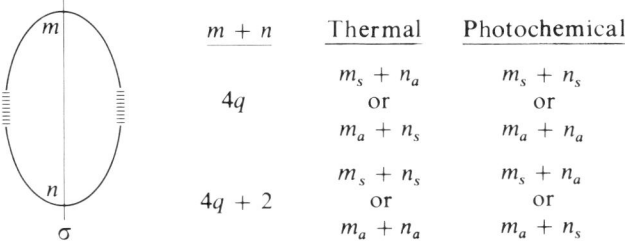

FIGURE 3.18 *The Woodward-Hoffmann rules for allowed cycloaddition reactions.*[1,2]

Graphical representations may be used to illustrate the symmetry-allowed processes relating various structural types, including relevant stereochemical features. A superb example of these representations may be seen in Figure 26 of Reference N, which interrelates bicyclobutanes, cyclobutenes, and butadienes in the ground states and excited states. Similar treatment of hexatrienes would include the well-known photochemical conversion of *cis*-hexatrienes to bicyclo[3.1.0]hexenes, the possible stereochemical consequences of which[1] are shown in Eq. 3.12. These

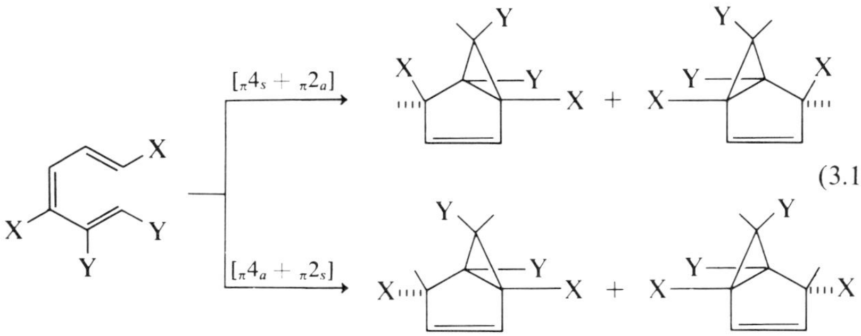

(3.12)

stereochemical predictions[1] have been verified by Padwa and Clough[12a] using 1,2,6-triphenyl-(E,Z,E)-hexa-1,3,5-triene (**I**) and 1,2,6-triphenyl-(E,Z,Z)-hexa-1,3,5-triene (**II**).[13] These photolyses were found to be stereospecific, with the cyclizations occurring only by the $[_\pi 4_s + _\pi 2_a]$ route (Eq. 3.13 and 3.14), possibly for steric reasons or because of the nodal

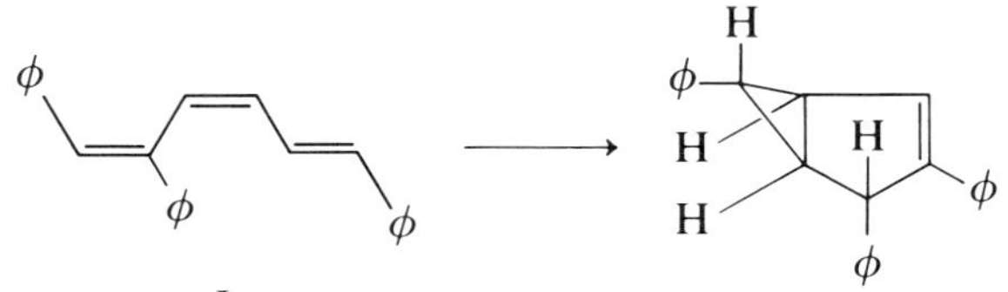

(3.13)

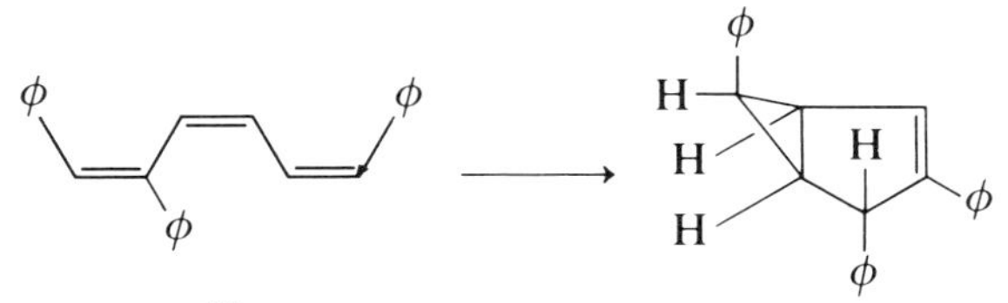

(3.14)

---

12. (a) A. Padwa and S. Clough, *J. Amer. Chem. Soc.* **92**, 5803 (1970), and A. Padwa, L. Brodsky, and S. Clough, ibid., **94**, 6767 (1972). (b) D. A. Seeley, ibid., **94**, 4378 (1972).
13. For a discussion of the E-Z stereochemical designations, see Chapter 10-3.

structure of the lowest π^* level of the hexatrienes.[12a] The concerted nature of these photocyclizations has been questioned by Seeley,[12b] who claims that the cyclization is nonconcerted, fast, and controlled by steric hindrance at the termini rather than by orbital symmetry (see p. 83 also).

The area of 1,3-dipolar cycloadditions (e.g., Eq. 3.15), for which Huisgen has proposed a concerted mechanism,[14] may also be analyzed by the Woodward-Hoffmann techniques[1,2e,15] since this is merely a special type of (2 + 4) cycloaddition. However, in some instances the reactions

$$a \overset{\oplus}{\underset{d=e}{\overset{b}{\diagdown}}} c^{\ominus} \longrightarrow a \overset{b}{\underset{d-e}{\diagdown\diagup}} c$$

$$\underset{R_2}{\overset{R_1}{\diagdown}}C \overset{+}{=} \underset{O^-}{\overset{R_3}{\diagdown}} \underset{}{N} + \diagdown C = C \diagup \longrightarrow \underset{R_2}{\overset{R_1}{\diagdown}} \underset{-C-C-}{\overset{R_3}{\underset{|}{\overset{|}{\diagdown}}\underset{C}{\overset{N}{\diagdown}}O}} \quad (3.15)$$

may not be concerted[16] or the polar character of the reactants may be extreme enough to favor a nonconcerted process with a relatively stable dipolar intermediate.[1]

Selection rules have also been derived for multicenter cycloadditions[1,2] involving concerted reactions between more than two π systems. Such reactions are fairly rare, because of the extremely low probability of multicenter collisions, an entropy effect which can be overcome with special geometrical constraints, such as placing several of the π systems in the same molecule.

3-4 Secondary Effects

It is fairly obvious that the best model for any type of analysis will be that which provides the greatest amount of useful correct information. The orbital symmetry approach shines in this respect. Not only does the Woodward-Hoffmann approach[1,2] predict that the Diels-Alder reaction will be thermally allowed (see discussion of Fig. 3.17), but it also supports

---

14. R. Huisgen, *Angew. Chem. Intern. Ed. Engl.* **2**, 565, 633 (1963), and **7**, 321 (1968); *J. Org. Chem.* **33**, 2291 (1968).
15. R. Hoffmann, *J. Amer. Chem. Soc.* **90**, 1475 (1968); A. Eckell, R. Huisgen, R. Sustmann, G. Wallbillich, D. Grashey, and E. Spindler, *Chem. Ber.* **100**, 2192 (1967).
16. R. A. Firestone, *J. Org. Chem.* **33**, 2285 (1968).

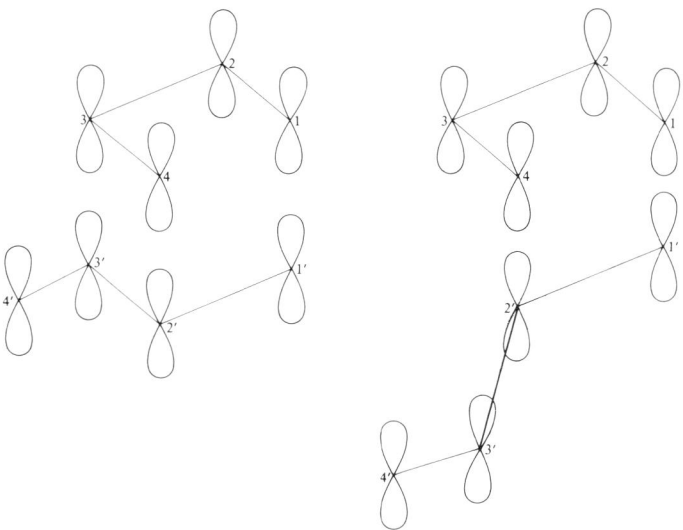

the *cis–endo* stereochemistry which is usually observed.[17] The *cis* union of diene and dienophile (Eq. 3.16) agrees with the $[_\pi 4_s + {}_\pi 2_s]$ nature of the symmetry-allowed reaction (p. 77), but the predominant *endo* stereochemistry cannot be determined from the relevant correlation diagram. The major electronic difference between the *endo* transition state and the *exo* transition state is the possibility of an interaction in the *endo* situation involving lobes on atoms 3 and 3'. If occupied MO's of both reactants are involved in such an interaction, the result would be to lower the energy of some orbitals and raise the energy of others, providing little net energy

---

17. A. Wasserman, *The Diels-Alder Reaction*, Elsevier, New York, 1966; J. Sauer, *Angew. Chem. Intern. Ed. Engl.* **5**, 211 (1966), and **6**, 16 (1967); J. G. Martin and R. K. Hill, *Chem. Rev.* **61**, 537 (1961).

stabilization. However, symmetry-allowed mixing of unoccupied orbitals of one reactant with occupied orbitals of the other reactant can produce significant energy stabilization. As shown in Fig. 3.19, the conformations resulting from interaction between diene HOMO and dienophile LUMO and

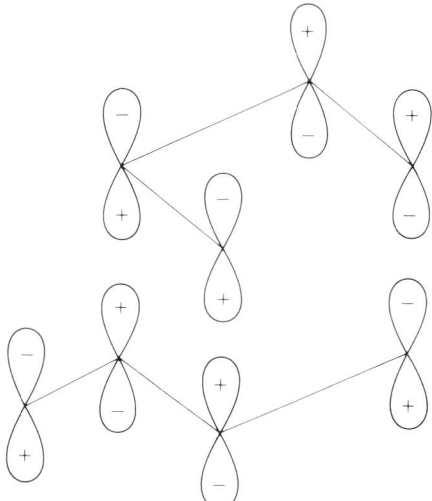

HOMO of diene and LUMO of dienophile

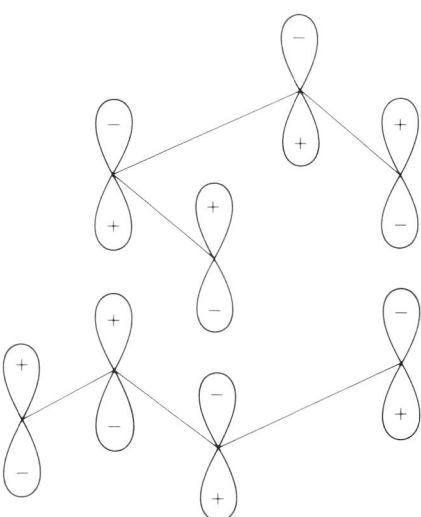

LUMO of diene and HOMO of dienophile

FIGURE 3.19 *Secondary interactions in the endo transition state for the Diels-Alder reaction.*[1,2]

between diene LUMO and dienophile HOMO (using butadiene as both diene and dienophile) both possess the proper symmetry for favorable secondary interactions in the *endo* transition state, while no secondary interactions are possible for the *exo* process because the orbitals are not in close enough proximity.[2h,18]

Because of the importance of symmetry in such secondary interactions, any preference for either the *exo* or *endo* orientation will be a function of the values of m and n, the numbers of π electrons involved in the desired combination of the two reactants.[1,2] For example, while secondary interactions stabilize the *endo* transition state for $[_\pi 4_s + {}_\pi 2_s]$ and $[_\pi 8_s + {}_\pi 2_s]$ cycloadditions, the same type of secondary interactions destabilize the *endo* transition state and produce predominant *exo* product in $[_\pi 6_s + {}_\pi 4_s]$ reactions.

Similar secondary interactions[1,2] favor the chairlike transition state

Chair form   Boat form

over the boatlike transition state in the Cope rearrangement[19]—a [3,3] sigmatropic rearrangement of 1,5-hexadienes (Eq. 3.17)—and related processes

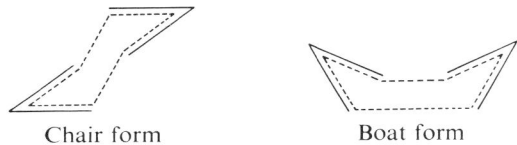

(3.17)

		trans,trans	cis,cis	cis,trans
Racemic	$\xrightarrow{225°}$	90%	10%	0%
meso	$\xrightarrow{180°}$	0%	0%	100%

---

18. The role of other secondary interactions has been stressed by L. Salem, *J. Amer. Chem. Soc.* **90**, 543, 553 (1968), but the conclusions are the same.
19. W. von E. Doering and W. R. Roth, *Angew. Chem. Intern. Ed. Engl.* **2**, 115 (1963); M. Simonetta, G. Favini, C. Mariani, and P. Gramaccioni, *J. Amer. Chem. Soc.* **90**, 1280 (1968); J. E. Baldwin and M. S. Kaplan, *Chem. Commun.* 1354 (1969).

such as the aromatic Claisen rearrangement.[20] However, because of the importance of stereochemical factors in such rearrangements,[19,21] the process involving the boatlike state is still sometimes predominant.

3-5 Summary

The principle of the conservation of orbital symmetry in pericyclic reactions is so fundamental that there can be no exceptions. Any experimental observation which appears to violate this principle must result either because of the existence of discrete intermediates[1,2] or because of geometric, steric, or entropic constraints of various kinds.[1,2] Ionic and neutral reactants are treated similarly since the major variable is the number of electrons involved in the process under consideration.

Some questions have been raised[2,22] as to how a photochemical process occurring through an excited triplet state could ever be concerted and thereby included in the Woodward-Hoffmann treatment. Woodward and Hoffmann have responded[1,2] that there is no requirement that a reaction involving a triplet excited state of a reactant proceed to an excited state of the product with or without spin-flipping. In other words, such a reaction could proceed directly to the ground state of the product in a concerted process, even though the physical rationale for such a radiationless transition has not yet been developed.[23] Other questions have been raised[22] about the proper choice of orbitals for photochemical processes. While arguments can be presented for a frontier-orbital approach involving the HOMO in thermal reactions,[1,2] it is more difficult to justify primary use of the antibonding orbital in the excited-state processes. Nevertheless, Ullman[24] has used orbital symmetry considerations to analyze some excited-state energy transfers.

The ideas inherent in the conservation of orbital symmetry have been applied and extended to many reaction types. Typical examples are application to group transfers and eliminations,[1,2] such as diimide reductions[25] (Eq. 3.18) or hydrogen eliminations; to *cheletropic reactions*,[1,2]

---

20. H. J. Hansen and H. Schmid, *Chem. Brit.* **5**, 111 (1969); G. Frater, A. Habich, H. J. Hansen, and H. Schmid, *Helv. Chim. Acta* **52**, 335 (1969); A. Jefferson and F. Scheinmann, *Quart. Rev.* **22**, 391 (1968).
21. C. L. Perrin and D. J. Faulkner, *Tetrahedron Lett.* 2783 (1969).
22. (a) W. Th. A. M. van der Lugt and L. J. Oosterhoff, *Chem. Commun.* 1235 (1968), and *J. Amer. Chem. Soc.* **91**, 6042 (1969). (b) For a review of orbital symmetry in photochemical transformations, see sources given in notes 2j and 21.
23. For approaches to this and related problems, see R. C. Dougherty, *J. Amer. Chem. Soc.* **93**, 7187 (1971).
24. E. F. Ullman, *J. Amer. Chem. Soc.* **90**, 4158 (1968).
25. J. Jacobus, *J. Chem. Educ.* **49**, 349 (1972).

where two σ bonds to a single atom are made or broken (Eq. 3.19); and to metal-catalyzed and inorganic reactions.[26]

$$\begin{array}{c}\text{H}\\ \text{N}\\ \parallel\\ \text{N}\\ \text{H}\end{array} + \begin{array}{c}\diagdown\diagup\\ \text{C}\\ \parallel\\ \text{C}\\ \diagup\diagdown\end{array} \longrightarrow \text{N}_2 + \begin{array}{c}\text{H}\diagdown\diagup\\ \text{C}\\ \vert\\ \text{C}\\ \diagup\diagdown\text{H}\end{array} \qquad (3.18)$$

$$\underset{\text{H}^{\prime\prime\prime}}{\overset{\text{H}_3\text{C}}{\diagup\hspace{-2pt}\diagdown}}\underset{\text{SO}_2}{\diagdown\hspace{-2pt}\diagup}\underset{\text{CH}_3}{\overset{\text{H}}{\diagdown\hspace{-2pt}\diagup}} \underset{}{\overset{\Delta}{\rightleftarrows}} \text{SO}_2 + \text{\Large$\diagdown\hspace{-4pt}\diagup\hspace{-4pt}\diagdown\hspace{-4pt}\diagup$} \qquad (3.19)$$

cis,trans

Various parallel or alternate theoretical approaches have been presented in the literature. Noteworthy are those of Fukui,[2c] Dewar,[2k,27] Salem,[18,28] Oosterhoff,[22,29] Zimmerman,[30] Trindle,[31] Goddard,[32] and Epiotis[33] (see Chapter 8-8). The Epiotis approach is especially significant because it illustrates the importance of configuration interaction on stereoselectivity and on forbidden character and emphasizes that orbital-symmetry-forbidden processes may be concerted and stereoselective. A perturbation frontier orbital method of analysis[33] also stresses polar influences and substituent effects in pericyclic reactions. Berson and others[34] support the idea that forbidden processes can be concerted and stereoselective and have used this idea to analyze pertinent experimental data with considerable success.

Because of the ease of application of the Dewar approach, it merits further scrutiny. In this approach the transition state is examined directly

---

26. F. D. Mango and J. H. Schachtschneider, *J. Amer. Chem. Soc.* **89**, 2484 (1967), and **91**, 1030 (1969); F. D. Mango, *Tetrahedron Lett.* 4813 (1969); D. R. Eaton, *J. Amer. Chem. Soc.* **90**, 4272 (1968); R. G. Pearson, *J. Amer. Chem. Soc.* **91**, 1252, 4947 (1969), and **94**, 8287 (1972), and *Accounts Chem. Res.* **4**, 152 (1971), and *Chem. & Eng. News* 66 (Sept. 28, 1970); T. H. Whitesides, *J. Amer. Chem. Soc.* **91**, 2395 (1969), W. Th. A. M. van der Lugt, *Tetrahedron Lett.* 2281 (1970).
27. Reference K, Chapter 8; M. J. S. Dewar, *Angew. Chem. Intern. Ed. Engl.* **10**, 761 (1971); C. L. Perrin, *Chem. Brit.* **8**, 163 (1972).
28. L. Salem and J. S. Wright, *J. Amer. Chem. Soc.* **91**, 5947 (1969).
29. J. J. C. Mulder and L. J. Oosterhoff, *Chem. Commun.* 305, 307 (1970).
30. H. E. Zimmerman, *J. Amer. Chem. Soc.* **88**, 1563, 1566 (1966), and *Accounts Chem. Res.* **4**, 272 (1971); H. E. Zimmerman and L. R. Sousa, *J. Amer. Chem. Soc.* **94**, 834 (1972); K. Shen, *J. Chem. Educ.* **50**, 238 (1973).
31. C. Trindle, *J. Amer. Chem. Soc.* **91**, 4936 (1969), and **92**, 3251, 3255 (1970).
32. W. A. Goddard, III, *J. Amer. Chem. Soc.* **94**, 793 (1972), and references therein.
33. N. D. Epiotis, *J. Amer. Chem. Soc.* **94**, 1924, 1935, 1941, 1946 (1972), and **95**, 1191, 1200, 1206, 1214 (1973).
34. J. A. Berson, *Accounts Chem. Res.* **5**, 406 (1972); J. A. Berson and L. Salem, *J. Amer. Chem. Soc.* **94**, 8917 (1972); J. E. Baldwin, A. H. Andrist, and R. K. Pinschmidt, Jr., *Accounts Chem. Res.* **5**, 402 (1972); W. Schmidt, *Tetrahedron Lett.* 581 (1972).

by a perturbational molecular orbital (PMO) treatment.[2k,27] The two crucial properties are the number of mobile electrons participating in the reaction and the number of out-of-phase overlaps present in the transition state. The usual procedure is to consider reaction in or between the lowest occupied molecular orbitals in the acyclic system. Assume that the termini of the π system are combined to form a continuous π-electron system. In the Hückel MO system, the lowest molecular orbitals have no nodes (p. 30), so the phases at the termini being combined are the same. The reaction will occur thermally if the transition-state species resembles an aromatic system, i.e., if it contains $4n + 2$ π electrons interacting in a planar monocyclic system. If the transition-state species resembles an antiaromatic system—if it has $4n$ π electrons—, the reaction will occur photochemically. In both of these cases, the reaction will take place with a stereochemistry dictated by the creation of maximum overlap in the transition state between the similarly-phased termini. For example, the cyclization of hexatriene to cyclohexadiene involves six electrons and would occur thermally in a disrotatory manner; the cyclization of butadiene to cyclobutene, a four-electron process, would occur photochemically in a disrotatory sense; and the Diels-Alder reaction of ethylene and butadiene would be a thermal $(4_s + 2_s)$ process.

An antiaromatic number of mobile electrons can be considered an aromatic number if an odd number of out-of-phase overlaps (where a positive lobe interacts with a negative lobe) is generated. The transition states would then be Möbius[30] rather than Hückel and the rules for aromaticity would be reversed. Thermal cyclization of butadiene to cyclobutene, a four-electron process, would occur through an aromatic transition state if the termini were combined so that a positive lobe would interact with a negative lobe. A conrotatory process would generate this type of interaction, and would be an allowed thermal reaction. Similarly, conrotatory photochemical cyclization of hexatriene to cyclohexadiene, and photochemical $(4_a + 2_s)$ and $(4_s + 2_a)$ Diels-Alder reactions, are permitted.

The Dewar method, therefore, becomes a search for aromatic transition states, which involve $4n + 2$ mobile electrons if they are Hückel (with zero or an even number of out-of-phase overlaps) and $4n$ mobile electrons if they are Möbius (with an odd number of out-of-phase overlaps). The activation-energy differences between allowed and forbidden processes may be estimated using PMO theory.[27] The major advantages offered by this approach are ease of analysis and freedom from symmetry considerations. The Dewar method tells nothing about the details of the reaction coordinate and suffers from the limitations of the $4n + 2$ rule itself (Chapter 2-2). In particular, it cannot be used for most heterocyclic and polycyclic systems. Nevertheless, it offers the promise of providing useful information in an extremely quick way.

General References PART I

A. Liberles, A. *Introduction to Theoretical Organic Chemistry*. New York: Macmillan, 1968.

B. Wiberg, K. B. *Physical Organic Chemistry*. New York: Wiley, 1964.

C. March, J. *Advanced Organic Chemistry: Reactions, Mechanisms, and Structure*. New York: McGraw-Hill, 1968.

D. Gould, E. S. *Mechanism and Structure in Organic Chemistry*. New York: Holt, Rinehart and Winston, 1959.

E. Hine, J. *Physical Organic Chemistry*, 2nd ed. New York: McGraw-Hill, 1962.

F. Coulson, C. A. *Valence*, 2nd ed. Oxford, England: Oxford University Press, 1961.

G. Streitwieser, A., Jr. *Molecular Orbital Theory for Organic Chemists*. New York: Wiley, 1961.

H. Roberts, J. D. *Notes on Molecular Orbital Calculations*. New York: W. A. Benjamin, 1962.

I. Liberles, A. *Introduction to Molecular Orbital Theory*. New York: Holt, Rinehart and Winston, 1966.

J. Riggs, N. V. *Quantum Chemistry*. New York: Macmillan, 1969.

K. Dewar, M. J. S. *The Molecular Orbital Theory of Organic Chemistry*. New York: McGraw-Hill, 1969.

L. Gilliom, R. D. *Introduction to Physical Organic Chemistry*. Reading, Mass.: Addison-Wesley, 1970.

M. Mislow, K. *Introduction to Stereochemistry*. New York: W. A. Benjamin, 1966.

N. Woodward, R. B., and R. Hoffmann. *The Conservation of Orbital Symmetry*. New York: Academic Press, 1970.

PART II

Structure and Reactivity

Nothing is more central to chemical thinking than the concept that each chemical compound or conformation has a three-dimensional structure different from that of any other chemical compound. Each molecule therefore will exhibit a unique set of chemical and physical properties. The chemist must look for general, recognizable features, and then use these features to further his understanding of the intimate details of molecular structure, both as such and as the cause of the various chemical and physical properties. A typical example is the way the organic chemist uses the concept of functional groups. All ketones and aldehydes have in common their carbonyl functionality, and therefore can be converted to derivatives such as oximes. Yet each ketone or aldehyde will react with hydroxylamine at a rate which is uniquely dependent on the exact molecular structure of the carbonyl compound under consideration. Part II will be concerned with the various ways in which structural changes within a given chemical environment (used in its broadest sense) manifest themselves in terms of measurable chemical and physical phenomena.
General references for Part II will be found on page 219.

Chapter 4

Substituent Effects

Within a particular gross molecular framework, such as benzoic acid and its variously substituted analogs, a substituent will probably have a different effect on ground-state phenomena than on excited-state or transition-state phenomena. The way in which a substituent affects those properties that are characteristic of the functional group under consideration should depend on the location of the substituent, a given substituent being able to interact with a given functional group by several different "mechanisms". The chemist who wishes to understand interaction phenomena must first qualitatively analyze such phenomena with respect to both the microscopic and macroscopic chemical systems with which he is dealing. Secondly, he must consider all of the possible ways whereby the substituent could interact with the functional group and thereby cause the observed changes relative to some standard system. Then he must devise experiments which would help to eliminate some or most of the interaction mechanisms he has considered. Lastly, he should attempt to quantitatively evaluate the importance of the various possible modes of interaction, thereby creating the basis for both the comparison of interaction mechanisms in significantly different chemical systems and the mathematical and physical descriptions of the interaction mechanisms.

4-1 Polarity and Polarizability

Under different conditions a molecule may behave as if its electrons were distributed differently. The ground-state electron distribution, or *polarity*, is usually evaluated by the gas-phase dipole moment or by microwave spectroscopy data. The dipole moment is symbolized by ↔, where the head of the arrow is the electron-rich end of the system. The term "polarity" assumes the absence of any perturbing electronic influence; otherwise, it could not be said to be a ground-state property. However, an electric field is imposed on the compound under consideration in order to measure its dipole moment. Therefore, the measured dipole moment μ is really composed of two terms: the permanent dipole moment of the system, and the additional induced dipole moment caused by the presence of the external electric field. This induced dipole moment involves the dynamic response of the system to an external influence, and is therefore an example of the *polarizability* of the compound. The situation is similar to the uncertainty principle in quantum mechanics (Eq. 1.5) in that the act of measurement changes the system being measured. Polarizability is a transition-state property of a molecule, and is usually measured by polarizability constants denoted by α. As will be discussed later (Chapter 8-2), polarizability is related to refractive index since refractive index may be defined as the response of the molecule to the electric vector of the light passing through the sample.

Polarity and polarizability may be cooperative or opposed in a directional sense. Vinyl chloride possesses a dipole moment with the electron cloud distorted toward the chlorine,

$$CH_2 = \overset{\longrightarrow}{CH-Cl}$$

yet it is attacked by electrophiles at the methylene carbon atom:

$$\overset{\frown}{CH_2} = CH \overset{\frown}{-} \overset{..}{Cl} + H^{\oplus} \longrightarrow CH_3 \overset{\oplus}{C}HCl$$

In this example, then, polarity and polarizability are directed in an opposite sense to one another. Similarly, chlorobenzene is represented with the negative end of the dipole on the chlorine, yet undergoes electrophilic aromatic substitution on the carbons *ortho* and *para* to the chlorine substituent.

The direction of the dipole moment of a molecule is based on the relative electronegativities of the component atoms. The total dipole moment of a molecule may be considered as the vector sum of the dipole moments of each bond in the molecule.[1] However, the concept of a bond dipole moment or a group dipole moment really has no direct physical significance, since the only experimental quantity is the dipole moment of the molecule as a whole. For example, while it is usually assumed that C—H bonds or C—C bonds between similarly hybridized carbon atoms

---

[1]. See Reference D, Chapter 3, Part I; Reference A, Sections 9.1–9.10; and Reference J, Sections 8 and 10.

do not possess bond dipole moments, quantum mechanical calculations at almost all levels of sophistication suggest that this assumption is not valid. Nevertheless, this discrepancy does not interfere with most analyses in terms of bond or group moments.

The dipole moment μ is equal to the product of the magnitude of the charges e and the distance between the centers of the charges r:

$$\mu = e \times r \quad (4.1)$$

Since it is impossible to evaluate e or r directly, the bond dipole–group dipole approach is the only practical way to predict dipole moments.

Bromobenzene has an experimental dipole moment of 1.54 D (for debyes, or 10^{-18} esu-cm). Since bromine is more electronegative than carbon, the negative end of the dipole should be toward the bromine. Treating this dipole moment as an aromatic ring–bromine moment, it is

$$\mu = 1.54 \text{ D}$$

possible to calculate dipole moments for other bromobenzene systems using the principles of vector addition. *m*-Dibromobenzene would be predicted to have a dipole moment of 1.54 D pointing from the ring in the direction of the mutually *ortho* position, and this agrees with experiment.

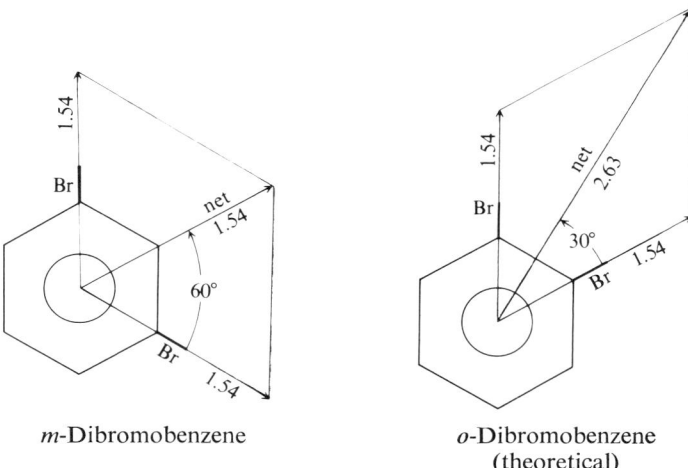

m-Dibromobenzene   *o*-Dibromobenzene
(theoretical)

By similar reasoning, *o*-dibromobenzene should have a 2.63 D dipole moment. However, the experimental value is 2.0 D. This discrepancy can be rationalized as being the result of the nonplanarity of the bromine atoms because of steric interactions between these two bulky groups when they are *ortho* to one another. If each bromine is allowed to lie at a 5° angle with the plane of the benzene ring, one bromine above the ring plane and the other below, then the experimental dipole moment may be calculated.

Such addition processes also permit the evaluation of the direction of a group moment. For example, toluene has an experimental dipole moment of 0.4 D. Since *p*-bromotoluene has an experimental moment of 1.9 D, the methyl-group moment must be directed so as to reinforce the bromine moment. The methyl-group moment therefore is directed toward the aromatic ring. Such a result is reasonable. An sp^2 carbon atom should

$$\mu = 0.4 \text{ D}$$

be more electronegative than an sp^3 carbon atom, because the sp^2 system, having greater *s* character, would hold its electrons closer to the nucleus (see discussion of Table 1.1).

4-2 Inductive and Field Effects

Given the fact that dipole moments may be predicted from group moments and bond moments using vector addition as described in the preceding section, the exact way in which an electronic distribution in one part of a molecule can affect or interact with the electronic distribution in another part of the same molecule remains to be defined. Chemists expect gaseous hydrogen chloride to be polar, because chlorine is more electronegative than hydrogen (p. 18).[2] This type of polarity is simply a question of the relative attractions of two atoms for the electrons in a bond formed between those atoms. It becomes slightly more difficult to predict just how an unequal distribution of electrons will affect an atom not directly bonded to either of the polar centers. Electrostatic interactions with nonbonded sites can occur either through the bonds or through space—either by an *inductive effect* or by a *field effect*, respectively.

Consider an alkyl halide. Because of the electronegativity difference between a carbon atom and a halogen atom, the halogen will act as an electron-withdrawing influence, inducing some semblance of a positive charge on the adjacent carbon atom. This somewhat positively charged

---

2. For a discussion of various approaches to the evaluation of electronegativity, see L. N. Ferguson, *The Modern Structural Theory of Organic Chemistry*, Prentice-Hall, Englewood Cliffs, N.J., 1963, Chapter II.

carbon atom will try to accommodate its induced charge by withdrawing electrons from the atoms to which it is bonded, creating an effect which will be relayed through successive bonds in the molecule:

$$\overset{+\delta\delta\delta\delta}{C} - \overset{+\delta\delta\delta}{C} - \overset{+\delta\delta}{C} - \overset{+\delta}{C} - \overset{-\delta}{X}$$

Because the magnitude of electrostatic effects depends on the size of the perturbing influence, this through-bond inductive effect rapidly decreases in intensity.

Branch and Calvin[3] were the earliest to develop a mathematical description of the inductive model. For inorganic acids derived from water (the —O—H acid group), they approximated acidity by Eq. 4.2:

$$\log K = \log K_{H_2O} + \sum_{\text{charges}} I_\alpha \alpha^i + \sum_{\text{atoms}} I_\alpha \alpha^i + \log \frac{n}{m} \quad (4.2)$$

Using a value of -16 for $\log K_{H_2O}$, i is the number of atoms between the substituent and the oxygen, α is the empirical evaluation of the inductive "fall-off" per bond and is assigned a value of $1/2.8$, I is the empirically determined inductive constant for each substituent atom in the acid (see Table 4.1), n is the number of equivalent ionizing hydrogens, and m is the

TABLE 4.1[3]

Substituent	Value of I^a
Charge (+ or −)	±12.3
Cl	+8.5
Br	+7.5
I	+6.0
O	+4.0
S	+3.4
N	+1.3
P	+1.1
C	−0.4

^{a} + denotes electron-withdrawing relative to hydrogen.

number of equivalent oxygen positions in the anion. In spite of successful calculations, such as that in Fig. 4.1, the Branch and Calvin approach fails miserably for many different types of systems. Resonance stabilization of the anion, such as in carboxylate anions, requires the addition of terms evaluating this resonance stabilization. Such resonance terms are dependent on the type of system involved and have little generality. Charged

---

3. G. E. K. Branch and M. Calvin, *The Theory of Organic Chemistry*, Prentice-Hall, New York, 1941, Chapter 6.

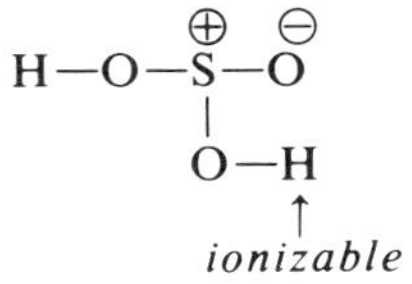

$$H-O-\overset{\oplus}{S}-\overset{\ominus}{O}$$
$$\underset{\underset{ionizable}{\uparrow}}{O-H}$$

From Eq. 4.2,

$$\log K = \underset{\underset{\log K_{H_2O}}{\uparrow}}{-16} + \underset{\underset{+charge}{\uparrow}}{12.3} - \underset{\underset{-charge}{\uparrow}}{\frac{12.3}{2.8}} + \underset{\underset{S}{\uparrow}}{3.4} + \underset{\underset{O}{\uparrow}}{\frac{4}{2.8}} + \underset{\underset{O}{\uparrow}}{\frac{4}{2.8}} + \underset{\underset{positions}{equivalent}}{\log \frac{2}{2}}$$

$\log K_{calc} = -1.8$
$\log K_{obs} = -1.8$

FIGURE 4.1 *Calculation of the acidity of H_2SO_3 by the Branch and Calvin approach.*[3]

Lewis structures create problems; e.g., should HNO_2 be treated as $O=N-O-H$, or as $\overset{-}{O}-\overset{+}{N}-O-H$? Substituent additivity is assumed, so the Branch and Calvin approach cannot accommodate the "saturation effect", which is the nonadditivity in the chemical or physical phenomenon being observed on successive substitution of the same atom or group in the same position. For example, chloroacetic acid is 80 times more acidic than acetic acid, dichloroacetic acid is 3000 times more acidic than acetic acid, and trichloroacetic acid is 13,000 times more acidic. On a relative basis, the second chlorine increases the acidity by a factor of less than the 80 associated with the first chlorine, and the third chlorine is acid-strengthening by a still smaller factor than the second chlorine. Adding each additional group is not as effective as adding the previous group with respect to the property being measured. Stereochemical considerations and hydrogen-bonding phenomena are also ignored in the Branch and Calvin approach: Maleic and fumaric acid would be expected in the terms of this approach to have identical second ionization constants, disagreeing with experiment.

In addition to the inductive effect, there is an electrostatic effect through "space", a field effect. The magnitude of such an effect depends on the magnitude, orientation, and shortest-path distance of the electrostatic perturbing influence, the substituent, from each atom under consideration. The transmission of the field effect within a given conformation therefore is a function of the dielectric constant of the medium through which the interacting lines of force must pass, acting through solvent molecules and/or portions of the molecule itself by dipolar or induced dipolar forces.

Kirkwood and Westheimer[4] attacked this problem of the medium through which the field effect operates by considering a model consisting of a spherical or ellipsoidal[4] cavity of low dielectric constant (the hydrocarbon portion of the molecule) surrounded by a medium of higher dielectric constant (the solvent) in order to calculate an "effective dielectric constant". The Kirkwood-Westheimer approach considers the field effect as a special case of the basic physical interactions between two poles: the substituent dipole (compared to the standard substituent hydrogen) and the "reaction center", the position where the effect is measured.

Consider the effect on K, the ionization constant of a carboxylic acid, of replacing a hydrogen in the acyl portion of the system with a substituent X. The dipole substituent effect may be calculated from Eq. 4.3,

$$\log \frac{K_X}{K_H} = \frac{e\mu \cos \theta}{2.3kTr^2 D_E} \tag{4.3}$$

where μ is the group or bond dipole moment of the substituent, r is the distance between the midpoint of the substituent dipole vector and the ionizable carboxyl proton, θ is the angle between the vector r and the dipole moment vector μ, and D_E is the effective dielectric constant as defined by the molecular geometry and an estimate of the molar volume. Because the substituent dipole vector cannot be specified in any meaningful manner, r and θ are not theoretically or experimentally accessible. For mathematical reasons, Kirkwood and Westheimer assumed a continuous internal dielectric constant for the molecular cavity, D_i, with a value of 2.0 derived from the liquid paraffin hydrocarbons. This internal dielectric constant of the molecular framework changes to some extent as a function of any through-bond inductive effects which might be operative, but the contribution of D_i to D_E should be relatively insensitive to the character of the hydrocarbon bonds within the cavity.[4] Therefore, while an accurate assessment of the field effect by the Kirkwood-Westheimer approach requires some knowledge of any inductive effect that is present, the uncertainties in r and θ are serious enough that D_E can be treated as an arbitrary parameter almost independent of the value of D_i. As we will see, the Kirkwood-Westheimer model underestimates by a factor of 2 the ΔpK_a values for acids with similar values of D_E, even for rigid bicyclic systems where r and θ are easiest to approximate.

Roberts and Moreland[5] used the pK_a's of 4-substituted bicyclo-[2.2.2]octane-1-carboxylic acids as a model for the evaluation of the relative importance of field and inductive effects. Because of the rigidity of

---

4. See (a) S. Ehrenson, *Progr. Phys. Org. Chem.* **2**, 195 (1964), (b) H. D. Holtz and L. M. Stock, *J. Amer. Chem. Soc.* **86**, 5188 (1964), and (c) J. T. Edward, P. G. Farrell, and J. L. Job, *J. Chem. Phys.* **57**, 5251 (1972).
5. J. D. Roberts and W. T. Moreland, Jr., *J. Amer. Chem. Soc.* **75**, 2167 (1953).

[Structure: bicyclic system with X substituent and COOH group]

this bicyclic system, the molecular geometry may be estimated with reasonable accuracy. Calculations based on the Kirkwood-Westheimer method give qualitative agreement with the experimental results, but quantitatively they lead to predictions of substituent effects which are roughly half of those found experimentally. Roberts and Moreland concluded that either the Kirkwood-Westheimer treatment is inadequate or an inductive effect exists independent of the field effect and roughly equal in magnitude.

Nevertheless, pK_a data in bicyclic systems (where conformational effects should be a minimum) or in other aliphatic systems permit a qualitative comparison of the electrostatic effects of various substituents relative to a hydrogen atom. Any substituent which is acid-strengthening (or base-weakening) relative to hydrogen in such aliphatic systems acts by an electrostatic electron-withdrawal process,[6] and is said to have a negative electrostatic effect, denoted by $-I$. Those substituents which are poorer electron-attractors than hydrogen have positive electrostatic effects, symbolized by $+I$. Note that this qualitative classification, shown in Table 4.2, is

TABLE 4.2 *Inductive Effects*

$-I$	$+I$
–Ammonium ions	$-R$
$-NO_2$	$-COO^{\ominus}$
$-C{\equiv}N$	
$-COOR$	
$-COR$	
–Halides	
$-OR$	
$-SR$	
$-C{=}C$	
$-C{\equiv}C$	
$-C_6H_5$	

---

6. R. T. Morrison and R. N. Boyd, *Organic Chemistry*, 3rd ed., Allyn and Bacon, Boston, 1973, Chapters 5.17, 18, and 22; Reference C, Chapter 1; and Reference D, Chapter 7.

independent of whether the electrostatic interaction mechanism is of the inductive type, the field type, or some combination of the two. Electrostatic interactions can be measured in the appropriate systems by acidity or basicity techniques, dipole moment measurements, intensities and band shifts in infrared spectra, nmr shielding parameters, or Taft polar substituent constants (Chapter 6-2), to name the most common methods.

The question of the existence of separate field and inductive effects as such might still be considered open at this point in our discussion. The significant experiments were performed by Roberts and Carboni[7] in 1955. These workers investigated the pK_a's of various ring-substituted phenylpropiolic acids,

$$R-C_6H_4-C\equiv C-CO_2H$$

For certain *ortho* substituents, such as halogens, these acids would be examples of "inverted" molecules,[8] systems wherein the inductive effect and the field effect should act in opposite directions because of the direction of the dipole of the substituents. Thus, for an *ortho* substituent such as a chlorine atom, the electron-withdrawing inductive effect should increase the acidity; and since the negative end of the C—Cl dipole is closer than the positive end to the carboxy proton, the field effect should decrease the acidity. The experimental result that *o*-chlorophenylpropiolic acid and related compounds are *less* acidic than the *meta* or *para* analogs provides an important piece of evidence for the existence and importance of a field effect. At the same time, the fact that the *o*-chloro compound is also *more* acidic than the unsubstituted species is evidence for an inductive effect. Roberts and Carboni were able to conclude that both the field effect and the inductive effect are present in systems of this type, and that these two electrostatic interaction mechanisms are of comparable importance. Extrapolation of this interpretation to the Roberts and Moreland results discussed above[5] suggests that the Kirkwood-Westheimer approach does have considerable validity for the calculation of field effects.

Dewar[9] has considered the relative importance of field and inductive effects in several systems. All of his evidence indicates that the field effect acting through space, including space occupied by bonds, accounts for all

---

7. J. D. Roberts and R. A. Carboni, *J. Amer. Chem. Soc.* **77**, 5554 (1955). For additional verification see K. Bowden, M. J. Price, and G. R. Taylor, *J. Chem. Soc., B* 1022 (1970), and earlier papers.
8. C. F. Wilcox and C. Leung, *J. Amer. Chem. Soc.* **90**, 336 (1968).
9. (a) M. J. S. Dewar and A. P. Marchand, *J. Amer. Chem. Soc.* **88**, 354 (1966), and previous papers. (b) For an analysis of the various types of effects in substituted benzenes and their relative importance, see A. R. Katritzky and R. D. Topsom, *J. Chem. Educ.* **48**, 427 (1971).

known results. He implies that the inductive effect through both σ and π bonds should be treated as part of the field effect. Sheppard and Henderson[10] present results supporting Dewar's point of view, including solvent effects which can best be explained by solute–solvent interactions creating a reduction in the magnitude of a field effect. However, several workers[11] feel that the inductive effect through π systems is more important than the field effect.

Baker, Parish, and Stock[12] have compared the 4-substituted bicyclo-[2.2.2]octane-1-carboxylic acids (**I**; investigated also by Roberts and Moreland, note 5), 4-substituted bicyclo[2.2.2]oct-2-ene-1-carboxylic acids (**II**), and 4-substituted dibenzobicyclo[2.2.2]octa-2,5-diene-1-carboxylic acids (**III**), in order to evaluate the effects of variation in the hybridization of the carbon–carbon bonds between the substituent and the reaction site;

and structure **I** and some cubanecarboxylic acids (**IV**), to measure the effects of variation in the number of paths between the substituent and the reaction site. The inductive effect should depend on both the hybridization of the bonds (greater transmission through the more polarizable π bonds) and the number of paths (transmission proportional to the number of paths). On the other hand, the field effect[4] should be more or less independent of these variables since minor changes in the nature of the cavity should not be too important. The substituent effects on the thermodynamic dissociation constants in any set of these compounds were found to be almost identical with those in every other set, results which are consistent only with the presence of a field effect. Kirkwood-Westheimer calculations reinforce this conclusion,[12] although there is some difficulty with the way in which alkyl and hydrogen substituents behave (see Chapter 4-4). Additional support for this preponderance of the field effect over the inductive effect has been presented, both from experimental results[13]

---
10. W. A. Sheppard and R. M. Henderson, *J. Amer. Chem. Soc.* **89**, 4446 (1967).
11. G. L. Anderson and L. M. Stock, *J. Amer. Chem. Soc.* **91**, 6804 (1969); Y. Nomura and Y. Takeuchi, *Tetrahedron Lett.* 5585 (1968).
12. F. W. Baker, R. C. Parish, and L. M. Stock, *J. Amer. Chem. Soc.* **89**, 5677 (1967).
13. Reviewed by L. M. Stock, *J. Chem. Educ.* **49**, 400 (1972); see also D. S. Noyce and coworkers, *J. Org. Chem.* **34**, 1247, 1252 (1969); C. L. Liotta, et al., *Chem. Commun.* 1251 (1969); *J. Amer. Chem. Soc.* **94**, 2129, 4891 (1972); R. Golden and L. M. Stock, ibid., **94**, 3080 (1972).

and from calculations.[14] Bowden[15] has suggested that the field effect predominates, an inductive effect through π bonds becoming important only wherever the substituent and reaction site are linked by a continuous π system.

Pople and Gordon[16] have used a molecular orbital approach to calculate charge distributions and dipole moments. In molecules containing polar substituents, their calculations suggest an alternation in magnitude of polarity in both saturated and unsaturated systems. While this alternation is well known in unsaturated systems, where it is attributed to resonance interactions (see Chapter 4-3), neither an inductive effect nor a field effect of the type thus far considered could account for such behavior in saturated molecules. The alternation calculated may be illustrated with n-butyl fluoride. The fluorine is a strong attractor of σ electrons from the entire alkyl group, but particularly from the carbon to which it is directly bonded (the α carbon). Electron-withdrawal from the β position is less than that from the α position, just as suggested for an inductive effect (p. 92). However, electron-withdrawal from the γ position is

$$\begin{array}{ccccc} +\delta\delta\delta\delta & +\delta\delta & +\delta\delta\delta & +\delta & -\delta \\ C & - C & - C & - C & - X \\ \delta & \gamma & \beta & \alpha & \end{array}$$

greater than that from the β position. The order of decreasing positive charge on carbon in such a saturated system would therefore be

$$\alpha > \gamma > \beta > \delta.$$

Pople and Gordon interpret their results in terms of a "back-donation" by a lone pair of electrons on the fluorine analogous to the polarization observed in vinyl chloride (p. 89). This alternation in magnitude of polarity is ascribed to an interaction of unspecified nature between σ electrons and nonbonding electrons (or π electrons).

Pople and Gordon have used their calculated results to propose a double classification scheme for substituent effects in saturated systems.[16] According to their proposal, the usual $+I$ or $-I$ electrostatic classification (Table 4.2) is applied to the substituent effect with respect to the hydrocarbon fragment as a whole. A $+$ or a $-$ superscript is added, denoting polarization of the σ electrons remaining in the hydrocarbon fragment by the n or π electrons of the substituent. Those substituents classified as $-I^+$, such as $-F$, $-OR$, and $-NR_2$, have the most electronegative atom directly attached to the hydrocarbon, while the $-I^-$ substituents $-CF_3$, $-COR$, $-CN$, $-NO_2$, and $-CO_2R$ have the electronegative atom one

---

14. R. B. Hermann, *J. Amer. Chem. Soc.* **91**, 3152 (1969).
15. K. Bowden and D. C. Parkin, *Can. J. Chem.* **46**, 3909 (1968), and earlier papers.
16. J. A. Pople and M. Gordon, *J. Amer. Chem. Soc.* **89**, 4253 (1967).

position removed. The superscript used corresponds to the sign assigned to the resonance effect of the given substituent (see Table 4.3).

Support for the Pople-Gordon alternation in magnitude of polarity may be found in nmr coupling constants,[17,18] e.g., the influence of electronegative substituents (X) on J_{AB} as a function of the number of intervening atoms[18] in systems such as

$$\begin{array}{ccc} \text{C—C—C} & \text{and} & \text{C—C—C—C} \\ |\ \ |\ \ | & & |\quad\ \ |\ \ | \\ \text{X}\ \ \text{H}_A\ \text{H}_B & & \text{X}\quad \text{H}_A\ \text{H}_B \end{array}$$

Calculations by Brownlee and Taft[19] and Hehre and Pople[20] using different calculation techniques support the Pople and Gordon results,[16] but another set of calculations disagree.[21] Wilcox and Leung,[8] in an analysis of substituent effects in several bicyclic systems, conclude that either field effects predominate in their systems or alternating effects play a role, but they comment that the latter possibility cannot be adequately evaluated yet.

The bulk of evidence at this time suggests that any effect within σ systems which drops off proportionately with distance is probably a field effect. It may be that any effect through the σ bonds themselves, any inductive effect, will show some sort of alternation in magnitude. However, Hammett has cautioned that the field effect and the inductive effect should never be treated as distinct physical phenomena since they are only empirical models.

4-3 Resonance Effects

Resonance effects[9b,22] (also called *conjugation effects* or *tautomeric effects*) are those substituent effects propagated by a substituent-induced polarization of a π system. Resonance effects are distributed throughout an unsaturated or aromatic system without the sharp decrease with distance noted for an inductive or field effect (Chapter 4-2) but are transmitted primarily to alternate atoms in the π system.[23]

---

17. S. Castellano and R. Kostelnik, *J. Amer. Chem. Soc.* **90**, 141 (1968).
18. A. D. Cohen and T. Schaefer, *Mol. Phys.* **10**, 209 (1966), as reviewed by S. Sternhell, *Quart. Rev.* **23**, 236 (1969).
19. R. T. C. Brownlee and R. W. Taft, *J. Amer. Chem. Soc.* **92**, 7007 (1970).
20. W. J. Hehre and J. A. Pople, *J. Amer. Chem. Soc.* **92**, 2191 (1970).
21. M. E. Schwartz, C. A. Coulson, and L. C. Allen, *J. Amer. Chem. Soc.* **92**, 447 (1970).
22. Reference A, Chapter 4; Reference C, pp. 231–233; Reference D, Chapters 3 and 7; Reference E, pp. 92–93; Reference I, Chapter 11; J. B. Hendrickson, D. J. Cram, and G. S. Hammond, *Organic Chemistry*, 3rd ed., McGraw-Hill, New York, 1970, Section 8.4.
23. The Pople-Gordon alternating substituent effect (p. 98) shows the characteristics of a resonance effect but is not one according to this complete definition.

It is often difficult to separate electrostatic (field and/or inductive) interactions from resonance interactions.[9b] For example, the pK_a of phenol is about 10 and the pK_a of ethanol is approximately 16. The direction of this acidity difference is expected, since a phenyl group is a stronger electron-attractor than a saturated alkyl group, but six powers of 10 is an unusually large electrostatic effect. Favorable resonance interactions in the phenoxide ion which are not present in phenol itself, in ethanol, or in the ethoxide ion provide a reasonable explanation for the magnitude of this acidity difference. Nevertheless, it is impossible to decide from arguments like these how much of this difference is an electrostatic effect and how much is a resonance effect.

The same problems and the same arguments are present in the basicity comparison of aniline and aliphatic amines. Aniline is 10^6 times less basic than an aliphatic amine because of the electron-withdrawing nature of the phenyl ring and a resonance interaction between the lone pair of electrons on nitrogen and the π electrons of the aromatic ring in aniline itself. However, here it is possible to roughly separate the electrostatic and resonance influences. The pK_a of quinuclidine is 10.58; that of benzoquinuclidine is 7.79. Conjugation effects should be minimal in the benzo-

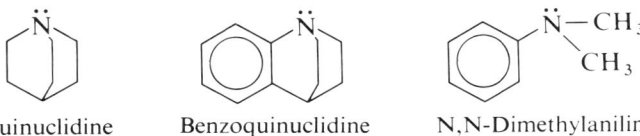

Quinuclidine   Benzoquinuclidine   N,N-Dimethylaniline

quinuclidine since the strain in the bicyclic system would prevent the lone-pair electrons from assuming the proper orientation for a resonance interaction with the aromatic ring (Fig. 4.2). The difference in basicity between quinuclidine and benzoquinuclidine should therefore be a measure only of the electrostatic effect of the aromatic ring. The difference in pK_a between an aniline and benzoquinuclidine, then, would measure the

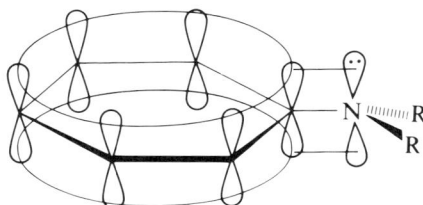

FIGURE 4.2 *The optimum orientation for a resonance interaction in an aniline system.*

resonance effect. Since the pK_a of N,N-dimethylaniline is 5.06, the resonance effect and the electrostatic effect in anilines are of comparable importance.

Resonance effects are often evident from dipole moment studies (Chapter 4-1). A dipole moment of 4.34 D would be expected for *p*-nitrophenol by addition of the dipole moments of nitrobenzene and phenol, each of which already possesses a resonance interaction between the substituent and the aromatic ring. The observed dipole moment of 5.04 D may be explained by invoking an additional resonance interaction involving the nitro group and the hydroxy group through the aromatic ring:

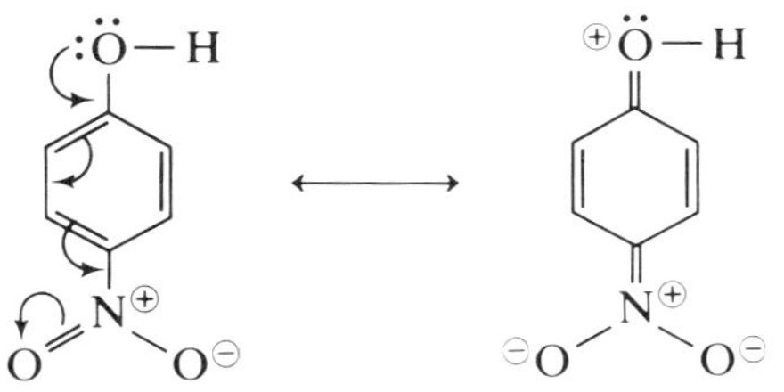

The behavior of such systems in the presence of additional substituents supports such an interpretation. For example, the calculated dipole moment of *p*-nitroaniline is 5.17 D, the observed dipole moment is 6.2 D, and the observed dipole moment of 2,3,5,6-tetramethyl-4-nitroaniline (**V**)

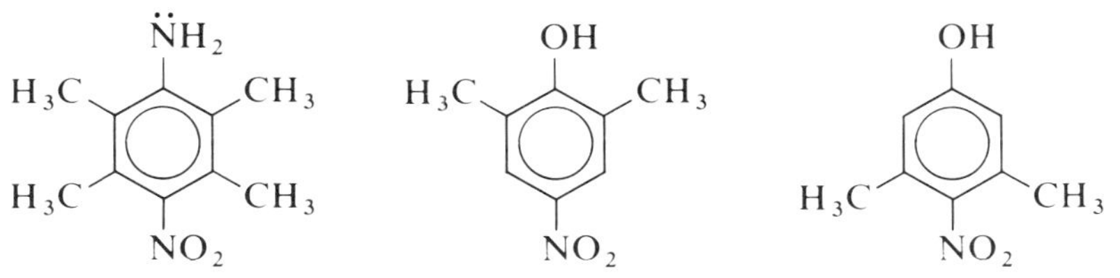

is 5.0 D. The latter effect cannot be the result of either electrostatic or resonance effects involving the methyl groups, because they are symmetrically disposed. The methyl groups must be introducing steric factors which prevent the coplanarity of the two relevant nitrogen orbitals and the aromatic ring π orbitals required for the through-ring resonance interaction. This steric effect is called *steric inhibition of resonance*.

Steric inhibition of resonance may also be illustrated using pK_a's. Phenol **VII** should be a stronger acid than phenol **VI**, because the electrostatic acid-weakening methyl groups operate over a shorter distance in **VI**. However, the pK_a of **VI** is 7.16 and that of **VII** is 8.24. Steric inhibition of resonance is possible in both phenols. Yet, precisely because it is a steric phenomenon, it assumes more importance with the larger NO_2, thereby preventing the acid-strengthening resonance interaction with the nitro group in **VII**.

Substituents may be classified as electron-donating by resonance ($+R$) or electron-withdrawing by resonance ($-R$) (Table 4.3) in the same

TABLE 4.3 *Resonance Effects*

$+R$	$-R$
–Halide	$-NO_2$
$-OR$	$-CN$
$-O-\overset{O}{\overset{\|}{C}}-R$	$-\overset{O}{\overset{\|}{C}}-R$
$-SR$	
$-NR_2$	$-\overset{O}{\overset{\|}{C}}-OR$
$-NH-\overset{O}{\overset{\|}{C}}-R$	$-\overset{O}{\overset{\|}{C}}-NR_2$
$-O^{\ominus}$	
$-S^{\ominus}$	$-CF_3$
–Alkyl	

way electrostatic effects were denoted (Table 4.2). Notice that each $+R$ group (except the alkyl groups) is attached to the π system by an atom possessing one or more unshared pairs of electrons, and that each $-R$ group is not attached to the π system by its most electronegative atom. In general, conjugation is more important for substituents which are first-row elements, because in these elements the $2p$ orbitals are most similar in size and energy to π orbitals, thereby producing better overlap and stronger bonds (see Chapter 1-2).

Any consideration of substituent effects through π systems must involve both electrostatic and resonance effects. For example, *m*-hydroxybenzoic acid is a stronger acid than benzoic acid, and *p*-hydroxybenzoic acid is a weaker acid than benzoic acid. In the *meta* compound, only the $-I$ effect of the hydroxy group is felt since one usually assumes that resonance interactions occur only between *ortho* or *para* substituents on a benzene ring (but see Chapter 6-3). This $-I$ effect stabilizes the carboxylate anion, thereby increasing the acidity of the compound. In the *para* compound, both the $-I$ and $+R$ effects of the hydroxy group are operative. Since the *para* acid is weaker than benzoic acid, the $+R$ effect of the hydroxy group must have greater magnitude than the $-I$ effect of the hydroxy group *in this series*.

4-4 Alkyl-group Effects: Hyperconjugation

As noted previously (p. 97), alkyl-group effects often do not fit into an otherwise reasonably consistent scheme of substituent effects. This difficulty may be considered to be threefold. First of all, exactly what are the electrostatic effects of alkyl groups? Secondly, how can an alkyl group exhibit a resonance effect when it possesses neither π nor nonbonding electrons for conjugative interactions? Thirdly, and possibly not independent of the first two problems, how are the steric effects of the larger alkyl groups adequately evaluated?

Historically, alkyl substituents connected to sp^2 carbon atoms were found to be electron-donating relative to hydrogen. Well-known examples are the order of carbonium ion stabilities ($3° > 2° > 1°$),[24] the activating effect of alkyl groups toward electrophilic aromatic substitutions,[25] and the magnitude and direction of the dipole moments of the alkylbenzenes:

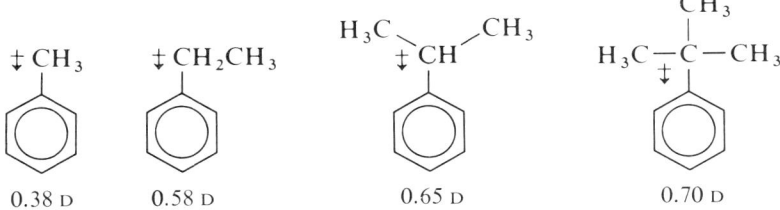

Some difficulties were encountered in terms of the so-called Baker-Nathan effect (p. 105) and in the order of basicity of the methylamines in aqueous media (Chapter 11-6):

$$NH_3 < CH_3NH_2 < (CH_3)_2NH > (CH_3)_3N$$

Nevertheless, the consensus of opinion among chemists was that alkyl groups were electron-releasing in all situations; e.g., methanol should be more acidic than ethanol, and a primary carbanion should be more stable than a tertiary carbanion.

While a controversy raged over the nature of the Baker-Nathan effect (to be discussed below), the order of amine basicities was shown to be solvent-dependent,[26] the gas-phase order indicating consistent electron

---

24. R. T. Morrison and R. N. Boyd, *Organic Chemistry*, 3rd ed., Allyn and Bacon, Boston, 1973, Chapters 5.18 and 8.21; Reference A, Section 9.13.
25. R. T. Morrison and R. N. Boyd, *Organic Chemistry*, 3rd ed., Allyn and Bacon, Boston, 1973, Chapters 11 and 12; Reference D, p. 431.
26. E. Grunwald and M. Cocivera, *Discussions Faraday Soc.* **39**, 105 (1965), and earlier papers; E. Grunwald and E. K. Ralph, *Accounts Chem. Res.* **4**, 107 (1971).

release by the alkyl groups:[27,28]

$$NH_3 < CH_3NH_2 < (CH_3)_2NH < (CH_3)_3N$$

Kwart[29a] and Laurie[29b] suggested that alkyl groups might be slightly electron-withdrawing when bonded to an sp^3 carbon atom, while Stock[12] argued that the electronic effect of a methyl group bonded to an sp^3 carbon atom would be little different from that of a hydrogen atom.

The bombshell arrived in 1968, when Brauman and Blair[28,30] used ion cyclotron resonance to measure the *gas-phase* acidities of alcohols and observed orders of acidity consistent only with electron-withdrawing alkyl groups:

$$(CH_3)_3CCH_2OH > (CH_3)_3COH > (CH_3)_2CHOH$$
$$> CH_3CH_2OH > CH_3OH > H_2O$$

$$(CH_3)_3COH \cong CH_3(CH_2)_3CH_2OH \gtrapprox CH_3(CH_2)_2CH_2OH$$
$$> CH_3CH_2CH_2OH > CH_3CH_2OH$$

Brauman concluded that alkyl-group effects result from the polarizability of the alkyl units and their proximity to a charged center. An alkyl group will stabilize either a positively charged center or a negatively charged center relative to a neutral species, with the stabilization increasing as the size or complexity of the alkyl group increases. This "internal solvation"[28,30] has been supported by molecular orbital calculations,[30-32] including calculated extrapolations to gas-phase alkane acidities[32] (tertiary carbanion stabilized relative to secondary carbanion, etc.), and by some nmr observations,[33] although the validity of the conclusions reached in the latter have been questioned.[34] Robinson[34] prefers that a steric argument presented originally by Schleyer and Woodworth[35] in a slightly different context be applied to the nmr data. When comparing alkyl-

---

27. M. S. B. Munson, *J. Amer. Chem. Soc.* **87**, 2332 (1965); J. I. Brauman, J. M. Riveros, and L. K. Blair, ibid., **93**, 3914 (1971); E. M. Arnett, F. M. Jones, III, M. Taagepera, W. G. Henderson, J. L. Beauchamp, D. Holtz, and R. W. Taft, ibid., **94**, 4724 (1972); D. H. Aue, H. M. Webb, and M. T. Bowers, ibid., **94**, 4726 (1972); J. P. Briggs, R. Yamdagni, and P. Kebarle, ibid., **94**, 5128 (1972); I. Dzidic, ibid., **94**, 8333 (1972).
28. J. I. Brauman and L. K. Blair, *J. Amer. Chem. Soc.* **90**, 6561 (1968).
29. (a) H. Kwart, et al., *J. Amer. Chem. Soc.* **83**, 4552 (1961), and **86**, 1161 (1964).
(b) V. W. Laurie and J. S. Muenter, *J. Amer. Chem. Soc.* **88**, 2883 (1966).
30. J. I. Brauman and L. K. Blair, *J. Amer. Chem. Soc.* **92**, 5986 (1970), and **93**, 3911, 4315 (1971). For supporting evidence from flowing afterglow techniques, see D. K. Bohme, E. Lee-Ruff, and L. B. Young, ibid., **94**, 5153 (1972).
31. R. B. Hermann, *J. Amer. Chem. Soc.* **92**, 5298 (1970); W. J. Hehre and J. A. Pople, *Tetrahedron Lett.* 2959 (1970).
32. T. P. Lewis, *Tetrahedron* **25**, 4117 (1969); P. H. Owens, R. A. Wolf, and A. Streitwieser, Jr., *Tetrahedron Lett.* 3385 (1970).
33. L. M. Jackman and D. P. Kelly, *J. Chem. Soc., B* 102 (1970).
34. P. M. E. Lewis and R. Robinson, *Tetrahedron Lett.* 2783 (1970).
35. P. von R. Schleyer and C. W. Woodworth, *J. Amer. Chem. Soc.* **90**, 6528 (1968).

substituent effects in several rigid systems, Schleyer[35] observed better correlations between substituent effects in different carbon frameworks when the hydrogen substituent, the unsubstituted compound, was omitted from the correlation. The introduction of a steric effect on replacing a hydrogen with an alkyl group may be seen in the change in C—C—C bond angle in going from isobutane (111.15°) to neopentane (109.5°). While the Schleyer-Woodworth work[35] does point out the importance of "substituent-induced structural changes",[36] it cannot provide an explanation for the Brauman-Blair results.[28,30]

The Baker-Nathan effect, while first reported for an S_N2 reaction,[37] is easier to discuss in the context of electrophilic aromatic substitutions. Rates of bromination of alkylbenzenes indicate a reactivity order of: toluene (340) > ethylbenzene (290) > cumene (180) > t-butylbenzene (110) > benzene (1). It is immediately apparent that, while all of the alkyl groups are electron-releasing, the order of electron release (and benzenonium ion stabilization) is: methyl > ethyl > isopropyl > t-butyl, the opposite of the "normal electrostatic order" discussed previously. Two general explanations have been presented to explain this order of electron release: the *hyperconjugation model* and the *solvation model*.

Hyperconjugation may be defined as the partial delocalization of σ electrons from a bonding C—H orbital,[38] a process for which there is spectroscopic evidence[39] but one for which the question of whether there is sufficient stabilization to affect a chemical process might be open to debate.[38] Assuming that this special conjugation effect occurs to a greater extent with C—H orbitals than with C—C orbitals,[40] toluene, with three C—H orbitals in the methyl group, should be able to donate electrons to the aromatic π system better than any other alkylbenzene, and the order observed should be merely a function of the number of C—H bonds present at the benzylic carbon atom.

In order to visualize the hyperconjugation model more clearly, let us take linear combinations of the three hydrogen atoms of the methyl group in toluene in such a way as to produce three molecular orbitals, one of which has σ and two of which have π symmetries.[41] The orbitals on the

---

36. G. L. Anderson and L. M. Stock, *J. Amer. Chem. Soc.* **90**, 212 (1968), and **91**, 6804 (1969).
37. J. W. Baker and W. S. Nathan, *J. Chem. Soc.* 1844 (1935).
38. E. M. Arnett and J. W. Larsen, *J. Amer. Chem. Soc.* **91**, 1438 (1969), and references cited therein.
39. T. Sorensen, *J. Amer. Chem. Soc.* **89**, 3782, 3794 (1967).
40. While it is usually assumed that C—C hyperconjugation is less than 75% as effective as C—H hyperconjugation, F. R. Jensen and B. E. Smart, *J. Amer. Chem. Soc.* **91**, 5686, 5688 (1969), have suggested that C—C hyperconjugation may be comparable to C—H whenever the C—C bonds contain a higher than usual amount of p character.
41. Reference B, Section 1.15.

benzylic carbon then could consist of two sp hybrids which could bond with the aromatic ring and the σ hydrogen orbital, and two p orbitals which could bond with the two π hydrogen orbitals while also interacting with the aromatic π system. The methyl group then would consist of two localized bonds (one σ and one π) in the plane of the aromatic ring, and one delocalized bond (π) perpendicular to the aromatic ring:

$$\langle \bigcirc \rangle - \overline{C = H_3}$$

Since the two π hydrogen orbitals produce a cylindrically symmetrical system (just as for the π electrons in acetylene), there should be no stereochemical requirement for hyperconjugation. Calculations for this model in the ground state using simple LCAO–MO give an electron distribution in good agreement with the observed dipole moments (p. 103). However, LCAO–MO exaggerates dipole moments, so the calculated dipole moment should be larger than the experimental dipole moment for the model to be considered adequate. This suggests that hyperconjugation is not important in the ground state.[42] Attempts to apply this model to transition states lead to results agreeing qualitatively with the Baker-Nathan order, suggesting that the hyperconjugation concept might be important in transition states.

Nonkinetic experimental support for the hyperconjugation model has been provided by thermodynamic measurements performed by Arnett's group.[38]

The hyperconjugation model is often couched in a valence-bond framework.[43] In this sense, hyperconjugation is essentially a "no-bond resonance". It is invoked to rationalize such phenomena as carbonium ion stabilities[43] ("isovalent hyperconjugation", Eq. 4.4), the greater stability of the more highly substituted alkene[43] ("sacrificial hyperconjugation", Eq. 4.5), and the electrophilic aromatic substitution reaction rates illustrating the Baker-Nathan effect (Eq. 4.6 for *para* attack by electrophile

$$\underset{H}{\overset{H}{\underset{|}{H-C-\overset{\oplus}{C}H_2}}} \longleftrightarrow \underset{H}{\overset{H}{H-C\overset{\oplus}{=}CH_2}} \longleftrightarrow \text{etc.} \qquad (4.4)$$

---

42. For a ground-state phenomenon for which hyperconjugation is presented as the possible cause, see G. W. Wahl, Jr., and M. R. Peterson, Jr., *J. Amer. Chem. Soc.* **92**, 7238 (1970).
43. R. T. Morrison and R. N. Boyd, *Organic Chemistry*, 3rd ed., Allyn and Bacon, Boston, 1973, Chapters 5.18, 6.28, and 8.21; Reference A, Sections 9.10 and 9.13; Reference C, pp. 56–59; Reference D, pp. 49–50; Reference E, Section 1-1d; and L. Radom, J. A. Pople, P. v. R. Schleyer, and coworkers, *J. Amer. Chem. Soc.* **94**, 5935, 6221 (1972).

$$\begin{array}{c}\text{H}\\\text{H}-\text{C}\\\text{H}\diagdown\\\text{H}\diagup\text{C}=\text{CH}_2\end{array}\quad\longleftrightarrow\quad\begin{array}{c}\text{H}\\\text{H}-\text{C}\diagdown\\\text{H}\diagup\text{C}-\dot{\text{C}}\text{H}_2\\\text{H}\end{array}\quad\longleftrightarrow\quad\text{etc.}\qquad(4.5)$$

(4.6)

E). In each situation the basic argument is that the more equivalent hyperconjugated resonance contributors that can be written, the more stable the resonance hybrid will be (p. 52). As discussed on page 103, the order of carbonium ion stabilities may be rationalized by electron-releasing alkyl groups, so Eq. 4.4 is unnecessary. Both alkene stabilities and arene dipole moments depend on the presence of sp^3-sp^2 carbon–carbon single bonds, which are shorter and more stable (Table 1.1) than sp^3-sp^3 single bonds. The only remaining possible role for hyperconjugation would be to explain the Baker-Nathan effect.

The second model to explain the Baker-Nathan order is the solvation model, originally presented by Schubert and Sweeney.[44,45a] The essence of this hypothesis is that the reaction medium plays a major role in the origin of the Baker-Nathan effect, particularly through alkyl-substituent steric hindrance to solvation of the transition state. The Baker-Nathan order, then, represents increasing ability of the alkyl group with increasing size to prevent solvation of the charged transition state. The Schubert-Sweeney model does not negate the possibility of an important C—H hyperconjugation, but does assume the "normal inductive order" regardless of electron demand. In other words, steric hindrance to transition-state solvation by the larger alkyl groups overcomes the "normal" order of alkyl electron release, which may or may not include C—H hyperconjugation. Striking support for the role of solvation in such alkyl-group phenomena is evident in the rates of alkaline hydrolysis of *para*-substituted ethyl benzoates. The *p*-methyl compound hydrolyzes faster than the *p*-*t*-butyl compound in 56% aqueous acetone, yet reacts slower in 85% aqueous ethanol. Further dramatic evidence for such a solvent effect is provided by Larsen and coworkers,[45b] who have observed the

---

44. W. M. Schubert and W. A. Sweeney, *J. Org. Chem.* **21**, 119 (1956).
45. (a) M. J. S. Dewar, *Hyperconjugation*, Ronald Press, New York, 1962; *Chem. & Eng. News* 86 (Jan. 11, 1965). (b) J. W. Larsen, P. A. Bovis, M. W. Grant, and C. A. Lane, *J. Amer. Chem. Soc.* **93**, 2067 (1971).

"normal inductive order" in the gas phase and the Baker-Nathan order in solution for the same system undergoing the same process.

However, Schubert[46] has pointed out that the Schubert-Sweeney hypothesis of steric hindrance to specific solvation near bulky alkyl groups[44] is not a unique explanation of the Baker-Nathan phenomenon. Equally applicable in many cases would be Shiner's suggestion[45,47] that solvation enhances C—H hyperconjugation over C—C hyperconjugation or the suggested importance of some type of solvation of the gegenion,[46] such as electrophilic solvation of the anion in an electrophilic aromatic substitution. Stock[48] has reported significant ground-state solvation effects and concluded, along with other workers,[46,49] that neither of the two major hypotheses advanced thus far for the Baker-Nathan effect is sophisticated enough to provide an adequate explanation, although the solvent must be an important factor. Both ionization potentials and charge-transfer spectra follow the "normal inductive order" and not the Baker-Nathan order, results which lead Traylor[50] to attribute the Baker-Nathan order to a "nonvertical" solvent effect[50] of some kind[40,47] and not to any kind of electron delocalization.

A satisfactory explanation of the Baker-Nathan effect apparently is not yet available, a problem which greatly complicates interpretation of β-deuterium isotope effects (Chapter 7-3). Hyperconjugation also provides the best explanation for several types of electron spin resonance hyperfine splitting phenomena. The observation of splitting involving hydrogens on an alkyl carbon directly bonded to an aromatic system and the observation that aliphatic hydrogens β to an electron site couple more strongly than those α to (or at) the radical site have only hyperconjugation as a reasonable explanation at the present time.

4-5 Hammett Equations for Equilibria

All of the substituent effects considered qualitatively thus far in this chapter may be viewed as perturbations on unsubstituted systems. If this perturbation is not large, there should be a linear relationship between

---

46. W. M. Schubert and D. F. Gurka, *J. Amer. Chem. Soc.* **91**, 1443 (1969).
47. V. J. Shiner, Jr., and C. J. Verbanic, *J. Amer. Chem. Soc.* **79**, 369, 373 (1957), and later papers.
48. A. Himoe and L. M. Stock, *J. Amer. Chem. Soc.* **91**, 1452 (1969).
49. C. Eaborn and R. Taylor, *J. Chem. Soc.* 247 (1961); H. C. Brown and R. A. Wirkkala, *J. Amer. Chem. Soc.* **88**, 1447, 1453 (1966).
50. (a) W. Hanstein, H. J. Berwin, and T. G. Traylor, *J. Amer. Chem. Soc.* **92**, 829 (1970), and earlier papers. (b) A "nonvertical" process is a process that is slow enough for nuclear motions to occur, which cannot be the case for any electron-delocalization or conjugation phenomenon.[45]

the change in electron density at the reaction site and the change in the reaction energetics caused by the substituent. For similar types of compounds undergoing similar reactions, the perturbation in energy caused by a given substituent in one of the compounds should be proportional to the perturbation in energy caused by the same substituent in one of the other compounds. For example, the logarithms of the thermodynamic ionization constants (in a given solvent at a given temperature) of a series of *meta*- and *para*-substituted phenylacetic acids relative to the ionization constant of phenylacetic acid should be proportional to the logarithms of the thermodynamic ionization constants (in the same solvent at the same temperature) of the similarly substituted benzoic acids relative to benzoic acid, because each type of carboxylic acid responds to the electronic effect of a particular substituent in a consistent manner. The correlation for these compounds (Eq. 4.7) turns out to be quite good, and the slope of the

$$\log\left(\frac{K_X}{K_H}\right)_{\text{phenylacetic acids}} = 0.49 \log\left(\frac{K_X}{K_H}\right)_{\text{benzoic acids}} \quad (4.7)$$

line is found to be 0.49. This slope is reasonable, since the intervening methylene group in the phenylacetic acids should attenuate the electrostatic effects of the substituents in these compounds.

This type of relationship was recognized by Hammett,[51] and is known as the *Hammett equation*. The comparison involves logarithms since

$$\Delta G° = -RT \ln K$$

thereby placing the relationship of Eq. 4.7 in the framework of a comparison of free energies:

$$(\Delta G_X° - \Delta G_H°)_{\text{phenylacetic acids}} \propto (\Delta G_X° - \Delta G_H°)_{\text{benzoic acids}} \quad (4.8)$$

This type of proportionality is called a *linear free-energy relationship*.[52] Since the quantities being compared in a linear free-energy relationship are thermodynamic quantities yet the relationships themselves are not part of the formal structure of thermodynamics, such comparisons are called *extrathermodynamic relationships*.

The standard chosen for the Hammett equation was the thermodynamic ionization constants of the *meta*- and *para*-substituted benzoic acids in water at 25°C—primarily because they were available with high

---

51. L. P. Hammett, *Physical Organic Chemistry*, 1st ed., McGraw-Hill, New York, 1940.
52. Reference B, Section 2-8 and pp. 403–417; Reference C, pp. 238–245; Reference D, Chapter 7; Reference E, Chapter 4; Reference G, Chapters 6 and 7; Reference I, Chapter 11; P. R. Wells, *Linear Free Energy Relationships*, Academic Press, New York, 1968; H. H. Jaffé, *Chem. Rev.* **53**, 191 (1953); J. Shorter, *Chem. Brit.* **5**, 269 (1969); P. D. Bolton and L. G. Hepler, *Quart. Rev.* **25**, 521 (1971).

accuracy. A *substituent constant* for a given substituent, σ_X, is defined by

$$\sigma_X = \log\left(\frac{K_X}{K_H}\right)_{\text{benzoic acids, H}_2\text{O, 25°C}} \quad (4.9)$$

Placing this into Eq. 4.7, and defining the slope as ρ (the *reaction constant* for the particular reaction at a given temperature in a given solvent), we can write the general form of the Hammett equation:

$$\log \frac{K_X}{K_0} = \rho\sigma \quad (4.10)$$

where the subscript 0 denotes the unsubstituted or reference compound.

In effect, the ionization of benzoic acids is the standard reaction type ($\rho = 1$), and σ is defined so that $\sigma = 0$ for benzoic acid itself. Since an electron-attracting *meta* or *para* substituent is acid-strengthening in the benzoic acid series, a positive σ value denotes a substituent which withdraws electrons relative to hydrogen. Similarly, a negative σ value means that the substituent donates electrons relative to hydrogen. Since a substituent in the *meta* position would interact differently with the ionization site from the same substituent in the *para* position, two sets of σ values must be defined (σ_m and σ_p), typical values of which are shown in Table 4.4. The

TABLE 4.4 *Hammett Substituent Constants*:[53] σ_m and σ_p

Substituent	σ_m	σ_p
H_3C-	-0.069	-0.170
CH_3CH_2-	-0.07	-0.151
$(CH_3)_3C-$	-0.10	-0.197
C_6H_5-	$+0.06$	-0.01
$H_3C-\overset{O}{\overset{\|}{C}}-$	$+0.376$	$+0.502$
$N\equiv C-$	$+0.56$	$+0.660$
F_3C-	$+0.43$	$+0.54$
H_2N-	-0.16	-0.66
O_2N-	$+0.710$	$+0.778$
CH_3O-	$+0.115$	-0.268
$HO-$	$+0.121$	-0.37
$F-$	$+0.337$	$+0.062$
$Cl-$	$+0.373$	$+0.227$
$Br-$	$+0.391$	$+0.232$
$I-$	$+0.352$	$+0.18$

53. D. H. McDaniel and H. C. Brown, *J. Org. Chem.* **23**, 420 (1958).

substituent constant is, in theory, independent of the nature of the reaction being considered, as long as the interaction mechanisms between a given substituent and the reaction sites do not change character. Those σ's based on measured thermodynamic ionization of the appropriate benzoic acids in water at 25°C are called *primary* values, and are necessarily the most reliable. Those σ's obtained by comparison with another set of compounds or reaction conditions are called *secondary* values, and are more susceptible to error since they depend on the accuracy of a series of measurements and their regression to a line determined using statistical methods. When two or more substituents are simultaneously present in *meta* or *para* positions, their σ values may usually be added together. Deviations from additivity will occur if the Hammett plot is not linear or if the substituents interact with one another, particularly if this interaction depends on the nature of the reaction taking place. Notice that *ortho* substituents have thus far not been considered. They will be discussed in Chapter 6-2.

Both σ_m and σ_p may include electrostatic and resonance effects. It would be extremely useful to obtain substituent constants reflecting only electrostatic effects and only resonance effects. The former are readily available from the 4-substituted bicyclo[2.2.2]octane-1-carboxylic acids.[5] A set of σ' values which evaluate substituent electrostatic effects is shown in Table 4.5. In general,

$$\sigma_m - \sigma' = \pm 0.1$$

suggesting that electrostatic interactions are the major constituents of *meta*-substituent effects. Values of $\sigma_p - \sigma'$ are negative for the substituents shown in the table other than the nitro group. Since all of these substituents other than the nitro group are electron-donating in resonance interactions (Table 4.3), the significance of both resonance and electrostatic interactions in *para*-substituent effects is indicated. This does not mean that the value $\sigma_p - \sigma'$ should be used as a quantitative measure of resonance

TABLE 4.5 *Electrostatic Substituent Constants:*[5] σ'

Substituent	σ'
HO—	+0.25
CH_3O—	+0.23
Cl—	+0.47
Br—	+0.45
H_3C—	−0.05
O_2N—	+0.63

effects. Electrostatic effects are distance-dependent (Eq. 4.3) and therefore should be different in the *meta* and *para* positions of a benzene ring.

The reaction constant ρ measures the susceptibility of a reaction to the substituent effects. Reactions with positive ρ values are aided by electron withdrawal from the reaction site, while those with negative ρ values are assisted by electron donation into the reaction site. The magnitude and sign of ρ are therefore a measure both of the amount of charge developing at the reaction site and of the extent to which the reaction site can interact with the substituents in the particular compounds being investigated in the given solvent at the given temperature. Table 4.6 presents

TABLE 4.6 *Reaction Constants for Ionizations at* $25°C$[52]

Reaction	ρ
Benzoic acids in H_2O	+1.00
Benzoic acids in 50% EtOH	+1.57
Benzoic acids in MeOH	+1.54
Phenylacetic acids in H_2O	+0.49
β-Phenylpropionic acids in H_2O	+0.21
Cinnamic acids in H_2O	+0.47

a few examples of the solvent and substrate sensitivities of ρ. The predicted[52] inverse temperature dependence of ρ has yet to be conclusively established.[52,54] The relationship of ρ and the reaction mechanism will be explored in greater depth in Chapter 6.

The validity of linear free-energy relationships has often been questioned.[52,55] As shown in Eq. 4.8, the Hammett equation is a free-energy comparison. Substituents may affect both the enthalpy and the entropy terms in a given reaction series. A linear relationship between free-energy changes in two different reaction series therefore would occur only if each series is isoentropic (results in no substituent effect on entropy), if each series is isoenthalpic (results in no substituent effect on enthalpy), or if enthalpy change is linearly related to entropy change ($\Delta\Delta H° = \beta \, \Delta\Delta S°$). Recent evidence[56] solves a major problem by finally establishing that the enthalpy changes are linearly related to the entropy changes in the standard series for the Hammett equation, the thermodynamic ionization of the substituted benzoic acids. Nevertheless, in spite of the fact that many attempts[52,55] have been made to understand why linear free-energy

---

54. T. E. Bitterwolf, R. E. Linder, and A. C. Ling, *J. Chem. Soc.*, B 1673 (1970).
55. C. D. Ritchie and W. F. Sager, *Progr. Phys. Org. Chem.* 2, 323 (1964).
56. P. D. Bolton, K. A. Fleming, and F. M. Hall, *J. Amer. Chem. Soc.* 94, 1033 (1972).

relationships such as the Hammett equation have broad scope or exist at all, this whole area remains somewhat ambiguous.

The relationship

$$\Delta\Delta H° = \beta \, \Delta\Delta S° \qquad (4.11)$$

was first recognized by Leffler,[52,57] who collated examples of experimental results illustrating this idea. If $\beta \geq 0$, this is termed an *isoequilibrium relationship*,[52,58] with β, in units of absolute temperature, being identified as the temperature at which all of the substituent effects will disappear ($\Delta\Delta G° = 0$). The implications of the possible existence of an isoequilibrium relationship are manifold. (1) A reaction studied at the temperature $T = \beta$ does not exhibit substituent effects. A result indicating lack of a substituent effect requires that the reaction series be studied at a distinctly different temperature to find out if substituent effects are actually possible in this series and the temperature originally used was the isoequilibrium temperature for the reaction. (2) The sign of ρ is different above and below the isoequilibrium temperature. Therefore, little significance should be attributed to the sign of ρ if the reaction constant is small in magnitude and an isoequilibrium relationship is indicated. (3) The reaction constant should be an inverse function of the temperature.[52,54]

Considerable discussion of the validity of isoequilibrium relationships has been presented in the literature,[52,55,59] but little common ground has been established thus far between the believers and the dissenters.

---

57. J. E. Leffler, *J. Org. Chem.* **20**, 1202 (1955).
58. For rate processes, if $\Delta\Delta H^{\ddagger} = \beta \, \Delta\Delta S^{\ddagger}$, the process is termed an *isokinetic relationship*. The discussion presented here applies equally to such relationships.
59. R. C. Petersen, *J. Org. Chem.* **29**, 3133 (1964).

Chapter 5

The Rate of a Reaction

One of the major areas of chemistry is that of chemical kinetics,[1-3] which is the study of reaction rates, the influence of variables on reaction rates, and the relationship of these results to the detailed mechanism of the reaction. No mechanism can ever be fully determined without a detailed study of its kinetics.

5-1 Kinetics

The reaction rate is defined as the rate of change in concentration of a reactant or product as a function of time. Most reactions fit a rate expression of the form

$$\text{rate} = -\frac{d[A]}{dt} = k[A]^m[B]^n[C]^o \cdots \qquad (5.1)$$

where k is the specific rate constant, and brackets denote concentration. The exponents, which must be obtained experimentally, define the *order* of the reaction with respect to each reactant. Similarly, the sum of the exponents defines the order of the total reaction. There is no necessary

---

1. For treatments equivalent to that presented herein, see Reference B, Section 3-1; Reference D, Chapter 6; Reference E, Chapter 3; Reference H; Reference I, Chapter 4; and K. J. Laidler, *Chem. Brit.* **3**, 475 (1967).
2. For background to this chapter, see W. J. Moore, *Physical Chemistry*, 3rd ed., Prentice-Hall, Englewood Cliffs, N.J., 1962.
3. For more detailed treatment of kinetics, see Reference H; K. J. Laidler, *Chemical Kinetics*, 2nd ed., McGraw-Hill, New York, 1965.

relationship between the stoichiometrically balanced equation for the reaction and the order of the reaction, nor is there a necessary relationship between the order and the *molecularity* of the reaction. The molecularity is a description of the number of individual molecules taking part in a collision in a given step in an assumed reaction mechanism involving the breaking and/or forming of bonds. In a multistep reaction mechanism, the molecularity most often is used with respect to the slowest step in the mechanism, the so-called *rate-determining step* (rds). For a multistep reaction, the rate equation (Eq. 5.1) describes the relationship between the rate and those species involved in or before the rate-determining step, thereby permitting some evaluation of the mechanism up to and including the rate-determining step.[4]

The normal procedure in a kinetic investigation is to determine first the order of the reaction with respect to each reactant by observing change in concentration with time. For this data, integrated forms of the rate expression may be used to calculate the value of the rate constant k. If the concentration of a reactant does not change or changes to such a small extent that the change cannot be accurately measured, this reactant either is involved in the mechanism after the rate-determining step, is involved as a catalyst, or is present in such a large amount that the amount consumed in the reaction is negligible, in which case the concentration of this reactant is included in the rate constant (Eq. 5.2) and the reaction appears to be of lower order than it actually is. The choice of the proper technique for

$$k' = k[A]^m$$
$$\text{rate} = k'[B]^n[C]^o \cdots \qquad (5.2)$$

determining the rate equation or the various rate constants depends on the magnitude of the speed required for the measurement: The act of measuring must be faster than whatever is being measured.

Following is a brief review of some of the more common rate expressions.[2,3]

First-order.
 EXAMPLE: An S_N1 process.

$$\text{rate} = -\frac{d[A]}{dt} = k_1[A]$$

Let x = amount of A reacted at time t
 $[A]_0$ = initial concentration of A, at $t = 0$

---

4. J. O. Edwards, E. F. Greene, and J. Ross, *J. Chem. Educ.* **45**, 381 (1968).

Then

$$-\frac{dx}{[A]_0 - x} = k_1 \, dt$$

Integrating,

$$\ln([A]_0 - x) - \ln[A]_0 = -k_1 t$$

Plot $\log([A]_0 - x)$ vs. t; slope $= -k_1/2.303$.

Second-order.
EXAMPLE: An $S_N 2$ process.

If 1:1 stoichiometry,

$$-\frac{d[A]}{dt} = k_2[A][B]$$

$$\frac{dx}{dt} = k_2([A]_0 - x)([B]_0 - x)$$

Integrating,

$$\frac{1}{[A]_0 - [B]_0} \ln\left(\frac{[B]_0([A]_0 - x)}{[A]_0([B]_0 - x)}\right) = k_2 t \tag{5.3}$$

Plot the logarithm vs. t; slope $= k_2([A]_0 - [B]_0)$.

If $[A]_0 = [B]_0$ or if the reaction is second-order, such as a dimerization, Eq. 5.3 becomes:

$$\frac{1}{[A]_0 - x} - \frac{1}{[A]_0} = k_2 t$$

Reversible.

$$A \underset{k_r}{\overset{k_f}{\rightleftarrows}} B$$

Start with only A at concentration $[A]_0$:

$$-\frac{d[A]}{dt} = k_f[A] - k_r[B] = k_f[A] - k_r([A]_0 - [A])$$

Integrating,

$$\ln \frac{k_f[A]_0}{k_f[A]_0 - (k_f + k_r)([A]_0 - [A])} = (k_f + k_r)t$$

At equilibrium, $k_f = k_r$, $[A] = [A]_e$, and

$$\ln \frac{[A]_0 - [A]_e}{[A] - [A]_e} = (k_f + k_r)t$$

Therefore, the reaction appears first-order with the rate constant being the sum of the rate constants for the forward and reverse reactions. For k_f or k_r, this sum and $K_{eq} = k_f/k_r$ are used as simultaneous equations.

Consecutive.

EXAMPLE: S_N1 processes of same substrate with different nucleophiles.

For the special case of different reactions with the same rds, both reactions will have the same rate constant and rate equation.

Competitive.

EXAMPLE: Competing first-order and second-order processes such as S_N1 and S_N2 with the same stoichiometry in the common substrate.

$$\frac{dx}{dt} = k_1([A]_0 - x) + k_2([A]_0 - x)([B]_0 - x)$$

At $[A]_0 = [B]_0$,

$$\frac{dx}{dt} = k_1([A]_0 - x) + k_2([A]_0 - x)^2$$

Integrating and simplifying,

$$\frac{k_2}{k_1} = \frac{1}{e^{k_1 t} - 1} \left(\frac{1}{[A]_0 - x} - \frac{e^{k_1 t}}{[A]_0} \right)$$

For further examples, see Chapter 5-2.

Steady-state Approximation.

$$A + B \underset{k_{-1}}{\overset{k_1}{\rightleftarrows}} C \overset{k_2}{\longrightarrow} \text{products}$$

$$-\frac{d[B]}{dt} = -\frac{d[A]}{dt} = k_1[A][B] - k_{-1}[C] \quad (5.4)$$

$$\frac{d[C]}{dt} = k_1[A][B] - k_{-1}[C] - k_2[C] \quad (5.5)$$

If [C] is very small, $d[C]/dt \ll -d[A]/dt$, so one can assume that $d[C]/dt = 0$, the steady-state approximation. From Eq. 5.5,

$$[C] = \frac{k_1[A][B]}{k_2 + k_{-1}}$$

Substituting into Eq. 5.4,

$$-\frac{d[B]}{dt} = k_1[A][B]\left(1 - \frac{k_{-1}}{k_2 + k_{-1}}\right)$$

Note: The steady-state approximation is not valid if [C] is ever appreciable.

5-2 Partial Rate Factors

One striking example of the effective use of kinetics data for primarily nonmechanistic purposes is the use of information gleaned from competitive electrophilic aromatic substitution reactions to obtain predictions about the isomer distributions from related substrates under the same reaction conditions. H. C. Brown's concept of partial rate factors[5] permits such predictions by measuring the ability of a reagent to discriminate both intramolecularly and intermolecularly.

For an electrophilic aromatic substitution in which a known mixture of two substrates is allowed to react with an insufficient amount of electrophile, the relative rates of the competitive reactions may be determined from the proportions of the products, provided that the reaction is (a) first-order in each substrate, (b) the same order in the electrophile for both substrates, (c) devoid of side reactions, and (d) subject to kinetic control (or irreversible) prior to the product analysis. If benzene is used as one of the substrates, the product analysis leads to the rate constant for the reaction of the second substrate relative to the rate constant for the reaction of benzene under the given set of reaction conditions.

The above type of analysis really underuses the data. The product analysis requires finding the amounts of all of the possible isomers formed from each substrate and summing them for each substrate. A far more useful result would be the relative rate constant for the formation of *each* isomer corrected for any statistical factors so that it may be ascertained on a per-reactive-site basis. These relative rate constants for a given position relative to a hydrogen in benzene are called *partial rate factors*.[5] For example, the partial rate factor for *para* substitution in toluene, $p_f^{CH_3}$, is defined as

$$p_f^{CH_3} = \frac{k_{C_6H_5CH_3}}{k_{C_6H_6}} \times \frac{\text{fraction } para\text{-substituted product}}{1/6} \quad (5.6)$$

where $k_{C_6H_5CH_3}/k_{C_6H_6}$ is the relative rate for the given reaction of toluene compared to benzene, and the factor of 1/6 provides the statistical balance between the single *para* position in toluene and the six equivalent positions in benzene. The *ortho* and *meta* partial rate factors may be similarly defined[5] (Eq. 5.7 and 5.8, respectively).

$$o_f^{CH_3} = \frac{k_{C_6H_5CH_3}}{k_{C_6H_6}} \times \frac{(1/2) \text{ fraction } ortho\text{-substituted product}}{1/6} \quad (5.7)$$

$$m_f^{CH_3} = \frac{k_{C_6H_5CH_3}}{k_{C_6H_6}} \times \frac{(1/2) \text{ fraction } meta\text{-substituted product}}{1/6} \quad (5.8)$$

---

5. Reference C, p. 391; Reference G, pp. 196–203; L. M. Stock, *Aromatic Substitution Reactions*, Prentice-Hall, Englewood Cliffs, N.J., 1968, Chapters 3 and 4; L. M. Stock and H. C. Brown, *Advan. Phys. Org. Chem.* **1**, 35 (1963).

The magnitude of the partial rate factors provides a quantitative measure of the activating or deactivating effect of a given substituent toward electrophilic substitution at a given position.[5] For example,[5] the partial rate factors for one set of reaction conditions for the bromination of toluene are found to be $o_f^{CH_3} = 800$, $m_f^{CH_3} = 5.5$, and $p_f^{CH_3} = 2400$. Since all are greater than unity, the methyl group in toluene activates all of the ring positions toward electrophilic attack. The greater magnitude of the *ortho* and *para* values provides a quantitative assessment of the well-known *ortho–para* directing ability of the methyl group.[5,6]

Partial rate factors may be extrapolated to polysubstituted systems. Bromination of *meta*-xylene under the same reaction conditions as used to determine the partial rate factors mentioned above[5] would be predicted

$$o_f^{CH_3} \times p_f^{CH_3} \qquad\qquad o_f^{CH_3} \times o_f^{CH_3}$$

$$m_f^{CH_3} \times m_f^{CH_3}$$

$$o_f^{CH_3} \times p_f^{CH_3}$$

(structure: meta-xylene, with CH₃ groups shown)

to yield (800 × 800 × 100)/4,480,000, or 14.3%, of 2,6-dimethylbromobenzene, (800 × 2400 × 2 × 100)/4,480,000, or 85.7%, of 2,4-dimethylbromobenzene, and (5.5 × 5.5 × 100)/4,480,000, or 0%, of 3,5-dimethylbromobenzene. If anything, the amount of the 2,6-dimethylbromobenzene would be expected to be less than calculated, because of the steric problems associated with attack between two groups *meta* to one another.[5,6] Variations of this type are consistently obtained when the pertinent experiments are performed, so partial rate factors provide only semiquantitative information for systems such as this one.

Intramolecular comparisons of partial rate factors may be used to measure the *selectivity* of a given reagent as well as the difference between the free energies of the transition states leading to the two isomers. Brown's selectivity[5]

$$S = \log \frac{p_f}{m_f} \tag{5.9}$$

measures the intramolecular selectivity, while

$$-RT \ln \frac{p_f}{m_f}$$

measures the transition-state free-energy differences. For an *ortho–para* directing group,[6] the selectivity (Eq. 5.9) is inversely related to the reactivity

---

6. R. T. Morrison and R. N. Boyd, *Organic Chemistry*, 3rd ed., Allyn and Bacon, Boston, 1973, Chapters 11 and 12; J. B. Hendrickson, D. J. Cram, and G. S. Hammond, *Organic Chemistry*, 3rd ed., McGraw-Hill, New York, 1970, Chapter 16.

of the electrophile:[7] The more selective the electrophile, the less reactive it will be.[8]

5-3 Transition-state Theory

Three theories are used to describe reaction rates in terms of energies. The first, *collision theory*,[9] is a mechanical model based on the kinetic theory of gases, and ultimately leads to Eq. 5.10,

$$k_r = PZe^{-E_a/RT} \tag{5.10}$$

where k_r is the rate constant; P is the probability factor, or steric factor; Z is the collision factor, a temperature-dependent value; E_a is the energy of activation; R is the gas constant; and T is the absolute temperature. This formalism for gas-phase reactions closely resembles the second approach, the *Arrhenius theory*,[10] which leads to Eq. 5.11 as a first approximation. In this equation, A is called the Arrhenius pre-exponential factor

$$k_r = Ae^{-E_a/RT} \tag{5.11}$$

and is assumed to be temperature-independent. The third approach, developed primarily by Eyring and Polanyi, is called *transition-state theory*[11] (or absolute reaction rate theory). This theory attempts to place kinetics on a more or less thermodynamic basis, and therefore is most adaptable to reactions of complex molecules in solution. Transition-state theory and the use of a potential-energy surface to discuss reaction kinetics and mechanisms have become an integral part of organic chemical thinking, so it is worthwhile to look at the development of this theory for the case of a bimolecular process.

Transition-state theory rests on three sets of fundamental assumptions.[11] The first is that one of the requirements for a reaction to occur is that the reacting species combine and pass through some state having greater potential energy than either the reactants or the products. This state of higher potential energy is called the transition state, and it constitutes a saddle point on a potential-energy diagram (Fig. 5.1). If a reaction leads to several products, each product is formed from a different

---

7. Deviations from this statement are usually the result of temperature, solvent, or catalyst influences.
8. For further consideration of this statement and its relationship to the Hammett equation and Hammond's postulate, see Chapters 6-1, 6-4, and 7-8.
9. See Reference H, Chapter 4; W. J. Moore, *Physical Chemistry*, 3rd ed., Prentice-Hall, Englewood Cliffs, N.J., 1962, Chapter 8.
10. The Arrhenius theory is not a theory as such. It is a relationship between temperature and reaction rate deduced from experimental observations.
11. Reference G, pp. 62–74; Reference H, Chapter 5; Reference I, Chapter 5.

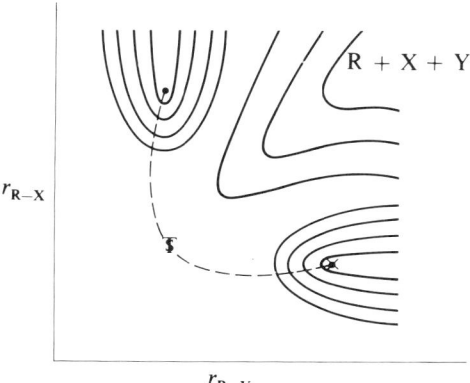

FIGURE 5.1 *The potential-energy surface for a reaction* $R-X + Y \to R-Y + X$ *(· denotes start of reaction, ‡ denotes transition state, and contours are constant potential-energy surfaces).*

transition state. The major product results from the transition state with the lowest energy. For a given transition state, the accumulation of atoms moving across this potential-energy barrier is called the *activated complex*. A double dagger ($\ddagger$) will be used to indicate the activated complex or any property associated with it. The activated complex has bonds, vibrations, and a well-defined composition, yet it differs from a molecule in that one degree of freedom, usually a vibration, causes decomposition to products. This vibration, which is transformed into a translation in one dimension, is called the *reaction coordinate*, and is assumed to be the most important degree of freedom for the given reaction.[12] The slice through the potential-energy surface for the reaction (Fig. 5.1) along the reaction coordinate is called the *reaction profile* (Fig. 5.2). Properties of the activated complex and parameters for the activation process may be calculated,[13] but such calculations are at best qualitatively accurate and applicable to only the simplest systems.

The second set of assumptions on which transition-state theory is based is that the populations of the transition state may be approximated by the Boltzmann distribution law and are in statistical equilibrium with the reactants. Mathematically, for a reaction

$$A + B \rightleftharpoons AB^{\ddagger} \longrightarrow \text{products}$$

---

12. See Chapters 7-2 and 7-8 for further discussion of this assumption.
13. Reference B, Section 3-3.

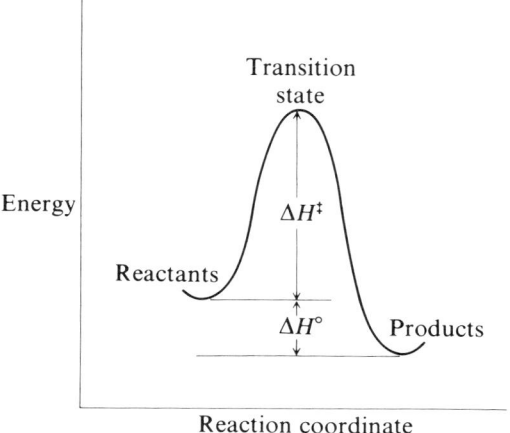

FIGURE 5.2 *The reaction profile.*

an equilibrium constant K is assumed (Eq. 5.12), and the activity coeffi-

$$K = \frac{[AB^\ddagger]\gamma_{AB^\ddagger}}{[A][B]\gamma_A\gamma_B} \tag{5.12}$$

cients are all assumed equal to unity. This equilibrium constant for the gaseous state also may be defined using statistical mechanics and partition functions,[14]

$$K = \frac{Q_\ddagger^0}{Q_A^0 Q_B^0} e^{-E_0^0/RT} = \frac{[AB^\ddagger]}{[A][B]} \tag{5.13}$$

where the superscript zeros indicate the standard state; the Q's are the partition functions, which are products of the contributions of the electronic, rotational, vibrational, and translational energies of each species;[14] and E_0^0 is the difference in energy between the reactants and products involved in the equilibrium at $0°K$. This assumption of an equilibrium situation permits the application of thermodynamic principles to kinetic phenomena.

The third set of fundamental assumptions centers on a definition of reaction rate as being proportional to the concentration of the activated complex. The average velocity, $\bar{v}$, for passage forward through a small distance δ at the potential energy of the transition state may be written as

$$\bar{v} = \left(\frac{kT}{2\pi m_\ddagger}\right)^{1/2}$$

---

14. Reference H, Chapter 5; Reference I, Chapters 3 and 5; W. J. Moore, *Physical Chemistry*, 3rd ed., Prentice-Hall, Englewood Cliffs, N.J., 1962, Chapter 15; F. Daniels and R. A. Alberty, *Physical Chemistry*, 3rd ed., Wiley, New York, 1966, Chapter 16.

where $m_\ddagger$ denotes the mass of the transition state, and k is the Boltzmann constant. The reaction rate therefore may be expressed as

$$\text{rate} = \frac{[AB^\ddagger]}{\delta}\left(\frac{kT}{2\pi m_\ddagger}\right)^{1/2} \qquad (5.14)$$

Since the rate may also be expressed as

$$\text{rate} = k_r[A][B] \qquad (5.15)$$

setting Eq. 5.14 and 5.15 equal leads to

$$k_r = \frac{[AB^\ddagger]}{[A][B]\,\delta}\left(\frac{kT}{2\pi m_\ddagger}\right)^{1/2} \qquad (5.16)$$

Using the earlier definition (Eq. 5.13) of an equilibrium constant in terms of partition functions, Eq. 5.16 may be written as

$$k_r = \frac{Q_\ddagger^0}{Q_A^0 Q_B^0\,\delta}\left(\frac{kT}{2\pi m_\ddagger}\right)^{1/2} e^{-E_0^0/RT} \qquad (5.17)$$

The partition function for the transition state, $Q_\ddagger^0$, may be factored to exclude the reaction coordinate[11,14]

$$Q_\ddagger^0 = Q_\ddagger Q_{RC} = Q_\ddagger \frac{\delta}{h}(2\pi m_\ddagger kT)^{1/2} \qquad (5.18)$$

leading to Eq. 5.19 on substitution into Eq. 5.17.

$$k_r = \frac{Q_\ddagger}{Q_A^0 Q_B^0} e^{-E_0^0/RT} \frac{kT}{h} = K^\ddagger \frac{kT}{h} \qquad (5.19)$$

Note that $K^\ddagger$ is defined as if it were a true equilibrium constant (Eq. 5.13), yet it is not one because one of the degrees of freedom of the activated complex, the reaction coordinate, has been excluded from this definition. Nevertheless, it is assumed that $K^\ddagger$ may be treated as if it were an equilibrium constant and that the normal thermodynamic equations may be utilized (Eq. 5.20 and 5.21).

$$\Delta G^\ddagger = -RT \ln K^\ddagger \qquad (5.20)$$

$$\Delta G^\ddagger = \Delta H^\ddagger - T\Delta S^\ddagger \qquad (5.21)$$

The symbol $\Delta G^\ddagger$ is known as the *free energy of activation*, $\Delta H^\ddagger$ is the *enthalpy of activation*, and $\Delta S^\ddagger$ is the *entropy of activation*. All of these quantities refer to the ground state of the reactants and the transition state. In order to be consistent with experimental results, a correction term κ, the transmission coefficient, must be introduced into Eq. 5.19, giving the "Eyring equation" (Eq. 5.22). The transmission coefficient usually has

$$k_r = \frac{\kappa k T K^\ddagger}{h} = \frac{\kappa k T}{h} e^{-\Delta G^\ddagger/RT} = \frac{\kappa k T}{h} e^{-\Delta H^\ddagger/RT} e^{\Delta S^\ddagger/R} \qquad (5.22)$$

a value of unity, and is defined as the fraction of forward-moving molecules of the activated complex that actually become products.

The Eyring equation (Eq. 5.22) may be compared to the Arrhenius equation (Eq. 5.11) or the collision-theory equation (Eq. 5.10). For a reaction occurring in solution, and assuming $\Delta S^{\ddagger}$ is not a function of temperature,

$$E_a = \Delta H^{\ddagger} + RT \qquad (5.23)$$

and

$$A = \frac{\kappa k T e}{h} e^{\Delta S^{\ddagger}/R} \qquad (5.24)$$

If A is expressed using seconds as the unit of time and the temperature is around 300°K, Eq. 5.24 simplifies to

$$\Delta S^{\ddagger} = 4.575 \log A - 60.53$$

which points out that a negative value of $\Delta S^{\ddagger}$ is related to a low value of A. The entropy and enthalpy of activation are usually obtained from a plot of $\log(k_r/T)$ with $1/T$. The slope equals $-\Delta H^{\ddagger}/2.303R$, and the intercept is approximately equal to $(\Delta S^{\ddagger}/2.303R) + 10.32$. Alternately, $\Delta S^{\ddagger}$ may be obtained from $\Delta H^{\ddagger}$ (Eq. 5.25), although the risk in this type of

$$\Delta S^{\ddagger} = 4.575(\log k_r - \log T - 10.32) + \Delta H^{\ddagger}/T \qquad (5.25)$$

approach is considerable because of the potential magnification of errors when a single experimental point or a closely related set of points is employed in the evaluation of experimental data.

The principle of *microscopic reversibility* is inherent in transition-state theory. According to this principle, a reaction and its reverse will follow the same path. A reaction profile (Fig. 5.2) supports this principle in that the path of lowest potential energy over a barrier would involve the same route and the same transition state for both the forward and reverse directions.

5-4 Entropy of Activation and Solvent Effects

The quantity from transition-state theory most commonly used in an analysis of reaction mechanism is the entropy of activation, $\Delta S^{\ddagger}$. As with any entropy term, $\Delta S^{\ddagger}$ measures changes in orderedness, including changes in the orientation of solvent molecules resulting from changes in solvation. A negative $\Delta S^{\ddagger}$ indicates a decrease in the total number of molecules on formation of the transition state, and/or a cyclic transition state, and/or some other increase in ordering in proceeding to the transition state. A typical example is a $\Delta S^{\ddagger}$ of -9.5 eu for an $S_N 2$ displacement on methyl iodide and a $\Delta S^{\ddagger}$ of -19.9 eu for a related displacement on neopentyl iodide (using the same nucleophile and reaction conditions), thereby reflecting the greater ordering in the more crowded transition state required with the bulkier neopentyl system.

The entropy of activation has a tendency to be less positive than anticipated for reactions in which neutral molecules form ions.[15] In the formation of ions, either the same number of species are formed as originally reacted, or more equivalents of products are generated than there were of reactants, leading to an anticipated zero or positive component to $\Delta S^{\ddagger}$. However, changes in solvent orientation impose a negative component on the overall $\Delta S^{\ddagger}$. This negative component is greatest for the least polar solvent, since in such a case more solvent ordering occurs in the presence of the ions than in the presence of the neutral species. A more polar solvent merely trades solvent–solvent ordering in the presence of the neutral reactants for solvent–solute ordering with the ions, leading to little or no effect on $\Delta S^{\ddagger}$. Similarly, there is a positive component in $\Delta S^{\ddagger}$ if the reaction involves the combination of two ions of opposite charge to form neutral products; and there is a negative component in $\Delta S^{\ddagger}$ for combination of two ions of the same charge, because the resulting species exhibits greater charge concentration and therefore greater solvent ordering than the original ions.

An inspection of Eq. 5.21 indicates that a highly negative entropy of activation may be just as effective in slowing down a reaction as a highly positive enthalpy of activation, and emphasizes the importance of the choice of solvent[15] for many reactions.

Solvent effects on reaction rates usually parallel those on equilibria. Reactions in which charge is created proceed most effectively in polar solvents. As discussed above, the more polar solvents disperse the charge separation created in the activated complex more efficiently by solvation than do less polar solvents, while they themselves undergo the least reorganization. Conversely, the less polar solvents favor the destruction of charge or the combination of a neutral species with an ionic species. Solvent effects are usually insignificant for reactions between neutral molecules not involving charged, charge-separated, or dipolar intermediates and for intramolecular rearrangements. Solvent effects would be minor for pericyclic reactions of the type discussed in Chapter 3.

Changes in solvent occasionally induce changes in mechanism and changes in order, but these are fairly unusual occurrences. However, it is fair to say that extrapolation of mechanistic information from one solvent to another solvent significantly different in polarity or hydrogen-bonding ability (in both a donor or acceptor sense) is often risky.

---

15. For a more detailed discussion of solvent effects on various types of reactions, see Reference I, Chapter 8, and Chapters 8-5 and 9 herein.

5-5 Salt Effects

If kinetic data are not evaluated under conditions of constant ionic strength, the *salt effects* resulting from changes in ionic strength must be considered.[16] Salt effects are deviations from ideality in solution behavior, and therefore are best treated as effects on activity coefficients.

Ionic strength, μ, is defined as

$$\mu = \tfrac{1}{2} \sum c_i z_i^2 \tag{5.26}$$

where c_i is the concentration of ion i, and z_i is its charge. According to the Debye-Hückel theory,[16] the activity coefficient of an ion in dilute solution at the distance of closest approach of another ion is related to the ionic strength by

$$-\log \gamma_i \cong z_i^2 \alpha \sqrt{\mu} \tag{5.27}$$

where α is a constant whose value in dilute aqueous solution at 25°C is 0.509. Brønsted[16] and Bjerrum[16] utilized this approach to evaluate the influence of ionic strength on reaction rate. The equation they arrived at (Eq. 5.28) states that an increase in ionic strength will accelerate a reaction

$$\log \frac{k}{k_0} = 2z_A z_B \alpha \sqrt{\mu} \tag{5.28}$$

between like charges and decelerate a reaction between unlike charges in a dilute aqueous medium where interionic attractions are minimal. If one reactant is neutral, Eq. 5.28 predicts no influence of ionic strength, in disagreement with results from many systems. In solvents where ion-pair formation is extensive, such as glacial acetic acid, the effect of added salts is often accounted for by:

$$k = k_0 + b \,[\text{salt}]$$

an empirical equation with no theoretical basis, where k_0 is the rate constant in the absence of salt, and b is an empirical parameter characteristic of the salt. In low-dielectric media, ionic strengths often are influential because of correction terms required to compensate for the nondilute nature of the solution.

All of the above phenomena are examples of *primary salt effects*. For reactions catalyzed by acids or bases or involving ionic equilibria prior to the rate-determining step, *secondary salt effects*[16] are possible. The ionic strength influences the extent of dissociation of weak acids or bases, thus altering hydrogen ion or hydroxide ion concentrations at fixed concentrations of acids or bases.

---

16. Reference B, pp. 390–396; Reference F, Section 2.9; Reference G, Chapter 2 and pp. 306–308; Reference H, pp. 150–157; and Reference I, Chapter 7.

Ambiguities in the application of salt effects to analyses of reaction mechanisms most often arise because secondary salt effects are mistakenly ignored or because the dilute solution limitation on the equations for primary salt effects is forgotten. In analysis for general acid catalysis, changes in buffer concentrations (Chapter 7-1) must be compensated by changes in concentrations of inert salts so that constant ionic strength can be maintained. Otherwise, the salt effect is often mistaken for the general acid catalysis being investigated.

Chapter 6

Substituent Effects on Rates

Kinetic processes are studied considerably more often than equilibrium processes, simply because the chemist performing a reaction would prefer to produce the desired product in quantitative yield and as fast as possible. It is therefore extremely useful to attempt to systematize substituent effects on kinetic phenomena.

6-1 Hammett-type Equations for Rate Processes; Separation of Interaction Mechanisms

The Hammett equation[1] (Eq. 6.1) applies as well to kinetic phenomena as to equilibria (Chapter 4-5). The benzoic acid thermodynamic ionization

$$\log \frac{k}{k_0} = \sigma\rho \tag{6.1}$$

constants are still used as the reference process to obtain values of σ_m and σ_p (Table 4.4). Essentially it is assumed that

$$\Delta G^{\ddagger} = c \, \Delta G^0$$

---

1. Reference B, Section 2-8 and pp. 403–417; Reference C, pp. 238–245; Reference D, Chapter 7; Reference E, Chapter 4; Reference G, Chapters 6 and 7; Reference I, Chapter 11; P. R. Wells, *Linear Free Energy Relationships*, Academic Press, New York, 1968; H. H. Jaffé, *Chem. Rev.* **53**, 191 (1953); J. Shorter, *Chem. Brit.* **5**, 269 (1969).

where c is a constant. This use of transition-state theory (Chapter 5-3) extends the scope of the Hammett equation to all types of kinetic processes.[1] In fact, the Hammett equation is even used to analyze substituent effects on physical properties such as infrared absorption frequencies and nuclear magnetic resonance chemical shifts[1] (see below).

The success of the Hammett equation depends on the relatively constant relationship between electrostatic effects and resonance effects (Chapter 4-5) in reactions occurring on aromatic side-chains. If a positive or negative charge is generated in a position whereby direct resonance interaction is possible between the charge site and the ring or a substituent on the ring through a conjugated system, the resonance effect may become predominant and the usual proportionality between the electrostatic effects and the resonance effects destroyed. If, for example, a positive charge is being generated on the aromatic ring, as in electrophilic aromatic substitution (Eq. 6.2), on the benzylic carbon atom (Eq. 6.3), or on a position which is stabilized by phenyl participation (Eq. 6.4), the relative

$$R-C_6H_5 + E^{\oplus} \longrightarrow R-C_6H_5^{\oplus}\begin{smallmatrix}E\\H\end{smallmatrix} \qquad (6.2)$$

$$R-C_6H_4-CH_2-X \xrightarrow{S_N 1} R-C_6H_4^{\oplus}=CH_2 \qquad (6.3)$$

$$R-C_6H_4-CH_2-CH_2-X \xrightarrow{S_N 1} R-C_6H_4^{\oplus}\triangleleft \qquad (6.4)$$

importance of the electrostatic and resonance effects is constant for these processes, but the constant differs from that involved in the usual Hammett equation. Different σ values therefore are required for such reactions.

Since the simplest prototype of these positively charged processes is electrophilic aromatic substitution, the new σ values are called *electrophilic substituent constants*[1,2] and designated by σ^+. The standard process adopted was the solvolysis of 2-phenyl-2-propyl chlorides in 90% acetone at 25°C (Eq. 6.5). Since the positive charge cannot effectively interact with

$$X-C_6H_4-\underset{CH_3}{\underset{|}{\overset{CH_3}{\overset{|}{C}}}}-Cl \longrightarrow X-C_6H_4^{\oplus}=\underset{CH_3}{\underset{|}{\overset{CH_3}{\overset{|}{C}}}} \qquad (6.5)$$

---

2. H. C. Brown and Y. Okamoto, *J. Amer. Chem. Soc.* **80**, 4979 (1958), and earlier papers.

meta substituents by the enhanced resonance mode, it was assumed that σ_m has the same value as $\sigma_m{}^+$, thereby generating ρ values on the same scale as the usual Hammett ρ values. The $\sigma_p{}^+$ values are then obtained using this value of ρ for the standard reaction, Eq. 6.5. Because $\sigma_p{}^+$ values reflect increased electron donation by substituents (for better interaction with a positive charge), they are always more negative (or less positive) than the corresponding σ_p values. Differences between σ_m and $\sigma_m{}^+$ indicate additional resonance effects with certain *meta* substituents in the electrophilic reactions. Since positive charges are being generated in the electrophilic reactions, ρ inevitably is negative and large.

In a similar vein, additional resonance interaction with a negative charge leads to *nucleophilic substituent constants*,[1] σ^-, which are most useful for nucleophilic aromatic substitutions and many reactions of phenols and anilines and their derivatives.

A different type of substituent constants is the group designated by σ^0 and called *normal substituent constants*.[1] The concept of σ^0's grew out of attempts to find a set of substituent constants which would not be affected by changes in mode of interaction with reactive site or changes in solvation. Two approaches were utilized. The first searched for "well-behaved" *meta* substituents which give good correlations at all times, such as halides and the methyl, acetyl, and nitro groups. The second sought substituent effects in systems where the reaction site was insulated from the ring well enough to minimize resonance interactions effectively, as in the arylpropionic acids and their derivatives. The number of atoms involved in the insulation could not be too great or ρ would become so small that experimental error would become too significant. In either approach, the normal substituent constants would be used to obtain ρ values. Deviations from the $\sigma^0 \rho$ plot would indicate increased resonance interaction or solvation phenomena for deviating substituents. The amount of deviation could lead to a sliding scale of "enhanced" σ values as advocated by Wepster.[1]

Few studies have been performed on solvent influences on substituent effects (as distinct from solvent effects on ρ). Nevertheless, it is evident that solvation can cause occasional significant changes in substituent effects.[1,3]

Deviations from normal substituent constants resulting from enhanced resonance interactions could be used in attempts to quantitatively evaluate both substituent electrostatic effects and substituent resonance effects. Separation of substituent effects into the individual components is complicated by the multiplicity of possible interactions, particularly whenever

---

3. T. Yokoyama, G. R. Wiley, and S. I. Miller, *J. Org. Chem.* **34**, 1859 (1969).

the substituents are attached to an aromatic ring.[4] The grouping of these interactions into electrostatic, resonance, and steric phenomena, while somewhat risky, greatly simplifies the analysis. The electrostatic component in substituent effects is best obtained from σ' values (Chapter 4-5 and Table 4.5), since σ_m^0 values often include a slight resonance component (Chapter 6-3). The resonance component, σ_R, which is often reaction-dependent (see above), then may be obtained in unenhanced situations from equations of the type[1]

$$\sigma_p = \sigma' + \sigma_R$$

since the electrostatic effect operates over similar distances in the bicyclo-[2.2.2]octanes used for σ' and in *para*-substituted benzoic acids.

An alternate approach to the separation of electrostatic and resonance effects (in the absence of steric effects) is the use of the ^{19}F chemical shifts of *meta*- and *para*-substituted fluorobenzenes.[1,5] Since the chemical shift is a measure of the electron density about the fluorine, the difference in the chemical shift for a given substituent in the *meta* position and in the *para* position (relative to fluorobenzene as a standard) should be proportional to the resonance interaction of the *para* substituent with the fluorine, assuming that there is little or no resonance interaction with the *meta* substituent and that the electrostatic effects are the same for the *meta* and *para* positions. The shift for the *meta*-substituted compound relative to fluorobenzene is a measure of the electrostatic interaction. It therefore is possible to separate and quantify the electrostatic and resonance effects using this nuclear magnetic resonance approach. Taft has also applied ^{19}F magnetic resonance as (a) a probe for saturation effects[6] (p. 93), supporting the idea that there never can be a set of resonance-enhanced substituent parameters such as σ^+ or σ^- within the framework of the Hammett equation which are truly general, and (b) a probe for substituent effects in reactants and products and, therefore, on the thermodynamic parameters of Lewis acid–Lewis base reactions.[7] Unfortunately, several groups[3,8] have questioned the use of nuclear magnetic resonance chemical shifts or the use of fluorobenzene derivatives for the evaluation or separation of substituent interaction mechanisms. The usefulness of these magnetic resonance data must therefore be considered questionable.

---

4. A. R. Katritzky and R. D. Topsom, *J. Chem. Educ.* **48**, 427 (1971); *Angew. Chem. Intern. Ed. Engl.* **9**, 87 (1970).
5. R. W. Taft, et al., *J. Amer. Chem. Soc.* **81**, 5352 (1959); **82**, 756 (1960); **85**, 3146 (1963); **91**, 4794 (1969); **92**, 7007 (1970); and other papers in this series.
6. L. D. McKeever and R. W. Taft, *J. Amer. Chem. Soc.* **88**, 4544 (1966).
7. R. W. Taft, et al., *J. Amer. Chem. Soc.* **89**, 2391, 2397 (1967).
8. M. J. Hogben and W. A. G. Graham, *J. Amer. Chem. Soc.* **91**, 283, 291 (1969); K. Bowden, J. G. Irving, and M. J. Price, *Can. J. Chem.* **46**, 3903 (1968); G. R. Wiley and S. I. Miller, *J. Org. Chem.* **37**, 767 (1972).

6-2 The Taft Equation; *ortho* and Steric Effects

Aliphatic systems other than rigid systems such as the bicyclo-[2.2.2]octane series (p. 111) and *ortho*-substituted aromatic systems present problems which are apparent from lack of correlations with the Hammett equation. Even when seemingly related nonrigid aliphatic systems are compared to one another, correlations are relatively unsatisfactory. For example, rates of saponification of ethyl alkanoates do not show any correlation with ionization constants of the corresponding alkanoic acids.

Following a suggestion made by Ingold in 1930, Taft[1,9] proposed a procedure for evaluating electrostatic and steric/conformational factors in aliphatic systems. The essence of this procedure is to combine data from related reactions in such a way as to isolate the electrostatic contribution. This electrostatic component could then be applied to one of the reference reactions so as to obtain the steric component by difference. The reactions chosen were the acid- and base-catalyzed hydrolyses of aliphatic esters, RCO_2R', relative to the corresponding acetates, CH_3CO_2R'. Assuming that free energies of activation could be factored into independent electrostatic, resonance, and steric components[10] (Chapter 6-1), Taft argued that steric and resonance effects should be similar in the acid-catalyzed and base-catalyzed hydrolyses because of the similarities in the transition states leading to **I** and **II**, respectively. Although resonance effects should be

$$\begin{bmatrix} & \text{OH} & \\ & | & \\ \text{R}-&\text{C}-\text{O}-&\text{R}' \\ & | \quad | & \\ & \text{HO} \quad \text{H} & \end{bmatrix}^{\oplus} \quad \begin{bmatrix} & \text{OH} & \\ & | & \\ \text{R}-&\text{C}-\text{O}-&\text{R}' \\ & | & \\ & \text{O} & \end{bmatrix}^{\ominus}$$

$$\qquad \textbf{I} \qquad\qquad\qquad\qquad \textbf{II}$$

similar in these two transition states because of the tetrahedral carboxylate carbons, and although the transition states differ only by fewer than two protons, which are themselves small enough to mitigate against most steric effects, the transition states leading to **I** and **II** differ significantly in charge type and number of electron pairs. Therefore, while Taft's procedure might effectively cancel resonance and steric effects, solvation effects might lead to failure in some correlations, thereby producing new problems. In addition, solvation phenomena could act cooperatively with steric effects, leading to potential doubt as to whether steric effects were really canceled by such reaction comparisons. Even if we accept Taft's pro-

---

9. R. W. Taft, *Steric Effects in Organic Chemistry*, ed. by M. S. Newman, Wiley, New York, 1956, Chapter 13.
10. J. Shorter, *Quart. Rev.* **24**, 433 (1970).

TABLE 6.1 *Electrostatic Parameters for Aliphatic Systems*[9]

R (in RCO_2R')	σ^*	R (in RCO_2R')	σ^*
CH_3-	0.00	Cl_3C-	+2.65
$H-$	+0.49	CH_3OCH_2-	+0.52
CH_3CH_2-	−0.10	$HOCH_2-$	+0.55
i-Pr−	−0.19	$C_6H_5CH_2-$	+0.22
t-Bu−	−0.30	C_6H_5-	+0.60
$ClCH_2-$	+1.05	$CH_2=CH-$	+0.65
Cl_2CH-	+1.94	$C_6H_5CH=CH-$	+0.41

cedure, the method of obtaining the electrostatic components is still not obvious. Basing his reasoning on Hammett ρ values for hydrolyses of *meta*- and *para*-substituted benzoates,[1] Taft assumed that the electrostatic effects reside almost entirely in the base-catalyzed reaction. The ρ values for the acid-catalyzed hydrolyses of the benzoates and related systems range from -0.2 to $+0.5$ and may be considered to be zero. The ρ values for the base-catalyzed benzoate series range from $+2.2$ to $+2.8$ and average at $+2.48$. Taft's *polar substituent constants*, σ^* (Table 6.1), may therefore be defined as

$$\sigma^* = \left(\frac{1}{2.48}\right)\left[\log\left(\frac{k}{k_0}\right)_B - \log\left(\frac{k}{k_0}\right)_A\right]$$
$$= \left(\frac{1}{2.48}\right)\left[\log\left(\frac{k_B}{k_A}\right) - \log\left(\frac{k_{0B}}{k_{0A}}\right)\right] \quad (6.6)$$

where k_0 represents the rate constant for the acetate standard, B indicates a base-catalyzed process, A indicates an acid-catalyzed process, and the 1/2.48 factor is used to make the σ^* values similar to the scale of the σ values. The base-catalyzed hydrolysis and the acid-catalyzed hydrolysis are carried out with the same alcohol portion of the ester, R', at the same temperature in the same solvent. Because of difficulties with solvents and temperatures, most σ^* values were obtained in different systems and transferred to the primary system by $\sigma^*\rho^*$ correlations, with the resulting inherent errors.

If the σ^* values are true measures of the electrostatic component, they should show excellent correlation with the σ' scale (Table 4.5), where no interaction other than solvation effects could be present. In fact, σ^* and σ' show an excellent linear correlation,[1,9] supporting the electrostatic interpretation of σ^*. Inspection of the σ^* scale indicates an attenuation of 0.34 ± 0.05 per intervening methylene group (XCH_2CH_2- vs. XCH_2-), a result surprisingly consonant with the 1/2.8, or 0.36, attenuation per intervening atom used by Branch and Calvin (see discussion of Eq. 4.2). The increase in the negative magnitude of σ^* with increasing size of alkyl groups corresponds to the normal electrostatic order (p. 103).

The general Taft equation[1,9,10] may therefore be written as

$$\log \frac{k}{k_0} = \sigma^* \rho^* + sE_s \qquad (6.7)$$

where s is the sensitivity of the reaction to steric effects, and E_s is the steric effect of each substituent relative to a methyl group. The *steric substituent constants* E_s are obtained most easily from the acid-catalyzed hydrolyses of the aliphatic esters. Since ρ is assumed to be zero for this reaction, Eq. 6.7 would reduce to

$$\log \left(\frac{k}{k_0}\right)_A = sE_s \qquad (6.8)$$

Assigning s a value of 1.00 for this reaction leads directly to the required E_s values (Table 6.2), which, as expected for steric phenomena, increase faster than additivity. These steric parameters involve the carboxyl portion of the ester. Different values are required for changes in the alcohol portion of the ester, indicating different steric interaction mechanisms. Charton[11] has analyzed the E_s values obtained from Taft's aliphatic systems and found them to be a linear function of the van der Waals radii of the substituents and independent of either electrostatic or resonance effects, supporting Taft's procedure for separating electrostatic and steric components in aliphatic compounds.

A useful device for predicting whether or not two systems will depend on the same steric interaction mechanism is the principle of *isosterism*. If the transition states are the same size and shape, the steric interaction

TABLE 6.2 *Steric Parameters for Aliphatic Systems*[9]

R (in RCO_2R')	E_s	R (in RCO_2R')	E_s
H—	+1.24	$ClCH_2$—	−0.24
CH_3—	0.00	Cl_2CH—	−1.54
CH_3CH_2—	−0.07	Cl_3C—	−2.06
n-Pr—	−0.36	$C_6H_5CH_2$—	−0.38
i-Pr—	−0.47	$C_6H_5CH_2CH_2$—	−0.38
-Bu—	−0.93	$C_6H_5CH(CH_3)$—	−1.19
t-Bu—	−1.54	$(C_6H_5)_2CH$—	−1.76
CH_3OCH_2—	−0.19		

11. M. Charton, *Progr. Phys. Org. Chem.* **8**, 235 (1971), and previous papers referenced therein.

mechanism will be similar regardless of the chemical elements involved. For example, reactions of 2-alkylpyridines with methyl iodide show a correlation with Eq. 6.7[1] with $s = 2.06$, while reaction of the same pyridines with boron trifluoride or trimethylboron gives equally good correlations[1] ($s = 5.49$ and 6.36, respectively). Such systems are called *isosteres*.

After establishing E_s values for the aliphatic system, Taft[9] then applied the σ^*–E_s approach (Eq. 6.7) to *ortho*-substituted aromatic systems. The new scales were called σ_o^* and E_s^o. Since $\sigma_{o-X}^* - \sigma_{o-CH_3}^*$ was often equal to $\sigma_{p-X} - \sigma_{p-CH_3}$, Taft concluded that the σ_o^* scale was free from steric effects and that the electrostatic and resonance effects operated equally from the *ortho* and *para* positions. While this latter equality might be reasonable for resonance interactions, all models for electrostatic interactions include some sort of distance dependence, be it the number of bonds as in the inductive effect or the distance per se as in the field effect (Chapter 4-2). In addition, correlations using a $\sigma_o^*\rho^*$ approach lead to values of ρ^* different from the values of ρ obtained for the same systems using the *meta*- and *para*-substituted analogs.[10,11]

Charton's[11] analysis of E_s^o finds these values to be independent of the van der Waals radii and definitely dependent on the electrostatic and resonance interactions. Charton concludes that the data for *ortho*-substituted aromatic systems are generally independent of steric effects. He postulates the existence of a "resonance proximity effect" and probably an "electrostatic proximity effect", both of which are in addition to the usual aromatic resonance and electrostatic phenomena. Charton also suggests that the unsubstituted compounds (hydrogen as *ortho* substituent) usually do not fall on any correlation line. This exclusion of the unsubstituted system is supported by work by Schleyer and Woodworth[12] in completely different systems (p. 105). Charton,[11] Shorter,[10] and others[1,13] seem to support the point of view that no single purely electrostatic–resonance *ortho*-substituent scale is possible because of the range of relative dependence on the resonance and electrostatic effects for seemingly related reactions, the large solvent dependence,[14] and the dependence on the number and type of atoms intervening between the aromatic ring and the reaction site. Even Smith's[15] attempt to use gas-phase data to eliminate solvent effects,[14] and thereby derive a scale of useful *ortho* constants, seems doomed[10,11] to failure when applied to solution data, in spite of his conclusion that proximity effects are minimal or nonexistent in the gas-

---

12. P. von R. Schleyer and C. W. Woodworth, *J. Amer. Chem. Soc.* **90**, 6528 (1968).
13. A. Buckley, N. B. Chapman, and J. Shorter, *J. Chem. Soc.*, B 195 (1969); K. Bowden and G. E. Manser, *Can. J. Chem.* **46**, 2941 (1968).
14. J. Steigman and D. Sussman, *J. Amer. Chem. Soc.* **89**, 6406 (1967).
15. G. G. Smith, K. K. Lum, J. A. Kirby, and J. Posposil, *J. Org. Chem.* **34**, 2090 (1969), and previous papers in this series.

phase reactions. In general, the Taft approach[9] to both aliphatic and *ortho*-substituted aromatic systems is in need of refinement,[10,11] particularly in the latter case.

6-3 The Swain-Lupton Approach[16]

All of the data analyzed in terms of the Hammett equation (Chapters 4-5, 6-1, and 6-2), including all of the various types of special sets of substituent constants, have been recently reanalyzed by Swain and Lupton[16] in an attempt to use computer-performed correlations to derive physically significant independent variables for the resonance and electrostatic contributions for each substituent regardless of its environment. These new substituent constants are called $\mathscr{F}$ and $\mathscr{R}$ for the electrostatic and resonance constants, respectively. Each set of experimental substituent constants, such as σ, σ', and σ^+, is expressed as

$$\text{``}\sigma\text{''} = f\mathscr{F} + r\mathscr{R}$$

The signs and magnitudes of $\mathscr{F}$ and $\mathscr{R}$ evaluate the directions and magnitudes of the substituent effects in the same sense as the composite substituent constants discussed earlier.[1] The ratio $\mathscr{R}/\mathscr{F}$ expresses the relative importance of the resonance and electrostatic effects for each substituent were these two interactions equally important in influencing the experimental situation. The sensitivity of the experimental or analytical technique to $\mathscr{F}$ and $\mathscr{R}$ is expressed by f and r, which then can be used to calculate the percentage composition of the technique in terms of the individual electrostatic and resonance components. The $\mathscr{F}$ and $\mathscr{R}$ scales are derived by assuming that $r = 0.00$ for the σ' series (p. 111), a system where resonance effects must be absent because of the rigid aliphatic framework; and that $\mathscr{R} = 0.00$ for the $-\overset{\oplus}{N}(CH_3)_3$ substituent, since the quaternary nitrogen effectively prevents resonance interactions involving this group.

Typical $\mathscr{F}$ and $\mathscr{R}$ values are shown in Table 6.3. If the qualitative sign conventions are ignored, perfect correlation is found to exist between these numbers and the qualitative evaluation of substituent effects (Tables 4.2 and 4.3). Deviations in $\mathscr{F}$ from σ' values (Table 4.5) result from the need to place all results on a similar basis. The electrostatic effects of the alkyl groups follow the normal order (p. 103). Resonance effects are larger for the first-row elements than for those lower in the periodic table on a group basis, supporting the idea of better overlap in $2p-\pi$ conjugation (p. 102).

The only surprise lies in the percent resonance derived from f and r. As expected, σ' shows less than a 4% resonance component and σ^* shows

---

16. C. G. Swain and E. C. Lupton, Jr., *J. Amer. Chem. Soc.* **90**, 4328 (1968).

TABLE 6.3 *Swain-Lupton Electrostatic and Resonance Parameters*[16]

Substituent	$\mathscr{F}$	$\mathscr{R}$
H	0.000	0.000
I	0.672	−0.197
Br	0.727	−0.176
Cl	0.690	−0.161
F	0.708	−0.336
NH_2	0.037	−0.681
CN	0.847	0.184
OH	0.487	−0.643
OCH_3	0.413	−0.500
SH	0.464	−0.111
NO_2	1.109	0.155
CF_3	0.631	0.186
CH_3	−0.052	−0.141
CH_2CH_3	−0.065	−0.114
$C(CH_3)_3$	−0.104	−0.138
C_6H_5	0.139	−0.088

$6 \pm 4\%$, suggesting that the Taft procedure for separating steric effects in aliphatic systems (Chapter 6-2) works reasonably well. The σ_p scale is 53% resonance and σ_p^+ shows $66 \pm 5\%$, the greater value reflecting the increase in resonance postulated for those systems where the electrophilic substituent constants are required. However, σ_m interactions have a 22% resonance component! If the Swain-Lupton approach proves valid, all interpretations based on the assumption that the σ_m values are purely electrostatic are probably erroneous, and the correlations justifying this assumption are probably fortuitous (pp. 111, 130, and 131). In addition, Swain and Lupton found that electrostatic effects were not equal for *meta* and *para* positions.

Godfrey[17] has presented an alternative to the Swain-Lupton approach, proposing a field and charge-transfer (FCT) approach. This approach, which is restricted to aromatic molecules, ignores electrostatic effects operating through σ electrons. While it is useful for spectroscopic phenomena and some aromatic reactions, the FCT approach is so limited in scope that Godfrey's contention that this constitutes an extension and improvement of the Swain-Lupton approach seems dubious.

6-4 ρ Values and Mechanisms

Our discussion of the Hammett equation and its various modifications has thus far concentrated on substituent effects. The ρ values and the types of correlations obtained are extremely useful in the analysis of

---
17. M. Godfrey, *J. Chem. Soc.*, B 1534, 1537, 1540, 1545 (1971), and previous papers in this series.

reaction mechanisms.[1] Positive ρ values indicate that a reaction is favored by electron withdrawal from the reaction site, while negative ρ values indicate acceleration by electron donation. If a mechanism includes several steps prior to and including the rate-determining step, the ρ value will be a composite of the ρ values for each individual step.

Nonlinear Hammett relationships[18] arise from a change in the nature of the substituent effect, a change in mechanism, a change in rate-determining step, or a change in transition-state structure. If a change in substituent interaction mechanism produces the nonlinearity, this often can be identified by improved correlation or linearity when σ is replaced with one of the other scales of substituent parameters (Chapter 6-1 and 6-2). In general, a change in mechanism or transition-state structure produces a concave upward plot. Assume that one mechanism is established for one linear portion of a concave upward Hammett plot. At the point corresponding to the change from linearity, this mechanism still is conceivable. The upward curvature is observed because the mechanism beginning to predominate must occur at a faster rate than the old mechanism or it would not be the predominant one. If the two mechanisms occur at comparable rates for all substituents, curvature will be noted and no linear segments will be apparent. If the mechanisms differ appreciably in substituent sensitivity, linear segments will be noticeable and the rates for each mechanism may be dissected. A change in rate-determining step usually causes a concave downward plot, since at least one of the slopes must result from a composite ρ value in which one of the components has become relatively unfavorable. Examples of these various possibilities will be noted in the following discussion.

The saponification of benzoates[1] provides a useful starting point (Eq. 6.9). The reaction is first-order in ester and in base. The tetrahedral intermediate possesses negative charge, so both the transition state leading

$$S\text{-}C_6H_4\text{-}C(=O)\text{-}OEt + {}^{\ominus}OH \rightleftharpoons S\text{-}C_6H_4\text{-}C(O^{\ominus})(OH)\text{-}OEt \tag{6.9}$$

$$\downarrow$$

$$S\text{-}C_6H_4\text{-}C(=O)\text{-}O^{\ominus} + HOEt \longleftarrow S\text{-}C_6H_4\text{-}C(=O)\text{-}OH + {}^{\ominus}OEt$$

---

18. J. O. Schreck, *J. Chem. Educ.* **48**, 103 (1971).

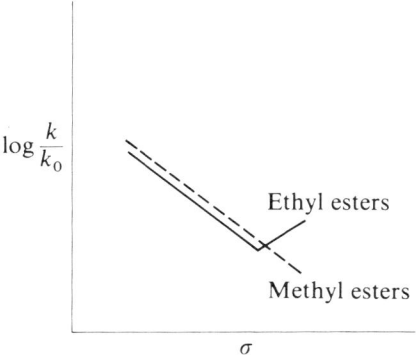

FIGURE 6.1 *The hydrolysis of alkyl benzoates in* 99.9% H_2SO_4.

to the intermediate and the transition state leading from intermediate to product should possess negative charge character. The negative charge would be stabilized by electron-withdrawing substituents, so a positive ρ value would support such a mechanism. Experimentally, $\rho = +2.5$.

Acid-catalyzed hydrolysis of methyl benzoates in 99.9% sulfuric acid[1] produces a linear $\sigma\rho$ plot (Fig. 6.1), which, along with other evidence,[19] supports a mechanism involving acyl-oxygen fission (Eq. 6.10).

---

19. Reference A, Chapter 17; Reference C, pp. 309–313; Reference D, Chapter 9; Reference E, Chapter 12; Reference I, Chapter 10.15; and Reference J, Section 58.

On the other hand, similar hydrolysis of the corresponding ethyl esters produces a concave upward $\sigma\rho$ plot (Fig. 6.1). The break in this plot corresponds to a change to alkyl-oxygen fission when strongly electron-attracting groups are present on the aromatic ring.[19] This change in mechanism is supported by nuclear magnetic resonance studies in strongly acidic media performed by Olah and coworkers.[20]

Nucleophilic substitution reactions provide an informative series. Reaction of substituted N,N-dimethylanilines with methyl iodide in 90% acetone proceeds with $\rho = -3.30$:[1]

$$S\text{-}C_6H_4\text{-}N(CH_3)_2 + CH_3I \longrightarrow S\text{-}C_6H_4\text{-}N^+(CH_3)_3 + I^-$$

Since increased electron density on the nitrogen should increase its effectiveness as a nucleophile, ρ should be negative and of reasonable magnitude if the mechanism is a bimolecular substitution. If the reaction rate is monitored as a function of change in substituent on the substrate, a similar ρ value would be expected for a unimolecular substitution. For example, solvolysis of benzyl chloride in water would be favored by electron-donating stabilization of the developing carbonium ion if the mechanism is S_N1 (Eq. 6.3). In addition, correlation with σ should be nonlinear and show significant improvement when σ^+ values are used. All of this is verified experimentally,[1] with $\rho = -1.875$ when correlated with σ^+. Reaction of the same series of benzyl chlorides with iodide ion in acetone produces a good linear $\sigma\rho$ plot with $\rho = +0.785$,[1] indicating a complete shift to an S_N2 mechanism, the intermediate or transition state of which is stabilized by electron-withdrawing groups yet demonstrates less sensitivity to substituent effects than that found for the closely related S_N1 process just discussed.[21] Reaction of various benzyl derivatives with various amines in nonhydroxylic solvents[1] produces curved $\sigma\rho$ plots, with the curvature being more pronounced for series of *para*-substituted compounds than for series of *meta*-substituted substrates. This type of result suggests the presence of both S_N1 and S_N2 processes or of a "merged" mechanism involving a slow change in transition-state structure from S_N1-like to S_N2-like.[21] This idea is reinforced by the greater curvature in the *para* series, since it is this series which should require σ^+ values for the

---

20. G. A. Olah, D. H. O'Brien, and A. M. White, *J. Amer. Chem. Soc.* **89**, 5694 (1967).
21. For a discussion of nucleophilic substitutions, see Reference A, Chapter 11; Reference C, Chapter 10; Reference D, Chapter 8; Reference E, Chapters 6 and 7; Reference J, Sections 27, 28, and 33; and A. Streitwieser, *Solvolytic Displacement Reactions*, McGraw-Hill, New York, 1962.

most strongly electron-donating substituents in an S_N1 mechanistic component, which would itself be most favorable relative to an S_N2 component in the presence of just such electron-donating groups.

The clear-cut distinction between S_N1 and S_N2 on the basis of the sign of ρ becomes somewhat confused when the solvolysis of aliphatic systems is investigated.[1,21] In 80% ethanol at 25°C, a series of tertiary halides shows $\rho^* = -3.29$, while reactions of primary tosylates in absolute ethanol at 100°C proceed with $\rho^* = -0.742$. These results suggest that (a) some carbonium-ion-like character is involved in both of these series (with the greater amount in the tertiary series as expected), (b) the Taft equation (Chapter 6-2) does not center about zero in the desired way and therefore does not provide an unambiguous description of substituent interactions, or (c) the change in solvent, temperature, and leaving group is sufficient to change the sign from positive to negative in the primary series. Discrimination between these three possibilities has yet to be clearly performed.[9-11,21]

In a somewhat similar vein, the acid-catalyzed rearrangement shown in Eq. 6.11 involves an allyl cation in which the π system of the cation is conjugated with the π system of the benzene ring. This conjugation[1] is supported by the fairly large negative ρ value (-4.67) obtained from a σ^+ correlation.

$$\text{S-C}_6\text{H}_4-\underset{\underset{\text{OH}}{|}}{\text{CH}}-\text{CH}=\text{CH}-\text{CH}_3 + \text{H}^\oplus$$

$$\downarrow \qquad\qquad\qquad\qquad\qquad\qquad (6.11)$$

$$\text{S-C}_6\text{H}_4-\text{CH}=\text{CH}-\underset{\underset{\text{OH}}{|}}{\text{CH}}-\text{CH}_3$$

Similar magnitudes of $\sigma\rho$ plots for related reactions provide support for similar trends in transition-state structures or electronic distributions in reactants and products. For example, ρ values differing by only 0.01 for the thermodynamic ionizations of anilinium salts and N,N-dimethylanilinium salts in 30% ethanol at 25°C suggest that the two methyls attached to the nitrogen in the latter series produce effects in the conjugate bases which cancel their effects in the acidic forms. Along the same lines, similar ρ values for the bromination of toluenes in carbon tetrachloride at 80°C by molecular bromine and by N-bromosuccinimide suggest that the hydrogen-abstracting species is atomic bromine in both of these reactions.

Hammett-type plots can often provide supportive evidence for multi-step mechanisms. For example, the acid-catalyzed bromination of acetophenones is believed to involve rate-determining formation of the enol[22] (Eq. 6.12). The ρ value of -0.45[1] supports this mechanism, which is

$$S\text{-}C_6H_4\text{-}\overset{O}{\underset{\|}{C}}\text{-}CH_3 + H_3O^{\oplus} \rightleftharpoons S\text{-}C_6H_4\text{-}\overset{OH}{\underset{|}{\overset{\oplus}{C}}}\text{-}CH_3 + H_2O$$

$$\text{rds} \updownarrow \quad (6.12)$$

$$\text{products} \xleftarrow{Br_2} S\text{-}C_6H_4\text{-}\overset{OH}{\underset{|}{C}}=CH_2 + H_3O^{\oplus}$$

kinetically independent of halogen concentration. The protonation of the ketone would be expected to be facilitated by electron-donating substituents, since this action would enhance the electronegativity of the carbonyl oxygen. The second step, formation of the enol from this protonated ketone, would probably be more facile in the presence of electron-withdrawing groups, since the carbonium-ion-like intermediate would thereby be destabilized. The composite ρ value would therefore be expected to be small and of unknown sign, which is in accord with the results.

The opposite situation exists for the Cannizzaro reaction[23] (Eq. 6.13). The first step, nucleophilic attack by hydroxide on the carbonyl carbon, would be favored by electron-withdrawing substituents (see the analogous saponification of benzoates, Eq. 6.9). The second step, hydride transfer, would be favored by electron withdrawal in the aldehyde and probably by electron donation in the hydride donor. The ρ value of $+3.63$ gives no information as to whether the first step or the second step in this mechanism is rate-determining. Experimentally,[23] the hydride transfer is thought to be rate-determining since the reaction is second-order in aldehyde. The problem is complicated further by a competing process which is second-order in both aldehyde and base.

A third situation is exemplified by the biochemically important reaction of aldehydes or ketones with amines or amine derivatives to form

---

22. Reference A, Chapter 16; Reference C, Chapter 12; Reference D, Chapter 10; Reference E, pp. 109–111; Reference I, Chapters 4.15, 4.18, 4.27, and 10.13; and Reference J, Section 47.
23. Reference A, pp. 554–556; Reference C, pp. 908–909; Reference D, pp. 546–548; Reference E, pp. 267–269; and Reference J, pp. 1029–1032.

$$\underset{S}{\overset{O}{\text{Ar}}}-\text{C}-\text{H} + {}^{\ominus}\text{OH} \longrightarrow \underset{S}{\overset{O^{\ominus}}{\text{Ar}}}-\underset{\text{OH}}{\overset{|}{\text{C}}}-\text{H}$$

$$\underset{S}{\text{Ar}}-\underset{\text{OH}}{\overset{O^{\ominus}}{\overset{|}{\text{C}}}}-\text{H} + \underset{S}{\overset{O}{\text{Ar}}}-\text{C}-\text{H}$$

$$\downarrow$$

$$\underset{S}{\overset{O}{\text{Ar}}}-\text{C}-\text{OH} + \underset{S}{\text{Ar}}-\underset{H}{\overset{O^{\ominus}}{\overset{|}{\text{C}}}}-\text{H} \qquad (6.13)$$

$$\updownarrow$$

$$\underset{S}{\overset{O}{\text{Ar}}}-\text{C}-\text{O}^{\ominus} + \underset{S}{\text{Ar}}-\underset{H}{\overset{\text{OH}}{\overset{|}{\text{C}}}}-\text{H}$$

imines (or Schiff bases) by dehydration of intermediate carbinolamines[24] (Eq. 6.14). In strong acid media, k_1 is rate-determining and ρ is greater than zero (0.91 with a σ^+ correlation) since the reaction would be accelera-

$$\underset{S}{\overset{O}{\text{Ar}}}-\text{C}-\text{H} + \text{H}_2\text{N}-\text{G} \underset{k_{-1}}{\overset{k_1}{\rightleftarrows}} \underset{S}{\text{Ar}}-\underset{H}{\overset{\text{OH}}{\overset{|}{\text{C}}}}-\text{NHG}$$

$$\text{H}^{\oplus} \downarrow k_2 \qquad (6.14)$$

$$\underset{S}{\text{Ar}}-\text{CH}=\text{NG} + \text{H}_2\text{O}$$

---

24. Reference E, pp. 254–257; Reference J, Section 54; and W. P. Jencks, *Progr. Phys. Org. Chem.* **2**, 63 (1964), and *Catalysis in Chemistry and Enzymology*, McGraw-Hill, New York, 1969, Chapter 10, Section A.

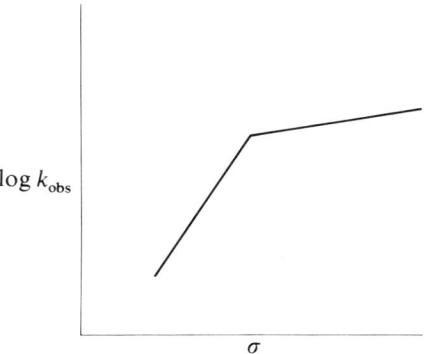

FIGURE 6.2 *The rate of reaction of substituted benzaldehydes with semicarbazide at pH 3.9 as a function of σ.*[24]

ted by electron-withdrawing groups, as are all nucleophilic additions to carbonyl groups (see above). In neutral media, k_2 is rate-determining, so the observed rate constant is a composite of the equilibrium constant for formation of the carbinolamine (k_1/k_{-1}) and the rate constant for dehydration of the carbinolamine (k_2). The dehydration would be facilitated by electron-donating substituents ($\rho = -1.74$), so the composite ρ (0.07) would be small and of unknown sign because of the opposite substituent effects of its two components. If the amine derivative is semicarbazide (G=NH—C(=O)—NH$_2$),[24] at pH 3.9[1,2,24] the $\sigma\rho$ plot exhibits a sharp break with an overall concave downward shape[24] (Fig. 6.2). At this pH, changing the substituents causes a fairly sudden change in the rate-determining step.

Extensive studies have been performed on electrophilic aromatic substitutions (Chapter 5-2). In general, ρ is very negative and correlation with σ^+ is preferred (Chapter 6-1). A discussion of the usefulness of these ρ values may be found in Chapter 7-8.

Free-radical reactions present a different type of problem.[1] Most free-radical processes either exhibit no sensitivity to substituents or correlate with some form of the Hammett equation with small values of ρ. For example, thermal cleavage of a series of symmetrical dibenzoyl peroxides (Eq. 6.15) at 80°C gives a reasonable $\sigma\rho$ plot with $\rho = -0.20$. This type of

$$S\text{-}C_6H_4\text{-}C(=O)\text{-}O\text{-}O\text{-}C(=O)\text{-}C_6H_4\text{-}S \xrightarrow{80°C} 2\ S\text{-}C_6H_4\text{-}CO_2\cdot \quad (6.15)$$

cleavage cannot involve formal charges or polar character in the transition state because of symmetry. Because ρ is temperature-dependent (Chapter 4-5), the negative sign obtained in this correlation probably has no significance.

Many free-radical reactions have a pronounced polar component. Hydrogen abstraction from a series of benzaldehydes by p-chlorobenzoyl-peroxy radicals (Eq. 6.16) exhibits $\rho = -1.67$, indicating that the reaction

$$\text{Cl}-\underset{}{\bigcirc}-\underset{\|}{\overset{\text{O}}{\text{C}}}-\text{O}-\text{O}\cdot + \text{H}-\underset{\|}{\overset{\text{O}}{\text{C}}}-\underset{\text{S}}{\bigcirc}$$

$$\downarrow \text{AC}_2\text{O} \qquad (6.16)$$

$$\text{Cl}-\underset{}{\bigcirc}-\underset{\|}{\overset{\text{O}}{\text{C}}}-\text{OOH} + \cdot\underset{\|}{\overset{\text{O}}{\text{C}}}-\underset{\text{S}}{\bigcirc}$$

is more sensitive to substituent effects than the standard completely ionic benzoic acid ionizations. This type of behavior results from the polar character in the transition state, **III**. Whenever a free-radical reaction possesses polar character in the transition state and shows a

$$\left[\text{Cl}-\underset{}{\bigcirc}-\underset{\|}{\overset{\text{O}}{\text{C}}}-\text{OO}^{\ominus}\cdots\dot{\text{H}}\cdots\overset{\oplus}{\underset{\|}{\overset{}{\text{C}}}}-\underset{\text{S}}{\bigcirc}\right]$$

III

Hammett-equation correlation (Eq. 6.17), (a) the nonpolar interactions

$$\log \frac{k}{k_0} = (\sigma\rho)_{\text{polar}} + (\sigma\rho)_{\text{nonpolar}} \qquad (6.17)$$

must be essentially constant, as in Eq. 6.16, (b) the polar interactions must be essentially constant, or (c) the polar interactions must be proportional to the nonpolar interactions. However, Zavitsas[25] argues that such polar character does not exist. He feels that these Hammett-equation correlations reflect the sensitivities of rates of hydrogen abstractions to the extent of bond breaking and to differences in bond dissociation energy.

Linear free-energy relationships of various types are used in the analysis of biochemical structure–activity relationships.[26] Various Hammett

---
25. A. A. Zavitsas and J. A. Pinto, *J. Amer. Chem. Soc.* **94**, 7390 (1972).
26. C. Hansch, *Accounts Chem. Res.* **2**, 232 (1969), and references therein.

and Taft equations often are useful, particularly in conjunction with additional parameters, such as the *partition coefficient factor*[26] π, which measures the "hydrophobic bonding" characteristics of a drug by using the partition coefficient between octanol and water for the given substituent in a standard system. Steric effects are often important, even when the substituent is remote from the site believed responsible for the drug activity. This type of steric effect is consistent with the "lock and key" hypothesis[26] for both enzyme and antigen action. Presumably, the entire structure of the drug is important for activity, including many fairly specific electrostatic and steric secondary interactions remote from the primary active site. This area of study is in need of further development, both from the point of view of developing linear free-energy relationships and from the point of view of better understanding the mechanism of drug action.[27]

---

27. Criticism has been leveled [C. D. Johnson and K. Schofield, *J. Amer. Chem. Soc.* **95**, 270 (1973)] at use of Hammett equations in structure–reactivity relationships (Chapter 7-8). Variable transition-state structures are implicit in most Hammett correlations derived from kinetic data. Therefore, values of ρ cannot be used to analyze transition-state structure by comparing structure–reactivity relationships in different reaction series.

Chapter 7

The Structure of the Transition State

Organic chemists of the present era—and, for that matter, inorganic chemists and a large group of biochemists—are heavily involved in questions related to reaction rates and mechanisms. For this reason, there is a large effort devoted to analysis of transition-state structure, for it is the energy of the transition state which is the most obscure and unpredictable factor involved in predictions of rates and evaluations of reaction mechanisms. Since the activated complex cannot be isolated and since calculations of its properties and/or parameters for the activation process are only qualitatively possible for even the most simple systems (Chapter 5-3), effort has been directed toward the search for techniques which might assist in predicting or evaluating transition-state properties. Even if one adopts the point of view that too much emphasis in current research is placed on the transition state and not enough emphasis on the ground state, one can hardly deny that the nature of the transition state deserves some degree of emphasis for the simple reason that it is one of the two components which must be understood in order for a reaction to be understood in the greatest possible detail.[1]

---

1. It is a basic tenet that a mechanism can never be proven. In order for a mechanism to be accepted, the experimental observations must be such as to rule out all other conceivable mechanisms. If we accept the premise that the current approach to the analysis of reaction mechanism is itself a cohesive and viable model, we recognize, therefore, that the limitation of this approach is the ability of human beings to conceive possible mechanisms.

7-1 Catalysis

Catalytic effects provide considerable information about the mechanism of a catalyzed reaction while simultaneously contributing little or no information about the uncatalyzed system. Fortunately, this does not constitute much of a problem, since most reactions which proceed efficiently when catalyzed occur slowly (if at all) when the catalyst is absent. The catalyst speeds up the reaction by changing the structure of the transition state or by changing the mechanism itself, leading to a lowering in energy of the transition state for the rate-determining step.

The most common catalysts are acids, HA, or bases, B. Several types of acid–base catalysis have been delineated. Because of the related nature of acid catalysis and base catalysis, only the former will be considered here. The proton will be represented as H^+ whenever the reactive catalytic species is either the proton itself or, the more common situation, the lyonium ion (e.g., the hydronium ion, $H(H_2O)_n^+$ or H_3O^+) in aqueous solution.

There are three major types of acid catalysis. In Type I, the reaction is catalyzed only by lyonium ion (Eq. 7.1),

$$S + H^+ \underset{k_{-1}}{\overset{k_1}{\rightleftharpoons}} SH^+ \qquad (K = k_1/k_{-1})$$
$$SH^+ \underset{\text{slow}}{\overset{k_2}{\longrightarrow}} \text{products} \tag{7.1}$$

with a rate law such as Eq. 7.2. This type of behavior is the simplest

$$\text{rate} = k_2[SH^+] = k_2 K[H^+][S] \tag{7.2}$$

example of what is called *specific acid catalysis*.[2] A typical reaction is the hydrolysis of many acetals (Eq. 7.3), where loss of alcohol, R'OH, occurs in

$$
\begin{array}{c}
\text{OR}' \\
| \\
R-C-OR' + H^+ \\
| \\
H
\end{array}
\overset{K}{\rightleftharpoons}
\begin{array}{c}
\overset{+}{H}OR' \\
| \\
R-C-OR' \\
| \\
H
\end{array}
$$

$$\Big\updownarrow \text{slow}$$

$$\text{products} \xleftarrow[\text{fast}]{H_2O} \begin{array}{c} \overset{+}{R-C-OR'} + R'OH \\ | \\ H \end{array} \tag{7.3}$$

---
2. Reference A, Chapter 2; Reference B, Section 3-7; Reference D, pp. 110–112 and 188–190; Reference E, Chapter 5-1 and 5-2; Reference H, Chapter 9; Reference I, Chapter 10; W. P. Jencks, *Catalysis in Chemistry and Enzymology*, McGraw-Hill, New York, 1969, Chapter 3.

the rate-determining step. If the reaction involves a similar equilibrium proton transfer from substrate to lyate ion prior to the rate-determining step, the phenomenon is called *specific base catalysis*, the prime examples of which are most base-catalyzed condensations

In Type II catalysis, the rate depends on the lyonium ion concentration and on the concentration of un-ionized acids which are present. There are several mechanistic schemes which correspond to this type of *general acid catalysis*.[2] In the simplest, the rate-determining step is transfer of a proton from an acid to the substrate (Eq. 7.4), a scheme analogous to Eq. 7.1

$$S + HA \xrightarrow[k]{slow} SH^+ + A^-$$
$$SH^+ \xrightarrow{fast} products \tag{7.4}$$

with k_1 rate-determining. The rate law, Eq. 7.5, indicates dependence on the concentration of acid HA. If more than one acid is present, the rate

$$\text{rate} = k[S][HA] \tag{7.5}$$

depends on two factors: the concentration of each acid, and the rate constant associated with proton transfer from each acid (Eq. 7.6, where n is the number of different acids present).

$$\text{rate} = [S] \sum_{i=1}^{n} k_i [HA_i] \tag{7.6}$$

A second scheme for Type II general acid catalysis is shown in Eq. 7.7, where B represents a base. The rate law, Eq. 7.8, indicates general acid

$$S + H^+ \xrightleftharpoons{K} SH^+$$
$$SH^+ + B \xrightarrow[slow]{k_2} products + BH^+ \tag{7.7}$$

catalysis by the conjugate acids of the bases involved. If the base B is

$$\text{rate} = k_2[B][SH^+] = k_2 K[S][B][H^+]$$
$$= k_2 K K_a^{BH^+}[S][BH^+] \tag{7.8}$$

the conjugate base A^- of an acid HA, Eq. 7.7 can be represented in slightly modified form as Eq. 7.9, with a rate equation of the form shown in Eq. 7.10. Reactions exemplifying this type of general acid catalysis are the enolization

$$HS + H^+ \xrightleftharpoons{K} HSH^+$$
$$HSH^+ + A^- \xrightarrow[slow]{k_2} HA + SH \tag{7.9}$$

$$\text{rate} = k_2 K[SH][H^+][A^-] = k_2 K K_a^{HA}[SH][HA] \tag{7.10}$$

of acetone (analogous to Eq. 6.12) and the acid-catalyzed mutarotation of glucose. The base-catalyzed halogenation of acetone provides an example of the corresponding general base catalysis.

A different type of general acid catalysis, Type III, involves formation of a complex between acid and substrate, followed by rate-determining decomposition of the complex[2] (Eq. 7.11). The rate expression (Eq. 7.12) indicates general acid catalysis whether or not the complex is hydrogen-

$$S + HA \xrightleftharpoons{K} (S \cdot HA)$$
$$(S \cdot HA) \xrightarrow[\text{slow}]{k} \text{products} \tag{7.11}$$
$$\text{rate} = k[(S \cdot HA)] = kK[S][HA] \tag{7.12}$$

bonded and whether or not proton transfer occurs in the rate-determining step. An example of this type of behavior is the acid-catalyzed decomposition of some diazo compounds.

General acid catalysis may be distinguished from specific acid catalysis by observing reaction rate as a function of buffer concentration. If the buffer solutions are at constant pH and constant ionic strength (Chapter 5-5) and differ only in concentration of acid HA, a plot of observed rate constant against buffer concentration would show zero slope for specific acid catalysis and some change in rate for general acid catalysis. In the latter case, the slope is the catalytic rate constant for the acid, k_{HA}, and the intercept is the rate constant for the reaction catalyzed by lyonium ion. Type II and Type III general acid catalysis are most commonly distinguished by use of solvent isotope effects (Chapter 7-4). Type III is also sometimes apparent from spectroscopic evidence for complex formation.

7-2 Primary Kinetic Isotope Effects

One of the most useful techniques for probing reaction mechanism or transition-state structure is kinetic isotope effects.[3] Isotopic substitution of some kind provides useful information while causing the least change in reactant and transition-state structure and in potential-energy surface. The two major limitations on the use of kinetic isotope effects are (1) the experimental problems associated with measuring small rate differences in a

---

3. L. Melander, *Isotope Effects on Reaction Rates*, Ronald Press, New York, 1960; F. H. Westheimer, *Chem. Rev.* **61**, 265 (1961); Reference A, Chapter 1.12; Reference B, Sections 2-7 and 3-4; Reference C, pp. 213–216; Reference D, pp. 192–193; Reference E, pp. 71–73; Reference I, Chapter 5.22 and 5.23; W. P. Jencks, *Catalysis in Chemistry and Enzymology*, McGraw-Hill, New York, 1969, Chapter 4; M. Wolfsberg, *Accounts Chem. Res.* **5**, 225 (1972); *Isotope Effects in Chemical Reactions*, ed. by C. J. Collins and N. S. Bowman, Van Nostrand Reinhold, New York, 1970.

reproducible manner, and (2) the usual kinetics problem that the information provided is most applicable to the rate-determining step, may often be a composite of results accumulated from several mechanistic steps, and cannot be used to analyze phenomena which occur after the rate-determining step.

Whenever the isotopic substitution involves an element at which a bond is formed or broken in or prior to the rate-determining step, the change observed as the result of the isotopic substitution is called a *primary isotope effect*. This is the type of isotopic substitution for which the largest effects are observed, since the isotope is most involved in the reaction sequence.

The simplest approach to understanding the basis for primary isotope effects is first to consider the effect of isotopic substitution of B' for B in a diatomic molecule A—B.[3] The zero-point energy is the lowest energy level for any molecule and may be approximated by $\frac{1}{2}h\nu$. Since 99% of all molecules possess the zero-point energy level at room temperature, any energy difference between A—B and A—B' must result from differences in zero-point energies. This would also be true for the dissociation process, provided that the bond is completely broken in the transition state and that the isotopic substitution has no effect on the potential-energy surface. Within the harmonic oscillator approximation for the vibration of the A—B bond, the difference in frequencies for this isotopic substitution may be expressed as

$$\frac{\nu'}{\nu} = \sqrt{\frac{m_B(m_A + m_{B'})}{m_{B'}(m_A + m_B)}} = M$$

(where the m's are the atomic masses) and the difference in zero-point energies as

$$E_{AB} - E_{AB'} = \tfrac{1}{2}h\nu(1 - M)$$

The most important conclusion is that the heavier isotope would lead to the lower zero-point energy (and the lower frequency of vibration)(Fig. 7.1).

Assuming no other effects are involved in primary kinetic isotope effects, the use of zero-point energy differences as the criterion for isotope effects could be extended to more complicated systems, as long as the bond under consideration is completely broken in the transition state. The energy difference may be translated into relative rates in terms of Eq. 7.13:

$$\frac{k_{AB}}{k_{AB'}} = e^{(E_a^{AB'} - E_a^{AB})/RT} = e^{h\nu(1-M)/2RT} \tag{7.13}$$

For a carbon–hydrogen bond with $\nu = 3000$ cm^{-1}, deuterium substitution would lead to a frequency of 2200 cm^{-1}, a zero-point energy difference of about 1.2 kcal/mole, and a calculated kinetic isotope effect of 6.9 at 25°C.

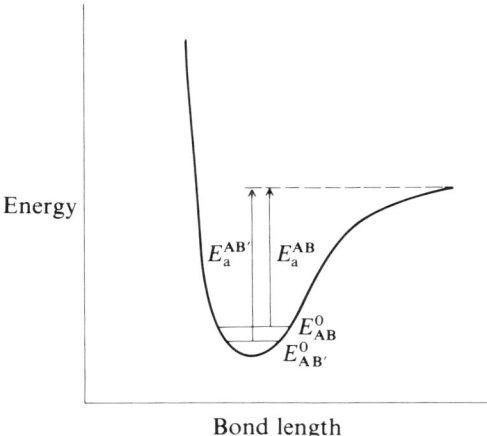

FIGURE 7.1 *Energy diagram for isotopic substitution in the dissociation of a diatomic molecule (B' heavier than B).*

This hydrogen–deuterium isotope effect is a function of temperature, exhibiting typical values of k_H/k_D of 8.2 at 0° and 2.1 at 500°C, and is the greatest of all isotope effects (other than tritium–hydrogen) because it involves the greatest mass difference between the isotopic species. A carbon 12–carbon 13 isotope effect would be calculated as exhibiting a kinetic ratio of 1.022 at 25°C, a small number which is often difficult to distinguish from experimental errors.

Unfortunately, the above extrapolation of the use of zero-point energy criteria to more complex molecules is only a rough approximation. A carbon atom must necessarily contain other substituents with vibrational frequencies of their own which would interact with the vibrational frequency of the bond being broken. Bending vibrations also become probable. In addition, bonds are usually not completely broken in the transition state, posing the problem of isotopic differences in transition-state energies.

A three-center transition state[3] provides a useful model, particularly because of the importance of proton-transfer reactions and catalysis. A model reaction would be

$$A-B + C \longrightarrow [A \cdots B \cdots C]^{\ddagger} \longrightarrow A + B-C$$

where the transition state could be treated first as a linear system independent of bending vibrations. If B were attached equally to A and C in such a transition state (that is, if force constants were equal), the asymmetric vibration would correspond to the reaction coordinate and the symmetric vibration would not involve movement of B, and would therefore be independent of the mass of B. Isotopic substitution of B' for B then would have no effect on the transition-state energy, and the isotope effect would correspond to that calculated from reactant zero-point energy differences (as in Eq. 7.13).

If the transition state is linear yet B is not attached equally to A and C, the force constants being unequal, the mass of B is important and the isotope effect will be reduced from the value calculated from Eq. 7.13. Just as the heavier isotope has the lower zero-point energy in the reactant, the heavier isotope also leads to a lower-energy transition-state species.

The conclusion of this type of analysis is really the expected one: The maximum primary kinetic isotope effect occurs when the bond is completely broken in the transition state or when the isotopic position is situated in a linear symmetrical orientation in the transition state; decreased isotope effects result when the isotopic position is unsymmetrically oriented in the transition state, in which instance both reactant and transition-state energy differences must be considered. In such linear unsymmetrical orientations, systematic variation of A and C components would cause the isotope effect to pass through a maximum value corresponding to the symmetrical orientation.[3-7] In the case of hydrogen and its isotopes, this variation in the magnitude of the isotope effect would be a function of the base strengths of the proton donor and proton acceptor, with the maximum occurring when the donor and acceptor have equal pK's. However, Bordwell[8] has presented evidence from the deprotonation of nitroalkanes suggesting that this maximum is poorly defined and approached with a gentle slope from each direction. He concludes that either the isotope effect in his system is relatively insensitive to the structure of the transition state or the symmetry of the transition state changes little over wide ranges of pK differences. Because of the existence of other systems with reasonably well-defined maxima and because of the unusual behavior of the nitroalkanes with respect to other probes of transition-state structure (see Chapter 7-5), the generality of Bordwell's results may be questioned.

If primary kinetic isotope effects for proton transfers depend on transition-state symmetry, there should be some relationship between isotope

---

4. Y. Pocker and J. H. Exner, *J. Amer. Chem. Soc.* **90**, 6764 (1968); S. B. Hanna, C. Jermini, and H. Zollinger, *Tetrahedron Lett.* 4415 (1969); J. E. Dixon and T. C. Bruice, *J. Amer. Chem. Soc.* **92**, 905 (1970); W. A. Pryor and K. G. Kniepp, ibid., **93**, 5584 (1971); R. P. Bell and B. G. Cox, *J. Chem. Soc., B* 194 (1970), and references therein.
5. A. J. Kresge and Y. Chiang, *J. Amer. Chem. Soc.* **91**, 1025 (1969); R. A. M. O'Ferrall, *J. Chem. Soc., B* 785 (1970).
6. R. A. M. O'Ferrall and J. Kouba, *J. Chem. Soc., B* 985 (1967); E. S. Lewis and J. K. Robinson, *J. Amer. Chem. Soc.* **90**, 4337 (1968); M. D. Harmony, *Chem. Soc. Rev.* **1**, 211 (1972).
7. A. M. Katz and W. H. Saunders, Jr., *J. Amer. Chem. Soc.* **91**, 4469 (1969).
8. F. G. Bordwell and W. J. Boyle, Jr., *J. Amer. Chem. Soc.* **93**, 512 (1971).

effects and relative reactivities[9] (assuming the latter can be used to evaluate transition-state structure—see Chapter 7-8). This type of correlation has been found for the acid-catalyzed hydrolysis of vinyl ethers,[9] a process involving rate-determining proton transfer to the carbon–carbon double bond.[9,10] However, the correlation exists only when the comparison is restricted to a single reaction type[9] (proton addition to a carbon–carbon double bond that is part of a vinyl ether system) brought about by a particular type of acid catalyst (such as carboxylic acids—see Chapter 7-5).

If the transition state is not linear, bending frequencies become involved in the energy considerations and the isotope effect is less than maximum.[5,6] Force constants to other substituents present at reaction sites must play some role in most kinetic isotope effects, because the entire electronic distribution changes in the vicinity of the reactive atoms. However, analysis of this effect has barely begun[7] (see Chapter 7-8). In a similar vein, the possibility of tunneling through the energy barrier is very real for the hydrogen isotope (and much less so for deuterium or tritium), yet it has been little investigated.[6] This possibility exists because the small size of the hydrogen atom requires a quantum mechanical treatment of its energy. Tunneling would lead to a larger isotope effect than otherwise anticipated. An additional potential complication for isotope effects involving hydrogen is that changes in electronic distribution and/or bond lengths, which are assumed to be insignificant, might be important for these species.

One example of the use of primary kinetic isotope effects may be seen in the benzylic bromination of alkylbenzenes by N-bromosuccinimide[3] at 77°C. The k_H/k_D values are 4.9 for toluene, 2.7 for ethylbenzene, and 1.8 for cumene (isopropylbenzene). This trend is interpreted as indicating the greatest transition-state linearity and symmetrical positioning of the hydrogen being abstracted in the toluene case, with increasing nonlinearity and/or loss of symmetry in ethylbenzene and cumene. Notice that the term "symmetry" refers to the transition-state force constants and not to the geometrical symmetry of the transition state as represented by various symmetry operations; the latter type of symmetry is probably identical in the toluene and cumene systems.

Another example of the use of primary kinetic isotope effects is the now-classic study of hydrogen isotope effects in electrophilic aromatic substitutions. Benzene and perdeuteriobenzene undergo electrophilic aromatic substitutions other than sulfonations (and desulfonations) at more or less the same rates. This absence of a primary hydrogen isotope effect indicates that the carbon–hydrogen bond in the aromatic ring is broken after the

---

9. A. J. Kresge, D. S. Sagatys, and H. L. Chen, *J. Amer. Chem. Soc.* **90**, 4174 (1968). For a review of isotope effects and their application to structure–reactivity relationships, see S. E. Scheppele, *Chem. Rev.* **72**, 511 (1972).
10. M. M. Kreevoy and R. Eliason, *J. Phys. Chem.* **72**, 1313 (1968).

rate-determining step, thereby supporting the commonly accepted mechanism involving rate-determining formation of either the σ-bonded benzenonium ion or the π-bonded complex.[11]

7-3 Secondary Kinetic Isotope Effects; Steric Kinetic Isotope Effects

Secondary kinetic isotope effects are rate differences resulting from isotopic substitution at a bond which is neither formed nor broken in or prior to the rate-determining step.[3,12] If this isotopic substitution involves an element at one end of a bond in which the element at the other end is involved in bond formation or bond scission, the isotope effect is termed an α effect. Removal of the isotopic substitution position one more atom from the reaction site leads to a β effect, and so forth. The types of possible isotope effects are designated for the S_N1 reaction of an alkyl halide by the subscripts in **I**. In general, only α and β secondary isotope effects are significant. Even these exhibit small magnitudes when maximal and are usually measurable only for hydrogen isotopes.

$$\begin{array}{cccc} H_\delta & H_\gamma & H_\beta & H_\alpha \\ | & | & | & | \\ C- & C- & C- & C-X \end{array}$$

I

Since α secondary isotope effects involve isotopic substitution on a bond which would be affected by changes in hybridization, electron density, and/or steric phenomena at the reaction site, these isotope effects must have their origins in changes in bond lengths, electron densities, or freedoms of motion. Nucleophilic aliphatic substitution reactions have been the primary source of information about α isotope effects. The overall conclusion is that the primary contributing factors are the out-of-plane bending vibrations in the ground state of the reactant.[3,12]

The magnitude of the α isotope effect may be used to discriminate between S_N1 and S_N2 processes. The S_N1 process exhibits the larger deuterium isotope effect ($k_H/k_D = 1.08$–1.25), with the magnitude a function of the nature of the leaving group, the solvent, and any ion pairing.[13,14]

---

11. H. Zollinger, *Advan. Phys. Org. Chem.* **2**, 163 (1964); E. Berliner, *Progr. Phys. Org. Chem.* **2**, 253 (1964); R. O. C. Norman and R. Taylor, *Electrophilic Substitution in Benzenoid Compounds*, Elsevier, New York, 1965; Reference C, Chapter 11.
12. E. A. Halevi, *Progr. Phys. Org. Chem.* **1**, 109 (1963).
13. V. J. Shiner, Jr., and W. Dowd, *J. Amer. Chem. Soc.* **93**, 1029 (1971), and references therein.
14. J. M. Harris, R. E. Hall, and P. v. R. Schleyer, *J. Amer. Chem. Soc.* **93**, 2551 (1971); V. J. Shiner, Jr., and R. D. Fisher, ibid., **93**, 2553 (1971).

Shiner[13] has emphasized the importance of rate-determining ion-pair formation and rate-determining ion-pair interconversion as different processes leading to different isotope effects. The sensitivity of these α isotope effects as a probe for detecting nucleophilic participation in a process approaching S_N1 character has been debated[14] with somewhat inconclusive results. The S_N2 process leads to lower α deuterium isotope effects (0.95–1.06), some of which are even inverse ($k_H/k_D < 1$).[15] These isotope effects increase with decreasing nucleophilic participation, and therefore are dependent on nucleophilic strength, solvent nucleophilicity, substrate sensitivity to nucleophilic attack, leaving group, and temperature.[15]

Application of α deuterium isotope effects to the Cope rearrangement[16] has led to the interesting conclusion that the isotope effect is a nonlinear function of the location of the transition state on the reaction coordinate. This means that these α isotope effects are not a sensitive probe for the extent of bond cleavage and/or bond formation in the transition state for the rate-determining step, bringing into question their usefulness in any studies other than the most extensive ones involving analysis of the isotope effect as a function of the many variables involved in the general reaction being studied.

The factors which must be considered in searching for the basis of β secondary isotope effects are many and varied.[12] Deuterium is more electropositive than hydrogen but less polarizable. The most accepted explanation invokes hyperconjugation or some other type of charge delocalization as the primary factor.[12] In solvolyses, β deuterium isotope effects exhibit a marked conformational dependence,[12] being greatest when the isotope is positioned *anti* to the leaving group. The operation of this specific *anti* stereoelectronic effect is paralleled by the behavior of alkyl groups when they are present in analogous compounds at the same position at which the isotopic substitution is being investigated.[17] Apparently, the most important relationship is the ability of the hydrogen or alkyl group to interact with the developing *p* orbital from the back side, a typical charge-delocalization phenomenon. An alternate explanation invoking steric effects has been shown to be of minor importance by several approaches.[18,19] As with α deuterium isotope effects, the β effects are larger in S_N1 processes than in S_N2 processes.[20] Somewhat surprisingly, the β isotope effects are

---

15. V. J. Shiner, Jr., M. W. Rapp, and H. R. Pinnick, Jr., *J. Amer. Chem. Soc.* **92**, 232 (1970), and references therein.
16. R. Humski, R. Malojcic, S. Borcic, and D. E. Sunko, *J. Amer. Chem. Soc.* **92**, 6534 (1970).
17. R. C. Bingham and P. v. R. Schleyer, *Tetrahedron Lett.* 23 (1971).
18. G. J. Karabatsos, G. C. Sonnichsen, C. G. Papaioannou, S. E. Scheppele, and R. L. Shone, *J. Amer. Chem. Soc.* **89**, 463 (1967).
19. B. L. Murr and J. A. Conkling, *J. Amer. Chem. Soc.* **92**, 3464 (1970).
20. V. J. Shiner, Jr., W. E. Buddenbaum, B. L. Murr, and G. Lamaty, *J. Amer. Chem. Soc.* **90**, 418 (1968), and references therein.

temperature-dependent for S_N1 reactions, yet temperature-independent for reactions involving significant nucleophilic participation.

An instructive example of the use of secondary isotope effects is an analysis of the retro-Diels-Alder reaction by Seltzer.[21] Decomposition of the adduct II formed from 2-methylfuran and maleic anhydride was studied using several different deuterated analogs. The equality of the isotope effects for deuteration at positions X and Y (IId and IIe) indicates that both bonds are broken more or less simultaneously in the transition state. The α isotope effect involving position Z (IIc) supports rate-determining bond scission at this carbon atom, while the β isotope effect (IIf) is too large for bond scission not to be occurring in the transition state at the carbon containing the methyl group. The overall picture supports the concerted or nearly concerted nature of this decomposition, and thereby supports a similar picture for the Diels-Alder reaction itself because of the principle of microscopic reversibility.

	Relative Rates
(a) all H	a/b = 1.16
(b) X = Y = D	a/c = 1.08
(c) Z = D	d/e = 1.00
(d) X = D	a/f = 1.03
(e) Y = D	
(f) R = D	

II

Extreme care must be taken in any study of secondary isotope effects to ensure that the isotope effect being observed is truly secondary. For example, β secondary isotope effects based on disappearance of reagent in a nucleophilic aliphatic substitution might actually reflect a combination of the desired β effect and a primary isotope effect on a concurrent elimination reaction. Ideally, product analysis should always accompany any analysis of secondary isotope effects (and most analyses of primary isotope effects as well).

Another group of secondary kinetic isotope effects are steric kinetic isotope effects, which involve isotopic differences in nonbonded nonelectronic interactions. Historically, part of the confusion regarding the existence of steric isotope effects centered on the problem of isolating steric effects from other secondary effects. Many systems in which steric isotope effects have been proposed have been subsequently clarified in terms of major electrostatic contributions and minor (if any) steric contributions.[18,22] Nevertheless, there are several instances in which the isotope effect must be unambiguously assigned a steric origin. Out-

---

21. S. Seltzer, *J. Amer. Chem. Soc.* **87**, 1534 (1965).
22. A. J. Kresge and R. J. Preto, *J. Amer. Chem. Soc.* **89**, 5510 (1967); A. J. Kresge and V. Nowlan, *Tetrahedron Lett.* 4297 (1971).

standing in this respect are isotope effects observed in racemizations of biphenyl systems and related structures, where the changes being observed are purely conformational and therefore not significantly sensitive to slight electrostatic phenomena. As with other secondary isotope effects, only the hydrogen isotopes exhibit differences large enough to be observable. Even with the hydrogen isotopes, highly crowded systems are required.[23]

In all of the biphenyl systems observed to date,[23,24] the steric isotope effect is measured by deuteration alpha to the biphenyl bridge (**III**) or in a methyl or methylene group alpha to the biphenyl bridge (**IV**), and the isotope effect is inverse ($k_H/k_D < 1$). The consistent magnitude of the inverse isotope effects for these biphenyl racemizations ($k_D/k_H = 1.05$–1.20) suggests that in sterically crowded situations deuterium has a smaller effective steric requirement than protium. Whether this difference in steric requirement reflects bond length differences, electron distribution differences, or nonbonded repulsion differences is not clear. The situation is slightly confused by the existence of a normal isotope effect ($k_H/k_D > 1$) for the racemization of the seemingly related bridged compound **V**.[25]

Systems in which a steric kinetic isotope effect is proposed and which do not involve in the transition state a clear-cut stereochemical influence independent of the isotope effect present a problem. The inverse isotope

---

23. K. Mislow, R. Graeve, A. J. Gordon, and G. H. Wahl, Jr., *J. Amer. Chem. Soc.* **86**, 1733 (1964); S. A. Sherrod and V. Boekelheide, ibid., **94**, 5513 (1972).
24. L. Melander and R. E. Carter, *J. Amer. Chem. Soc.* **86**, 295 (1964); R. E. Carter and L. Dahlgren, *Acta Chem. Scand.* **23**, 504 (1969).
25. K. Mislow, M. A. W. Glass, H. B. Hopps, E. Simon, and G. H. Wahl, Jr., *J. Amer. Chem. Soc.* **86**, 1710 (1964).

effects noted in systems capable of F-strain (Chapter 11-6) probably are legitimately steric in origin.[22,26] Other proposed steric isotope effects, whether inverse[27] or normal,[28] may be open to question.

7-4 Solvent Isotope Effects

Considerable effort has been directed toward the understanding of solvent isotope effects, with major emphasis on the comparison of rates of reactions in deuterium oxide and normal water.[29] Most of the systems studied involve acid or base catalysis, so the observed isotope effects are often a composite of primary and secondary isotope effects. For example, D_3O^+ is a stronger acid than H_3O^+ by a factor of 3. This isotope effect includes the primary isotope effect involving the transferred species (printed in boldface in **VI**) as well as the secondary isotope effect generated by the isotopic differences at the other sites bonded to the oxygen (underlined in **VI**). Calculations[30] support the idea that the primary isotope effect

$$S \cdots \mathbf{D} \cdots O \begin{matrix} \underline{D} \\ \underline{D} \end{matrix}$$

VI

(k_{H_2O}/k_{D_2O}) proceeds through a maximum (Chapter 7-2) as the transition state is varied from reactantlike character to productlike character and that the secondary isotope effect decreases monotonically throughout this same sequence. The two components may therefore reinforce each other or oppose each other. In addition, the magnitude of either one of the isotope effects would be increased by deuteration at the other type of position.[30] Nevertheless, the calculations suggest that this latter effect may be neglected for correlations with transition-state structure.

Not only is D_3O^+ a stronger acid than H_3O^+, but OD^- is a stronger base than OH^-. Solvent isotope effects (k_{H_2O}/k_{D_2O}) range from 0.5 to around 6, with the most common values falling between 1.5 and 2.8.[29] Such solvent isotope effects may be used to discriminate between the two major types of general acid catalysis, which were designated Type II

---
26. H. C. Brown and G. J. McDonald, *J. Amer. Chem. Soc.* **88**, 2514 (1966); H. C. Brown, M. E. Azzaro, J. G. Koelling, and G. J. McDonald, ibid., **88**, 2520 (1966).
27. J. G. Jewett and R. P. Dunlap, *J. Amer. Chem. Soc.* **90**, 809 (1968).
28. G. H. Cooper and J. McKenna, *Chem. Commun.* 734 (1966).
29. Reference B, Section 3-7; Reference E, Chapter 5-3c; Reference I, Chapter 5.23; W. P. Jencks, *Catalysis in Chemistry and Enzymology*, McGraw-Hill, New York, 1969, Chapter 4.
30. R. A. M. O'Ferrall, G. W. Koeppl, and A. J. Kresge, *J. Amer. Chem. Soc.* **93**, 9 (1971).

(p. 149) and Type III (p. 150). For Type II, the concentration of the conjugate acid of the substrate, SH^+ (or SD^+), would be about three times greater in D_2O than in H_2O because of the above-mentioned greater acidity of D_3O^+. Since there would be no isotope effect on the ensuing rate-determining step, k_{H_2O}/k_{D_2O} would definitely be less than 1. This result of an inverse isotope effect applies only when the proton removed by the anion or base does not exchange with the acid source or does so at a rate much slower than the reaction rate itself.[29] For Type III, the formation of the complex should not exhibit a primary isotope effect. However, the rate-determining step must involve a proton transfer, leading to a normal kinetic isotope effect, $k_{H_2O}/k_{D_2O} > 1$.

Even when the primary–secondary duality in solvent isotope effects is recognized, the question of their origins remains partially unsettled. A zero-point energy argument does not suffice to explain all of the available data, even when viewed from the broader point of view of differences in hydrogen bonding between solutes or transition states and the solvent. There are also so-called *nonspecific effects*,[29,31] which are small and tend to produce slower reaction rates in D_2O than in H_2O. Typical manifestations of these effects are the 23% greater viscosity of D_2O, which suggests that D_2O is more structured than H_2O,[31] and the fact that most ions are more soluble in H_2O, with the existence of some correlations between relative ionic solubilities and structure-making or structure-breaking tendencies. However, much of this is exceedingly speculative, since little is known about water structure. This lack of understanding is extremely unfortunate from the viewpoint of biochemistry, since aqueous solutions are crucial to the processes studied in that discipline.

Since proton transfer is an integral part of solvent isotope effects in acid- or base-catalyzed reactions, it is conceivable that a correlation exists between solvent isotope effects and transition-state structure similar to that anticipated for purely primary isotope effects (Chapter 7-2). In some slight parallel to these primary effects, correlations are observed for certain systems.[9] However, it has become more generally accepted in recent years that the search for such a correlation using a probe as complicated as solvent isotope effects is no more than a vain hope.[8,32]

7-5 The Brønsted Catalysis Law

In general acid catalysis, the rate constant k_a is often related to the strength of the catalyzing acid K_a by a linear free-energy relationship

---

31. See Reference F, Sections 2.1 and 2.5, for a discussion of solvent structure with emphasis on water structure.
32. Reference I, Chapter 5.23.

called the *Brønsted catalysis law*[33] (Eq. 7.14),

$$k_a = G_a K_a^\alpha \tag{7.14}$$

in which G_a and α are constants characteristic of the reaction, solvent, and temperature. Taking logarithms of Eq. 7.14 gives Eq. 7.15,

$$\ln k_a = \ln G_a + \alpha \ln K_a \tag{7.15}$$

which indicates that α may be determined as the slope of a plot of $\ln k_a$ with $\ln K_a$. Similar relationships (Eq. 7.16) may be observed for general base

$$k_b = G_b K_b^\beta \tag{7.16}$$

catalysis. While our discussion will be directed toward acid catalysis, it will not ignore base catalysis in that α and β behave in a completely analogous manner since both represent the same type of proportionality constant relating a free energy of activation to an equilibrium free energy change. In fact, the Brønsted law, proposed in 1924, was the first known linear free energy relationship. Since the Brønsted relationship compares a system in which proton transfer is incomplete (the catalysis rate) with another system in which the proton is dissociated (the equilibrium constant of the acid), values of α might be expected to range from 0 to 1 and to reflect the amount of proton transfer in the transition state for the catalytic process (see below).

The actual relationship between general acid catalysis and the Brønsted law may be analyzed in the framework of Type II catalysis. The rate law for Type II catalysis (Eq. 7.8) gives an observed rate constant as in Eq. 7.17,

$$k_{obs} = kKK_a \tag{7.17}$$

where k is the rate constant for the rate-determining step, K is a substrate constant, and K_a is the catalyst ionization equilibrium constant. Substitution of the Brønsted relationship (Eq. 7.14) into this equation and rearrangement gives

$$k = \frac{G_a}{K} \frac{K_a^\alpha}{K_a} = AK_a^{\alpha-1} \tag{7.18}$$

where A is a constant for a given set of substrate and reaction conditions other than the catalytic acid.

If α has a value of 1 in Eq. 7.18, the rate constant would be independent of the strength of the catalytic acid. This is general acid catalysis in its

---
33. Reference A, p. 107; Reference B, Section 3-7; Reference C, pp. 226–227; Reference D, pp. 113–115 and 226–227; Reference E, Chapter 5-3b; Reference G, pp. 156–161 165–167, 235–242, and 369–374; Reference H, Chapter 9; Reference I, Chapter 10.2–10.6 and 11.3; and W. P. Jencks, *Catalysis in Chemistry and Enzymology*, McGraw-Hill, New York, 1969, Chapter 3.

broadest sense and differs from specific acid catalysis. In this case the general acid catalysis as such cannot be observed, because the solvated proton is so effective a catalyst that its low concentration is not important. If α has a value of 0, Eq. 7.18 indicates that the reaction is catalyzed by the solvent, the highest concentration species, to such an extent that the effects of all other acids present are swamped out. Again, the general acid catalysis cannot be detected experimentally. For values of α between 0 and 1, k decreases as K_a increases. In a general sense, α decreases with increasing substrate reactivity, although the sensitivity of α to reactivity is not very great[34] (for a discussion of reactivity–selectivity relationships, see Chapter 7-8). This type of relationship is supported by primary deuterium isotope effects in a series of related catalytic carboxylic acids: The stronger the acid, the smaller the primary isotope effect.[3]

Four general limitations exist for the Brønsted catalysis law. First of all, the reaction must be subject to general acid catalysis. Secondly, if the catalytic acid contains more than one equivalent proton and/or more than one equivalent basic site after a proton has been lost, the rate constant must be corrected for the number of equivalent protons which could be transferred to the substrate, and the equilibrium constant for the catalyst must be corrected for both the number of equivalent protons in the acid and the number of equivalent basic sites in the conjugate base (see the analogous Branch and Calvin approach in Chapter 4-2). Thirdly, the law holds only for the catalytic action of similar acids. Even for the same reaction under the same conditions, a different linear relationship is usually observed for carboxylic acid catalysts than for phenol acid catalysts, etc. The α value is therefore a function of catalyst type as well as of substrate type.[9,10] The hydronium ion (and the hydroxide ion) usually does not fall on any Brønsted plot. This is because the catalysts other than the hydronium ion tend to be neutral species. Charge effects are therefore significant. The deviation of the hydronium ion does not occur when the catalysts are charged metal ions, particularly univalent ones. All of these differences reflect the importance of solvation on reaction rate phenomena. The fourth limitation, therefore, is that the reaction rate and acid equilibrium being compared must be measured in the same solvent at the same temperature. Since measurement under constant conditions is not the norm, curvature and erratic behavior are often the consequences. It is important to recognize that the Brønsted catalysis law must always be demonstrated experimentally and can never be assumed, even by extrapolation from an apparently analogous system.

In his original presentation, Brønsted realized that a linear rate–equilibrium correlation must be part of a longer curved line. The reasons are

---

34. A. J. Kresge, H. L. Chen, Y. Chiang, E. Murrill, M. A. Payne, and D. S. Sagatys, *J. Amer. Chem. Soc.* **93**, 413 (1971).

that reaction rates are limited by collisions or vibrations and are not thermodynamic properties. Work by Hine[35] provides experimental verification of this curvature, while Marcus[36] has provided the theoretical equations for this phenomenon.

The coefficient α is an experimental measure of the relative stabilization of the transition state for the reaction by acids of different strengths.[33] Since acids of different strength should transfer a proton to different degrees in the transition state, it has been traditional to associate α with the amount of proton transfer in the transition state and with the position of the transition state along the reaction coordinate (Chapter 5-3). Since α varies from 0 to 1 and since values of 1 indicate complete proton transfer to give protonated substrate while values of 0 indicate absence of proton transfer and resemblance to starting substrate, it is indeed attractive to associate α with fraction of progress along the reaction coordinate of the transition state. Values of α then could be used as probes for reaction mechanism and transition-state structure, bearing in mind the stringent restrictions on changes in substrate structure and catalyst structure discussed previously.

Values of α between 0 and 1 require that the position of equilibrium be more sensitive to structural changes than the reaction rate being considered. Consideration of the forward and reverse components of the acidity equilibrium[37] (Eq. 7.19) indicates that structural changes must affect k_1 and

$$HA + H_2O \underset{k_{-1}}{\overset{k_1}{\rightleftharpoons}} A^- + H_3O^+ \tag{7.19}$$

k_{-1} in opposite directions for this requirement to be met. However, for systems in which k_1 and k_{-1} are affected in the same direction (e.g., both increased by the structural change), α could be greater than 1 or less than 0. Investigations of base-catalyzed proton removal from carbon acids such as the nitroalkanes have indeed revealed values of β greater than 1 and less than 0.[37] Bordwell[37] has concluded that oxygen and nitrogen acids probably will produce α values between 0 and 1 while carbon acids very probably will not. The implication[38] is that either Brønsted coefficients are a poor guide to the extent of proton transfer in the transition state and/or the transition-state structures in certain systems are very insensitive to wide ranges in acidity or basicity.

A parallel approach to the question of the meaning of α is that developed to different degrees by Hine,[33,35] Jencks,[33] and, in particular,

---

35. J. Hine, *J. Amer. Chem. Soc.* **93**, 3703 (1971).
36. R. A. Marcus, *J. Amer. Chem. Soc.* **91**, 7224 (1969).
37. F. G. Bordwell, W. J. Boyle, Jr., J. A. Hautala, and K. C. Yee, *J. Amer. Chem. Soc.* **91**, 4002 (1969).
38. F. G. Bordwell and W. J. Boyle, Jr., *J. Amer. Chem. Soc.* **93**, 511 (1971), and **94**, 3901 (1972).

Kresge.[34,39] Potential-energy curves for proton transfer reactions need not have the same shapes for different types of acids or bases. Further, even if they have the same shapes, they may show deviations when comparing acids or bases of significantly different strengths. The Brønsted coefficient can, then, at best be no more than a measure of the extent to which substituents in the acid or base stabilize the transition state compared to the extent to which these substituents affect the equilibrium for complete proton transfer. Within the kinetic process, correlation of α with transition-state structure leaves no room for substituent effects in the transition state which do not have a counterpart in the reactants or the products. In every bimolecular proton transfer there must be forces present in the transition state which are new to the system. Kresge[39] terms these new transition-state interactions *intermolecular effects*. Values of α greater than 1 or less than 0 would occur whenever the intermolecular effects are greater than the intramolecular effects of the substituent within the catalyst or substrate reactant–product pair. This is most likely for carbon acids (or bases) which possess groups capable of removing negative charge from the vicinity of the site of proton transfer in a direction other than toward the substituent being varied. For example, ionization of a carboxylic acid (**VII**) produces an anion where charge delocalization can occur only toward the substituent R to a greater or lesser extent as the substituent is changed. On the other hand, ionization of a nitroalkane (**VIII**) or an analogous pseudoacid produces an anion where the charge delocalization occurs primarily toward the nitro group regardless of the substituent R, which must exert its effect in a different direction from the nitro group. Even

$$R-CO_2^{\ominus} \qquad R-\overset{\ominus}{C}-NO_2$$
$$\textbf{VII} \qquad\qquad \textbf{VIII}$$

in those "normal" situations where α has a value between 0 and 1, the intermolecular effect probably would exert sufficient influence for α not to be a valid measure of transition-state position, a position which might be otherwise deduced from the fraction of electronic charge transferred or the degree of bond formation or bond cleavage.

The overall ideas developed in these two approaches are supported by Marcus' theoretical work,[36] which leads to certain limitations on the ability of the Brønsted coefficient to even approximate the position of the transition state along the reaction coordinate. The overall conclusion is unmistakable. Even for a homogeneous set of catalysts, there is no hope for a simple general correlation of Brønsted coefficients with the extent of proton transfer in the transition state[8,38] (see Chapter 7-8 also).

---

39. A. J. Kresge, *J. Amer. Chem. Soc.* **92**, 3210 (1970).

7-6 Acidity Functions

The acidity of a dilute aqueous solution usually is identified with the activity of the hydrogen ion in that solution. However, since it is impossible to determine individual ion activities experimentally, hydrogen ion activities cannot be used directly to define acidity on some general scale.[40] The solution to this problem is to resort to an operational definition of a logarithmic scale of acidity based on emf measurements of series of standard solutions in electrochemical cells. This pH scale is itself significant only for dilute aqueous solutions of ionic strength less than 0.1. The problem of a general quantitative scale of acidity and basicity for any concentration of substrate in any given solvent requires a somewhat different approach.

Most chemists are used to dealing with acids in aqueous solution. It is accepted at an early stage of chemical development that mineral acids (such as nitric, hydrochloric, and sulfuric) are strong acids while carboxylic acids and phenols are weak acids. The strong acids are fully ionized in water, while the weak acids are ionized to a slight extent as measured by equilibrium constants for their ionizations. However, it must be recognized that these statements are applicable only to aqueous solutions. In liquid ammonia the carboxylic acids are fully ionized and are equal in strength to the mineral acids. The statement that an acid is strong means that its reaction with solvent (S) is virtually complete (Eq. 7.20). The apparent

$$HA + S \rightleftharpoons A^- + HS^+ \qquad (7.20)$$

strength of an acid therefore depends on the basicity of the solvent. The more basic the solvent (such as ammonia compared to water), the more the substrates will appear to be completely ionized. This is called a *leveling effect*. In order to evaluate relative acidities, solvents of basicity low enough to prevent complete ionization of the acids being compared must be utilized. For example, nitric acid and hydrochloric acid are not equally strong acids when present in the same concentrations in a solvent of low basicity such as glacial acetic acid. In general, when different acids are compared in the same solvent under identical conditions, the stronger acid is the most ionized one.

How about comparing acid strengths in different solvents? Which is more acidic, a given concentration of hydrogen chloride in benzene or the same concentration in water? Strange as it may seem, the most strongly acidic solution involving a single acid in different solvents is that in which

---

40. Reference B, Section 2-9; Reference C, pp. 222–226; Reference D, pp. 96–106; Reference E, Chapter 2-2 and 2-3; Reference G, pp. 269–281; Reference H, pp. 325–327; Reference I, Chapter 9; C. H. Rochester, *Acidity Functions*, Academic Press, New York, 1970, Chapters 1, 2, 3, 6, and 7.

the acid is the *least* ionized. The benzene solution of HCl in which little ionization has occurred is the stronger acid, because the water has promoted the ionization of the HCl and, acting as a base, has solvated the proton and (in a certain sense) neutralized it.

In order to quantitatively evaluate acidities or basicities, thermodynamic equilibrium constants using activities must be used. For the general dissociation reaction

$$HA \rightleftharpoons H^+ + A^-$$

the K_a is defined as

$$K_a = \frac{a_{H^+} a_{A^-}}{a_{HA}} = \frac{[H^+][A^-]}{[HA]} \frac{\gamma_{H^+} \gamma_{A^-}}{\gamma_{HA}} \quad (7.21)$$

where the a's are activities and the γ's are activity coefficients. This leads to the result that the acidity constant of an acid HA in any solvent S can be calculated from its acidity constant in some other solvent, the acidity constants of some other acid HA′ in the two solvents, and the activity coefficient product $(\gamma_{A'} \cdot \gamma_{HA})/(\gamma_A \cdot \gamma_{HA'})$, provided both acids are measurably but not fully ionized in the two solvents. Since

$$pK_a = - \log \frac{a_{H^+} a_{A^-}}{a_{HA}}$$

substitution into Eq. 7.21 gives

$$pK_a = \log \frac{[AH]}{[A^-]} + pH \quad (7.22)$$

in sufficiently dilute aqueous solution. The problem of defining pK_a in nondilute or nonaqueous media is still not solved because Eq. 7.22 and the above discussion require that one of the solvents be the dilute aqueous system.

For weak bases, fairly acidic media must be used to evaluate their basicity. For the general case

$$BH^+ \rightleftharpoons B + H^+$$

pK_{BH^+} is calculated as follows:

$$pK_{BH^+} = \log \frac{a_{BH^+}}{a_B} - \log a_{H^+} \quad (7.23)$$

If another base C is present in the same solution, $\log a_{H^+}$ in Eq. 7.23 must be identical for both bases, giving

$$\begin{aligned} pK_{CH^+} - pK_{BH^+} &= \log \frac{a_{CH^+}}{a_C} - \log \frac{a_{BH^+}}{a_B} \\ &= \log \frac{[CH^+]}{[C]} - \log \frac{[BH^+]}{[B]} + \log \frac{\gamma_{CH^+} \gamma_B}{\gamma_C \gamma_{BH^+}} \end{aligned} \quad (7.24)$$

If the concentration ratios in Eq. 7.24 may be determined, such as by spectrophotometric techniques, which usually require that each ratio be between 0.1 and 10 for high accuracy, the pK difference between B and C could be evaluated if the activity coefficient ratio could also be determined. Since the pK's are defined in terms of activities, they are medium-independent. If the concentration ratio terms as a unit were medium-independent, the activity coefficient ratio would of necessity be medium-independent. For similarly structured bases in various H_2SO_4–H_2O mixtures, Hammett and Deyrup[40] found in 1932 that the difference in the logarithms of the concentration ratios was indeed independent of the medium[41]—including dilute aqueous solutions. The activity coefficient ratio in Eq. 7.24 also then must be independent of the medium. Since the activity coefficient ratio for a single base, log (γ_{BH^+}/γ_B), is defined as zero in dilute aqueous media, the equality of Eq. 7.25 is found to exist,

$$\log \frac{\gamma_{CH^+}}{\gamma_C} = \log \frac{\gamma_{BH^+}}{\gamma_B} \tag{7.25}$$

and the activity coefficient logarithm is equal to zero in Eq. 7.24.

Bases for which Eq. 7.25 may be demonstrated are called *Hammett bases*. A pK_{BH^+} (Eq. 7.23) which can be determined in dilute aqueous medium could then be evaluated in various acidic solutions by observing concentration ratios as a function of the concentration of acid in the solvent system. Because of curvature in such extrapolations, pK's of weak bases are best determined by a stepwise procedure using Hammett bases of varying strengths. Differences in pK's are determined using Eq. 7.24 in dilute aqueous solution and in succeedingly more acidic media. The bases being compared are chosen so that they are successively weaker, yet are partially protonated in the medium in which they are being compared in a pairwise manner. By using these bases as overlapping indicators, the pK of a weak base can be determined in an acidic solution and extrapolated to the dilute aqueous system.[42] Examples of weak Hammett bases for which the pK's of the conjugate acids have been obtained in this way are shown in Table 7.1.

Since a series of Hammett bases are used as indicators in the overlap method discussed above, it is important to show that the indicator behavior is the desired proton-loss equilibrium. The simplest test is the use

---

41. Isosbestic points and behavior according to the Beer-Lambert law are used to test for spectral reliability.

42. The validity of this indicator-overlap method has been firmly established for bases of similar structure by P. D. Bolton, C. D. Johnson, A. R. Katritzky, and S. A. Shapiro, *J. Amer. Chem. Soc.* **92**, 1567 (1970); M. J. Kamlet and R. R. Minesinger, *J. Org. Chem.* **36**, 610 (1971); and E. M. Arnett, et al., *J. Amer. Chem. Soc.* **92**, 1260, 3977 (1970).

TABLE 7.1 pK_{BH^+} Values for Primary Amine Hammett Bases (H_2O, 25°C)[40]

Base	pK_{BH^+}
p-Nitroaniline	0.99
2,4-Dichloroaniline	2.02
o-Nitroaniline	-0.29
2,4-Dinitroaniline	-4.48
2,4,6-Trinitroaniline	-10.04

of the van't Hoff *i factor*[40] derived from cryoscopic measurements. Since the bases are fairly weak, the ideal solvent for the freezing-point depression studies is a fairly concentrated sulfuric acid solution, the most common of which freezes at 10.36°C and has a cryoscopic constant of 6.12. Small amounts of water are desirable in this solvent system to repress the self-ionization of the sulfuric acid. Surprisingly, the freezing-point depressions of these sulfuric acid solutions remain proportional to the concentration of ionic or nonionic solutes even to relatively high concentrations, indicating little fluctuation in the activity coefficients of the ions involved and extremely efficient ion-solvating ability on the part of the sulfuric acid solvent. The *i* factor is the number of moles of nonsolvent ions and/or molecules present per formula weight of added solute. Should the solute fail to ionize, *i* would be equal to 1 (this would be the case, for example, with trifluoroacetic acid or perchloric acid). The monobasic amines used as Hammett indicators should exhibit a value for *i* of 2 because of the following reaction,

$$B + H_2SO_4 \longrightarrow BH^+ + HSO_4^- \qquad (7.26)$$

as should ethers, ketones, carboxylic acids and their derivatives, and related systems. An *i* factor between 1 and 2 is often found with nitro compounds, sulfonic acids, and cyclic anhydrides, indicating incomplete conversion to the conjugate acid. Alcohols often exhibit $i = 3$ because of formation of alkyl hydrogen sulfates (Eq. 7.27), while triarylcarbinols lead to $i = 4$

$$ROH + 2H_2SO_4 \longrightarrow ROSO_3H + H_3O^+ + HSO_4^- \qquad (7.27)$$

(Eq. 7.28). All Hammett bases must therefore give *i* factors of 2 in sulfuric

$$\phi_3COH + 2H_2SO_4 \longrightarrow \phi_3C^+ + H_3O^+ + 2HSO_4^- \qquad (7.28)$$

acid solution studies.

If the pK's of a series of substrates are known, Eq. 7.21 or 7.23 may be used to obtain a measure of the ability of the solvent to donate a proton to a given type of base.[40] Since

$$K_{BH^+} = \frac{[B]a_{H^+}}{[BH^+]} \frac{\gamma_B}{\gamma_{BH^+}} \qquad (7.29)$$

it is possible to let h_0 equal the part of the equation which should characterize the solvent system and be independent of the nature of the Hammett base:

$$h_0 = a_{H^+} \frac{\gamma_B}{\gamma_{BH^+}} \qquad (7.30)$$

Substituting Eq. 7.30 into Eq. 7.29 and taking logarithms produces Eq. 7.31,

$$pK_{BH^+} = \log \frac{[BH^+]}{[B]} + H_0 \qquad (7.31)$$

where $H_0 = -\log h_0$ and is called the *Hammett acidity function*.[40] This H_0 is a quantitative measure of the ability of the solvent to donate protons to a Hammett base and has a definite value for every acidic solvent mixture. For a given pK_{BH^+}, the magnitude of H_0 tells how much of the base is present as its conjugate acid. The Hammett acidity function is referred to infinite dilution as the standard state and represents an extension of the pH scale into concentrated acids where activity coefficients are not equal to 1. In dilute aqueous solution, H_0 equals pH (giving an equation analogous to Eq. 7.22), as shown in Fig. 7.2. At higher acid concentrations,

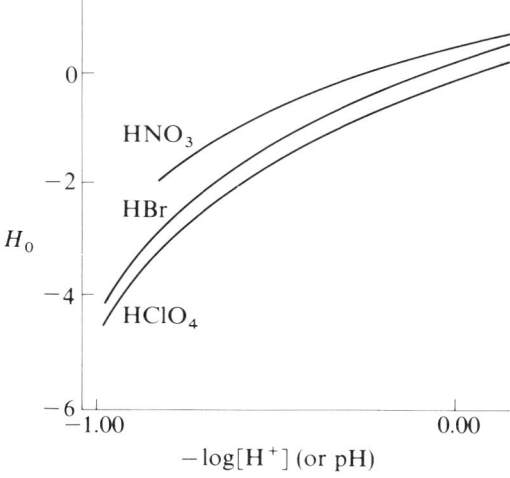

FIGURE 7.2 *A plot of H_0 with $-\log [H^+]$ for various acids.*

the slope becomes as much as 12. This deviation means that $\gamma_{H^+}(\gamma_B/\gamma_{BH^+})$ becomes increasingly greater than 1 as the acid concentration is raised. The most important role in this phenomenon appears to be played by solvation effects. The key concentration ratio seems to be $[H^+]/[H_2O]$. Accepting a solvation number of 4 for the proton in water, various calculations indicate insufficient water molecules to solvate all of the protons as the acid solvent becomes more concentrated beyond a certain point. The acidity function is therefore a function of the water activity as well as of the concentration of protons.

At the outset of the development of the H_0 concept, it was hoped that the H_0 scale would be found to be dependent only on the acid solution under a given set of reaction conditions using Hammett bases as indicators. The temperature variation of H_0 has been studied[43] and empirical equations developed to correlate the results. The importance of ionic strength in the development of an acidity function has been demonstrated by studies[40,44] of the effect of the addition of neutral salts to the acidic media. The changes in H_0 are linear in the concentration of the added salt, as are the changes in γ_B. All of the relevant activity coefficients and ratios of activity coefficients change on the addition of neutral salts, with the changes dependent on salt structure and concentration. The smaller the cation and the higher the charge on the cation, the larger the increase in acidity.[44] This type of behavior suggests that the greater the solvation requirements of the cation, the greater the effect on the acidity function.

The subscript zero in H_0 indicates an acidity scale based on neutral bases as indicators (Eq. 7.23). For mononegatively charged bases (Eq. 7.32),

$$\text{BH} \rightleftharpoons \text{B}^- + \text{H}^+ \tag{7.32}$$

the acidity function H_- is defined[40] as

$$H_- = pK_{BH} + \log \frac{[\text{B}^-]}{[\text{BH}]} = \log \frac{\gamma_{BH}}{\gamma_{B^-}} - \log a_{H^+} \tag{7.33}$$

The major difficulty in the development of the H_- scale has been the lack of availability of suitable negatively charged indicators which would be protonated only in fairly strongly acidic solutions. Other than for carbon acids,[45] for which a separate acidity function H_C has also been suggested,[46] most of the development of the H_- function has taken place through

---

43. P. Tickle, A. G. Briggs, and J. M. Wilson, *J. Chem. Soc.*, B 65 (1970); C. D. Johnson, A. R. Katritzky, and S. A. Shapiro, *J. Amer. Chem. Soc.* **91**, 6654 (1969).
44. J. P. H. Boyer, R. J. P. Corriu, and R. J. M. Perz, *Tetrahedron* **27**, 5255 (1971).
45. J. P. Jones, *Quart. Rev.* **25**, 365 (1971).
46. M. T. Reagan, *J. Amer. Chem. Soc.* **91**, 5506 (1969).

studies of strongly basic solutions,[40,47] where H_- measures the ability of the solvent to remove a proton from an electrically neutral weak acid indicator. The H_- scale is identical with pH in dilute aqueous solution (as for H_0). Typical acid indicators[47] used are phenols, anilines, hydrazones, alcohols, and cyano-, nitro-, and aromatically activated carbon acids. Basic media which have been evaluated[47] are aqueous alkali, alkoxides, aqueous amines, and dipolar aprotic solvents containing various added bases and cosolvents. Differences between H_0, H_- (Table 7.2), and other

TABLE 7.2 Acidity Functions for Aqueous Sulfuric Acid

H_2SO_4[a]	H_0	H_-	H_R
10%	−0.31	−0.09	−0.72
25%	−1.37	−1.46	−2.55
50%	−3.38	−3.91	−6.60
70%	−5.80	−6.21	−11.52
90%	−8.92	—	−16.72

[a] Weight % H_2SO_4 in H_2O.

SOURCE: C. J. O'Connor, J. Chem. Educ. **46**, 686 (1969).

acidity functions differing only in the charge on the indicator base are most likely solvation effects combined with electrostatic contributions to the various activity coefficients. From a comparison of H_- and H_{2-} scales, Bowden[47] has concluded that these functions are largely independent of the charge type of the indicator, provided similar indicators are being used and the charge in the H_{2-} series is situated on a different functional group from that being ionized in the indicator process.

An acidity function H_R has been defined[40] (Eq. 7.34) for the equilibria

$$H_R = -pK_{ROH} - \log \frac{[R^+]}{[ROH]} \qquad (7.34)$$

(Eq. 7.35) of substrates with van't Hoff i factors of 4 in sulfuric acid (see

$$ROH + H^+ \rightleftharpoons R^+ + H_2O \qquad (7.35)$$

Eq. 7.28). Since water is one of the components in the equilibrium mixture used to define the H_R scale, this acidity function is more sensitive to the effective water concentration than the others discussed herein and decreases more rapidly with increasing acid concentration (Table 7.2). Other than this effect related to a_{H_2O}, there are also solvent effects related

---

47. K. Bowden, Chem. Rev. **66**, 119 (1966); C. H. Rochester, Quart. Rev. **20**, 511 (1966).

to fundamental behavioral differences in the activity coefficients of carbonium ions and ammonium (and related) ions in concentrated acid solutions. From the simplest point of view, the ions produced in the H_R equilibria have unfilled bonding orbitals, while the others do not.[48]

Many of these acidity functions have also been studied in nonaqueous media.[40,47] Parallel behavior is observed except in solvents with low dielectric constants, in which cases ion-association phenomena become significant and prevent indicators from exhibiting Hammett-base behavior (Eq. 7.25).

Unfortunately, the hope that a given acidity function would be completely independent of indicator structure has been destroyed. Even though Hammett and Deyrup used primary amines, secondary amines, ketones, quinones, and other functional groups in generating the H_0 scale, later workers[40] have found significant deviations for indicators with different structures or different basic sites. In particular, primary amines, secondary amines, tertiary amines, ketones, esters, amides, and alkenes have all been observed to generate different acidity-function scales, although in some cases the differences become apparent only at extremely high acid concentrations. The symbol H_0 is reserved for primary amines and other substrates which exhibit the identical acidity-function behavior, and various other symbols are used for the other substrate classes. The most obvious difference between some of these structural classes, such as the various amines, is the amount of hydrogen bonding possible. Both the number and type of hydrogen bonds should have an effect on the activity coefficients, particularly on the cationic forms,[40,47] although an explanation in terms of hydration effects is far from simple. Nevertheless, it is obvious that γ_{BH^+}/γ_B is not a constant for all neutral bases in a particular acid solution. This activity coefficient ratio often may be a constant for structurally similar bases, but it is difficult at this time to predict which structural variations will cause significant changes. The implication of these results has been well stated by Hammett: There is no unique general operational definition of acidity or an inherent measure of the strength of an acid or base.[40] The observed behavior of an acid–base system involves the interplay of three variables:[40] (1) the degree to which a base is protonated under given reaction conditions; (2) the tendency of a reaction medium to transfer a proton to a base of given strength; and (3) the difference in the first and second variables for different bases of the same strength in the same medium ascribed to solvation phenomena (and possibly involving several variables since it is not even vaguely understood). As will be discussed in the next section, the use of acidity functions in analyses of reaction

---

48. Olah [*J. Amer. Chem. Soc.* **94**, 808 (1972)] has suggested a reclassification and renaming of positively charged carbon species to reflect this fundamental difference from (or similarity to) 'onium ions.

mechanisms requires either that a new acidity function be generated for the substrate under consideration or that an empirical factor be determined relating the behavior of the substrate to some previously generated acidity function.[49] In no instance can structural modification be freely assumed to cause no change in the behavior of the acidity function.

7-7 Acidity Functions and Transition-state Structure

Whenever a reaction mechanism contains a proton-transfer equilibrium prior to the rate-determining step (Eq. 7.1, 7.7, and 7.9), the rate depends on—among other factors—the equilibrium constant for this step. If the substrate is a weak base, as are most of the common functional groups, this equilibrium constant must be determined by the extrapolation of results obtained in concentrated acid solutions to the dilute aqueous media in which many reactions are most commonly performed. This extrapolation is performed most easily by use of the appropriate acidity function if one is available.[40,49] If correlation with a known acidity function is not warranted, either a new acidity function must be generated or an empirical factor must be used to improve correlation with a known acidity function.[49] Once either of these is done, it is then reasonable that acidity-function behavior might be a useful probe of reaction mechanism.[50]

The earliest example of such a rate–acidity mechanistic relationship was the *Zucker-Hammett hypothesis* (1939) for concentrated acid media.[50] This hypothesis grouped acid-catalyzed reactions into two categories. A reaction not involving water or another additional reactant in the transition state was expected to show a linear relationship between log (rate) and $-H_0$. For a Type I reaction (Eq. 7.1), the rate expression is most accurately

$$S + H^+ \underset{}{\overset{K}{\rightleftharpoons}} SH^+$$

$$SH^+ \xrightarrow[\text{slow}]{k_2} \text{products} \qquad (7.1)$$

written as:

$$\text{rate} = \frac{Kk_2[S][H^+]\gamma_S\gamma_{H^+}}{\gamma_\ddagger} \qquad (7.36)$$

where $\gamma_\ddagger$ represents the activity coefficient of the transition state involved in the rate-determining unimolecular decomposition of the protonated sub-

---

49. K. Yates, *Accounts Chem. Res.* **4**, 136 (1971).
50. Reference B, Section 3-8; Reference C, pp. 225–226; Reference D, pp. 190–192; Reference E, Chapter 5-3d; Reference G, pp. 281–289; Reference H, pp. 327–334; Reference I, Chapter 9; and C. H. Rochester, *Acidity Functions*, Academic Press, New York, 1970, Chapters 4 and 5.

strate. If the acidity is represented by h_0 (Eq. 7.30), Eq. 7.36 becomes:

$$\text{rate} = \frac{Kk_2[S]h_0 \gamma_S \gamma_{BH^+}}{\gamma_\ddagger \gamma_B} \tag{7.37}$$

If the transition-state activity coefficient is close to that of protonated substrate and if $\gamma_S/\gamma_{SH^+} \cong \gamma_B/\gamma_{BH^+}$, the rate would be equal to $Kk_2[S]h_0$ and therefore be linear in h_0. These requirements most likely would be satisfied if the substrate were structurally similar to the Hammett base indicators and if the transition state resembled the protonated substrate, SH^+, from which it was formed. Correlations of this type have been observed for the hydrolysis of sucrose, the hydrolysis of hindered esters, and the hydrolysis of β-propiolactone, processes for which other mechanistic criteria support the A1 mechanism (as in Eq. 7.1). While the slope of a plot of log (rate) against $-H_0$ usually is between 0.85 and 1.15 for these A1 processes, there exist other processes which show this type of linear relationship with much smaller slopes.[49,50] On the basis of the discussion in Chapter 7-6, it should not be surprising that acidity function correlations might show small or large deviations from unity, yet still be linear. In fact, any relationship at all might be considered fortuitous. Nevertheless, sufficient evidence has accumulated that linear correlation of log (rate) with $-H_0$ with a slope greater than 0.85 may be considered reasonable assurance of an A1 mechanism or of rate-determining protonation (Eq. 7.4) if the substrate is a nitrogen or oxygen base that such relationships must be considered significant.

The second Zucker-Hammett category is those reactions where water or another reactant combine with the protonated substrate in the transition state for the rate-determining step.[50] Evidence for this category, an A2 mechanism, would be a linear relationship with unit slope between log (rate) and $-\log[H^+]$ (or pH in the dilute acid region). For a reaction of the type represented by Eq. 7.7, if the base is water the rate law can be written as Eq. 7.38.

$$S + H^+ \xrightleftharpoons{K} SH^+$$

$$SH^+ + B \xrightarrow[\text{slow}]{k_2} \text{products} + BH^+ \tag{7.7}$$

$$\text{rate} = \frac{Kk_2[S][H^+][H_2O]\gamma_S \gamma_{H^+} \gamma_{H_2O}}{\gamma_\ddagger} \tag{7.38}$$

Substitution of h_0 (Eq. 7.30) gives

$$\text{rate} = \frac{Kk_2[S]h_0 a_{H_2O} \gamma_S \gamma_{BH^+}}{\gamma_\ddagger \gamma_B} \tag{7.39}$$

which differs in form from Eq. 7.37 by the presence of a term for the activity of water. Since the activity of water is believed to change with acid concentration (p. 171), it is unlikely that Eq. 7.39 would lead to a linear correlation between rate and h_0. In addition, it probably would not be expected that $\gamma_{\ddagger}$ would resemble γ_{SH^+} and be like a Hammett base, the requirement for the activity coefficient term to be medium-independent, since the transition state now contains protonated substrate and a water molecule. For this latter reason, the linear correlation between log (rate) and $[H^+]$ (Eq. 7.38) predicted by Zucker and Hammett would require an extremely fortuitous set of circumstances. Somewhat surprisingly, several reactions,[50] such as the enolization of acetophenone (similar to Eq. 6.12), the hydrolysis of unhindered esters, and the hydrolysis of γ-butyrolactone, do exhibit linear correlation between log (rate) and $-\log [H^+]$. Nevertheless, conclusions of A2 mechanisms based on this second Zucker-Hammett category require extreme caution and considerable external supporting evidence. The use of either Zucker-Hammett category is often frustrated by nonunit slopes or nonlinear behavior.

One category for which the Zucker-Hammett treatment has been completely inadequate is acid-catalyzed reactions involving carbon bases. The major reason lies in the difference between the H_0 and H_R acidity scales (p. 171). Reactions where the transition state for the rate-determining step includes substrate which has been protonated, yet has lost the elements of a molecule of water, often show excellent H_R correlations.[50] Reactions such as the acid-catalyzed hydrations of alkenes also often exhibit linear log (rate)–H_R plots.[50]

Hydration parameter treatments (w, w^*, ϕ, etc.) of the type originated by Bunnett[49,50] represent refinements of the Zucker-Hammett approach, showing improved mechanistic predictions and improved understanding of the detailed solution phenomena. Bunnett (1961) discovered an empirical relationship, Eq. 7.40, relating reaction rate, an acidity function, and

$$\log (\text{rate}) + H_0 = w \log a_{H_2O} + \text{constant} \tag{7.40}$$

water activity. When w is zero, the results agree with the first Zucker-Hammett category. On the basis of comparisons with reactions whose mechanisms were known, values of w were correlated with mechanistic processes. Values of w less than zero indicate that water is not present in the transition state, positive values less than $+3.3$ indicate water involvement as a nucleophile, and values greater than $+3.3$ implicate water acting as a proton-transfer agent in the transition state for a substrate protonated on oxygen or nitrogen (for substrates protonated on carbon, the w values are approximately zero). The theoretical interpretation given w is that it measures transition-state hydration requirements relative to the substrate and relative to the hydration effects in a Hammett base indicator pair. Unfortunately, different mineral acids lead to different w values for the same reaction and do not extrapolate to the same point in dilute media (where the activity coefficients should all be unity). Temperature changes also lead to marked effects on w values. The upshot has been the development of more complicated hydration parameters[49,50] such as Bunnett's linear free-energy relationship ϕ (1966), with negligible improvement in the understanding of transition-state structure or in the use of these parameters as mechanistic criteria.

Kresge and coworkers[51] have almost ended the hope that different mechanisms exhibit different acidity dependences by demonstrating large changes in acidity dependences within a single reaction mechanism on changing substrate and transition-state structures. However, they assert that this loss of acidity-function behavior as a mechanistic criterion is compensated by the rather detailed picture of transition-state bonding which acidity-function behavior might be able to supply. In addition, detailed understanding of acidity-function behavior will of necessity go hand-in-hand with improved understanding of solvation phenomena, an area sorely needing clarification (see Chapters 8-5 and 9).

7-8 Other Analyses of Transition-state Structure

As should be evident from the preceding sections of this chapter, the process of analyzing the structure of a transition state is very complicated (Chapter 5-3). The various probes proposed for quantitative evaluation of the position of the transition state along the reaction coordinate have either failed or become entangled in controversy. An alternative approach would be to search for principles firmly established for stable molecules which might legitimately be extrapolated to transition states. Most approaches of this sort involve a description of the changes in bond lengths and bond orders that occur on changing substituents.

The earliest commonly accepted approach is called the *Hammond postulate* (1955), which is also sometimes referred to as the *reactivity–selectivity principle*[52] or the *Bell-Evans-Polanyi principle*.[52] One statement

---

51. A. J. Kresge, S. G. Mylonakis, Y. Sato, and V. P. Vitullo, *J. Amer. Chem. Soc.* **93**, 6181 (1971).

52. (a) Reference G, pp. 162–168; R. T. Morrison and R. N. Boyd, *Organic Chemistry*, 3rd ed., Allyn and Bacon, Boston, 1973, Chapter 3.28; M. J. S. Dewar, *The Molecular Orbital Theory of Organic Chemistry*, McGraw-Hill, New York, 1969, Chapter 8.3; J. R. Murdoch, *J. Amer. Chem. Soc.* **94**, 4410 (1972). (b) Ritchie and coworkers [*Accounts Chem. Res.* **5**, 348 (1972), *J. Amer. Chem. Soc.* **94**, 4963, 4966 (1972) and **95**, 1882 (1973), and references therein] have concluded from studies of cation–anion recombination reactions that the entire reactivity–selectivity principle may be fallacious. They find that reactions of nucleophiles with cations are correlated by an equation of the form

$$\log \frac{k_N}{k_0} = N_+$$

where k_N is the rate of reaction of nucleophile N in a given solvent with the given cation, k_0 is a constant characteristic only of the given cation, and N_+ is a parameter of the nucleophile and solvent and independent of the nature of the cation. This requires that cation selectivities be constant in these reactions and, therefore, bear no relationship to any type of reactivity–selectivity relationship.

of this approach is that two states occurring consecutively along a reaction coordinate with similar energies will have similar geometries. An S_N1 reaction, for example, involves in the rate-determining step a sequence of: starting material – transition state – carbonium ion. The rate of such a reaction depends on the relative energies of the starting material and transition state, yet it is common to argue that a more stable carbonium ion corresponds to a faster reaction rate. The basis for this argument is the Hammond postulate. For a process involving a transition state possessing hybridization and geometry intermediate to those present in the starting material and an unstable intermediate (the carbonium ion), stabilization of the intermediate would produce stabilization in the "incipient intermediate" (the transition state) and make the transition state more like the starting material in geometry and energy. Stabilization of the reactant (or destabilization of the intermediate) would lead to a more intermediate-like transition state. The position of the transition state along the reaction coordinate, its energy, and its geometry are therefore interrelated and depend on the relative stabilities of the reactant and intermediate, or two intermediates, being considered. For a series of similar reactions, the later the transition state is reached along the reaction coordinate, the higher the energy of activation and the slower the reaction rate (Fig. 7.3).

Consideration of an S_N2 reaction provides additional insight into the Hammond postulate.[52] If a comparison is made of the relative bond lengths in the transition state for attack by two nucleophiles on a given alkyl halide, that nucleophile which leads to the more stable pentacoordinate system will be positioned farther from the carbon atom at which the

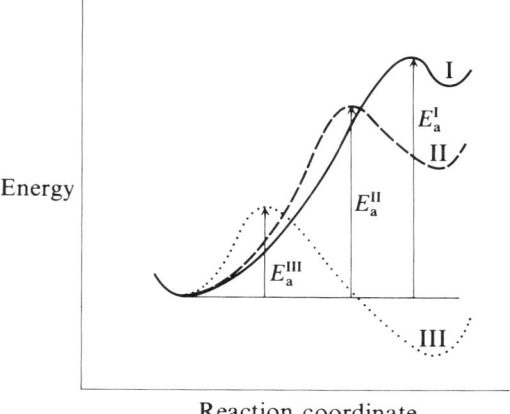

FIGURE 7.3 *The effect of product stabilization on the reaction profile. Product I, solid line, is less stable than product II, dashed line, which is less stable than product III, dotted line.*

substitution is occurring.[53] As before, the more stable intermediate corresponds to the more reactant-like geometry and the lower energy of activation when similar processes are being compared.

The dual concepts of partial rate factors and Brown's selectivity factors (Chapter 5-2) for electrophilic aromatic substitution reactions provide elegant instances of both the advantages and disadvantages in applications of the Hammond postulate to transition-state structures.[54] The partial rate factors (Eq. 5.6, 5.7, and 5.8) are quantitative measures of intermolecular selectivity on the part of the electrophile (relative to a benzene hydrogen), while the selectivity factors (Eq. 5.9) are quantitative measures of the intramolecular selectivity by the same electrophile.

For a given electrophile under a given set of reaction conditions, linear relationships have been observed between log (partial rate factor) and selectivity factors for many systems. The larger the partial rate factor, the larger the selectivity factor. This proportionality infers a similar mechanistic basis for intermolecular and intramolecular selectivities. For a given aromatic substrate, such as toluene, the electrophile exhibiting the largest selectivity factor is also usually the electrophile reacting at the slowest rate. Since the aromatic substrate is constant, this latter observation should be dependent on electrophile stability. Assuming a constant mechanistic scheme, the reaction involving the most selectivity and highest energy of activation should be that with the most stable electrophile, according to the Hammond postulate. The more stable and less reactive the electrophile, the more the transition state should resemble the benzenonium ion (or σ-complex or Wheland) intermediate (Eq. 7.41) and the

$$\text{ArH} + \text{E}^+ \longrightarrow \text{Ar}\begin{array}{c}\diagup \text{H}\\+\\\diagdown \text{E}\end{array} \qquad (7.41)$$

less it should resemble the starting aromatic system. A transition state resembling the benzenonium ion should involve significant positive charge development on the ring and should therefore exhibit high sensitivity to the electronic effects of substituents, leading to Hammett correlations with σ^+ and large negative ρ values.[54,55] The more negative the ρ value, the higher the selectivity and partial rate factors. On the other hand, a less stable, more reactive electrophile would lead to a transition state with little charge

---

53. No conclusion may be drawn regarding relative amounts of charge separation in the transition states for the reactions being compared.
54. Reference C, pp. 393–395; Reference G, pp. 196–203; L. M. Stock, *Aromatic Substitution Reactions*, Prentice-Hall, Englewood Cliffs, N.J., 1968, pp. 67–69; L. M. Stock and H. C. Brown, *Advan. Phys. Org. Chem.* **1**, 35 (1963).
55. G. A. Olah, M. Tashiro, and S. Kobayashi, *J. Amer. Chem. Soc.* **92**, 6369 (1970), and **94**, 7448 (1972); G. A. Olah and S. Kobayashi, ibid., **93**, 6964 (1971).

TABLE 7.3 *Typical Relationships for Electrophilic Aromatic Substitutions*[54]

	ρ	$m_f^{CH_3}$	$p_f^{CH_3}$	S
Br_2, HOAc (25°)	−12.1	5.5	2420	2.64
CH_3COCl, $AlCl_3$, ethylene dichloride (25°)	−9.1	4.8	750	2.19
$(CH_3)_2CHBr$, GaBr, benzene (25°)	−2.3	1.4	5	0.56

development on the aromatic ring (and probably of a π-complex nature),[55] Hammett-equation linearity with σ or σ^*, and a less negative ρ.[54,55]

The structure of the transition state, as expected, influences the selectivity factors, the partial rate factors, and the Hammett-type relationships, and depends on the electrophilicity of the electrophilic reagent and the nucleophilicity of the aromatic substrate (Table 7.3). Olah[55] has also argued that when the transition state resembles the starting materials, the *ortho : para* ratio will be approaching or greater than 1 because of the statistical effect, and that when the transition state resembles the benzenonium ion intermediate, the *ortho : para* ratio will be low because substituents in the σ complexes themselves are greater *para* than *ortho*. Deviations from any of these trends would most likely result from temperature effects, solvent effects, phenomena associated with the generation and nature of the electrophile, or changes in mechanism or rate-determining step.

The Hammond postulate deals with potential-energy relationships, yet often is extrapolated to free-energy relationships. Within the framework of additional assumptions, it can provide the basis for many linear free-energy relationships, such as the Hammett-type equations (Chapter 4-5 and 6) and the Brønsted catalysis law (Chapter 7-5). Many exceptions are known to the reactivity–selectivity principle,[52] yet many of these "exceptions" result from failure to recognize that the principle itself is applicable in a strict sense only to a potential-energy framework.

Because of the legitimate exceptions to the Hammond postulate which do exist, several alternative approaches have been developed. One of these is called the *reacting-bond rule*.[56] The central concept in this approach is the definition of a reacting bond as a bond present to some extent in the transition state, yet completely absent in either the reactants or the products. The reacting-bond rule states that an electron-donating substituent

---

56. C. G. Swain and E. R. Thornton, *J. Amer. Chem. Soc.* **84**, 817 (1962); Reference C, pp. 236–238.

will produce a transition state in which the nearest reacting bond is longer and the next nearest reacting bond shorter (and so forth in an alternating manner) than for the unsubstituted system. An electron-withdrawing substituent behaves in the opposite manner. The magnitude of the effect, in either instance, is dependent on the magnitude of the electronic effect (presumably, some other technique permits the electronic situation to be quantitatively evaluated in the ground state or transition state). The rule is based on the properties of individual bonds, and is therefore related to stretching-force constants in transition states of closely related processes.

The S_N1 reaction of substituted cumyl chlorides (Eq. 6.5, the standard system for the determination of σ^+ values) provides a simple example of the reacting-bond rule (ignoring solvation phenomena).[56] In the transition state for this reaction (**IX**), the carbon–chlorine bond is the only reacting

$$X-\underset{}{\underset{}{\bigcirc}}-\overset{CH_3}{\underset{CH_3}{\overset{|+\delta}{\underset{|}{C}}}}\cdots\overset{-\delta}{Cl}$$

IX

bond. Electron-donating X groups would lengthen this bond and give a picture of a transition state with more carbonium ion character. In terms of the orbitals involved, the electron donation would be expected to stabilize the developing carbonium ion so that the reacting bond could be stretched to a greater extent in the transition state, thereby also producing greater electron density on the chlorine leaving group in this reaction. If there were a second reacting bond, such as between the chlorine leaving group and a metal ion acting as an electrophilic catalyst, the increased charge density on the leaving group would permit greater overlap with the metal ion and produce a shorter second reacting bond.

Notice the difference from the Hammond postulate for this system. The Hammond postulate predicts that stabilization of the carbonium ion intermediate would stabilize the transition state and lead to a transition state earlier along the reaction coordinate. The reacting-bond rule predicts that stabilization of the carbonium ion intermediate would stabilize the transition state and lead to a transition state more like the intermediate and in which more bond breaking has occurred. Experimentally, the reaction is accelerated to a greater extent in a more polar solvent when electron-donating groups are present (Chapter 9-4), indicating greater charge separation and more carbonium ion character in the transition states for the systems involving the electron-donating substituents. This result could be predicted in a reasonably straightforward manner using the reacting-bond rule, but not by Hammond's postulate.

An extension of the reacting-bond rule is the *solvation rule*.[57] If a proton is being transferred from one atom containing one or more unshared electron pairs in the reactant state (e.g., oxygen, nitrogen, or halogen) to another similar atom in a reaction involving bond changes on a carbon atom in the rate-determining step, the proton should lie in a stable potential in the transition state and not form reacting bonds or give rise to a primary hydrogen isotope effect. The basic idea is to locate a species wherever it will provide maximum stabilization. This idea is applicable to any species involved in the transition state but not in the reaction coordinate. In other words, it applies to hydrogen bonds and to any kind of bonds to solvent molecules or added ionic species which can stabilize the transition state for a given reaction. The rule, then, determines the positions of all solvating solvent molecules and catalysts in the transition state. If a proton is being transferred, it is closer to the more basic atom in the transition state. Any substituent making an atom more basic would cause the proton to be closer to the site of that atom in the transition state. The solvent molecules and catalysts would position themselves so as to most effectively stabilize these preordained transition states as a function of the electron distributions but without the possibility of changing one of the sites at which an interaction is occurring.

An alternative approach to the reacting-bond–solvation rules is the *anthropomorphic rule*,[58] which considers the question,[57] "What would I do in that particular situation if I were an unshared pair of electrons?" The essential idea[58] is that catalysis will occur where it is most needed and the degree of stabilization achieved by the catalyst will be determined by the overall stabilization of the transition state and the substituent effects on the proton being transferred. Swain[57] has criticized the anthropomorphic approach as involving unsound reasoning and overdependence on intuition, while Jencks[58] has criticized the reacting-bond–solvation rules as neglecting the perturbations in the transition-state structure produced by the addition of the catalyst and overemphasizing the stability of the hydrogen bond as such.

The controversy between the Swain approach[56,57] and the Jencks approach[58] may be of little importance. Thornton[59] has pointed out that the reacting-bond rule really assumes only one major reacting-bond system and, therefore, is useless for reactions involving the simultaneous breaking of more than one bond at different parts of the molecule, such as bimolecular eliminations. The various approaches up to this point,

---

57. C. G. Swain, D. A. Kuhn, and R. L. Schowen, *J. Amer. Chem. Soc.* **87**, 1553 (1965); Reference C, pp. 236–238.
58. J. E. Reimann and W. P. Jencks, *J. Amer. Chem. Soc.* **88**, 3973 (1966); W. P. Jencks, ibid., **94**, 4731 (1972).
59. E. R. Thornton, *J. Amer. Chem. Soc.* **89**, 2915 (1967).

he remarks, consider substituent effects only in terms of changes in geometry along the reaction coordinate, while it is apparent[59] that substituents should linearly affect the vibrational potentials in the transition state of every normal mode of motion involving bonds being made or broken. Effects along the reaction coordinate should correspond to the Hammond postulate, while the effects perpendicular to the reaction coordinate would often be small but significant perturbations. The net effect of a substituent change would be obtained by summing the effects on all the normal modes, yet be dominated by those modes involving nuclei participating in reacting bonds, particularly those closest to the substituent. Analysis of the S_N1 mechanism by the reacting-bond rule[56] assumes that transition-state geometry and charge separation are parallel phenomena, which need not be true because of the intervention of substituent effects both parallel and perpendicular to the reaction coordinate. Solvation effects are incorporated into Thornton's rules by treating the solvent molecule as part of the transition state and analyzing proton motion (or lack of it) along the reaction coordinate, as opposed to analysis of the extent of proton transfer.

Because prediction of substituent effects by Thornton's rules is far from straightforward, the reader is referred to the original paper[59] for the rules (and examples). A later paper,[60] utilizing a bond-order approach for S_N2 transition states, avoids the necessity for summing the various normal modes while clarifying ambiguous predictions resulting from the Thornton rules. Substituent effects in S_N2 transition states have also been estimated from solvent effects on enthalpies of activation and enthalpies of transfer of transition states and reactants and products from one solvent to another that is significantly different.[61] Kurz[62] has also proposed that consideration of transition states as acids and bases with pK's will be useful for analysis of structural changes. Other approaches, progressively becoming more complex and less accessible to the average chemist, seem destined to follow in the near future. Hopefully, one of these will provide the sought-after key to the structure of transition states.

---

60. J. C. Harris and J. L. Kurz, *J. Amer. Chem. Soc.* **92**, 349 (1970).
61. P. Haberfield, *J. Amer. Chem. Soc.* **93**, 2091 (1971).
62. J. L. Kurz, *Accounts Chem. Res.* **5**, 1 (1972); J. L. Kurz and L. C. Kurz, *J. Amer. Chem. Soc.* **94**, 4451 (1972).

Chapter 8

Nucleophilic Character

One of the important aspects of any consideration of the interplay between structure and reactivity is an analysis of the detailed interactions between the reactants, including any solvent effects. One of the systems most intensively investigated in this regard is aliphatic nucleophilic substitution, for which many empirical criteria have been developed to correlate structure with reactivity.[1] Seven major factors[2] have been singled out as significantly influencing nucleophilic reactivity: the nature of the nucleophile, the nature of the electrophilic center (including the substituents), the nature of the solvent, the charges on the nucleophile and substrate, the bond strength in the product, the nature of the leaving group, and the "alpha effect." An investigation of these factors, besides being useful in its own right, points out many of the difficulties inherent in any quantitative analysis of relative reactivities.

8-1 Nucleophilicity

The first problem in analyzing structure–reactivity relationships in aliphatic nucleophilic substitutions is to define a nucleophile and to attempt to quantitatively evaluate the strength of a nucleophile on some scale. Ingold originally defined nucleophiles as reagents which donate electrons to or share electrons with another site. This definition is too broad for most applications since it includes all bases, all ligands, and all reducing

---
1. K. M. Ibne-Rasa, *J. Chem. Educ.* **44**, 89 (1967).
2. J. O. Edwards, *J. Chem. Educ.* **45**, 386 (1968).

agents, as well as those compounds which can be the attacking group in an aliphatic nucleophilic substitution.

A more generally accepted definition is that bases and nucleophiles have a tendency to form covalent bonds by sharing electron pairs, with basicity a thermodynamic property measured by an acid–base equilibrium (Eq. 8.1) and nucleophilicity a kinetic property measured by the rate

$$B^- + H_3O^+ \xrightleftharpoons{K} BH + H_2O \qquad (8.1)$$

constant for an S_N2 reaction (Eq. 8.2).

$$B^- + RX \xrightarrow{k} BR + X^- \qquad (8.2)$$

For a given group of nucleophiles containing the same nucleophilic atom and similar structural features in the immediate vicinity of the nucleophilic site, the rate constants for attack on a given substrate will often show excellent correlation with the basicities of the nucleophiles. Correlations with the Brønsted catalysis law and Hammett-type correlations are the usual manifestations of this type of behavior. Lack of parallel nucleophilicity–basicity behavior results when the nucleophilic atom is changed (thus, there are different Brønsted slopes for oxygen nucleophiles than for sulfur nucleophiles—see p. 162); when steric phenomena intrude (Chapter 6-2); when the nucleophile is very polarizable (see below); and when the nucleophilic site is bonded to another atom containing at least one nonbonded pair of electrons (the alpha effect—see Chapter 8-6). Nonlinear correlation with basicity is not unreasonable since basicity involves the proton, which is a unique electrophile because of its small size and concentrated positive charge.

Two alternative approaches which may be taken are to search for empirical criteria other than basicity as such or to use basicity as the first of several parameters to be combined in an empirical way. Hopefully, either of these approaches could be developed on a sound physical basis so as to increase understanding of the interactions occurring in the reaction process while providing predictive power as to reaction rate for a given unstudied system.

Only one single-criterion approach has been developed for nucleophilic aliphatic substitution reactions—the 1952 *Swain-Scott equation*,[3]

$$\log \frac{k}{k_0} = sn \qquad (8.3)$$

where s measures substrate discrimination among various nucleophiles, and n is characteristic of the nucleophile itself. Attack by water ($n = 0.0$) on

---
3. Reference B, pp. 423–429; Reference C, pp. 287–290; Reference D, pp. 299–302; Reference E, Section 7-2; Reference F, pp. 78–81; and Reference G, pp. 246–254.

methyl bromide ($s = 1.00$) was chosen as the standard system k_0. Values of n were obtained using methyl bromide in aqueous medium (Table 8.1); these values then were applied to other nucleophiles to obtain s values (Table 8.2). Linear correlations were satisfactory only for nucleophilic

TABLE 8.1 Swain-Scott Nucleophilic Constants[1,3]

Nucleophile	n	Nucleophile	n
H_2O	0.0	N_3^-	4.0
F^-	2.0	HO^-	4.2
$CH_3CO_2^-$	2.7	$C_6H_5NH_2$	4.5
Cl^-	3.0	NCS^-	4.8
Br^-	3.9	$S_2O_3^{2-}$	6.4

TABLE 8.2 Swain-Scott Substrate Constants[1,3]

Substrate	s
Methyl bromide	1.00
Epichlorohydrin	0.93
Mustard ion	0.95
Ethyl tosylate	0.66
Benzyl chloride	0.87
Benzoyl chloride	1.43[a]
Benzenesulfonyl chloride	1.25[a]

[a] Less-than-satisfactory linear correlation.

attack on sp^3 electrophilic sites with no more than minor variations in steric environment. Major changes in electrophilic site required definition of new sets of n values, a thoroughly unsatisfactory state of affairs. The conclusion reached was that nucleophilic attack must involve at least two interaction mechanisms whose relative importance is substrate-dependent, suggesting the use of linear combinations of two model processes (see Chapter 8-4 for Swain's approach to this problem and Chapter 7-8, note 52b, for a complicating factor).

8-2 The Edwards Equations

Whenever two model processes are to be combined in an attempt to predict a reaction rate or analyze the interactions involved in a reaction, the model processes should be chosen on the basis of the practical criteria that data are readily available for the model processes and that details of the model processes are fairly well understood. For this reason, basicity is an attractive choice as one of the model processes to be combined in an empirical manner. One of the principle differences between basicity and nucleophilicity (as discussed in the S_N2 framework) is the formation of a definite bond to the proton in the former and the formation of a partial bond to the electrophilic site in the latter.[3] The partial bond formed between the nucleophilic site and the electrophilic site in the transition state could be strengthened if the bonding orbitals would be polarized so as to increase the overlap between these sites. The Swain-Scott nucleophilic constants (Table 8.1) support the idea that polarizability is a meaningful factor since they are greatest for the most polarizable group in a given column of the periodic table (I > Br > Cl > F; S > O) even though base strength would increase in the opposite direction. In addition, the Swain-Scott substrate constants (Table 8.2) suggest that the more polarizable electrophilic sites (the last two entries) are the most sensitive to the nucleophile while showing the poorest linear correlations for Eq. 8.3. The deviations from Eq. 8.3 for these electrophilic sites are primarily in the direction of a nucleophilic reactivity sequence more closely paralleling base strengths. Apparently, the polarizability of the nucleophile becomes less important as the polarizability of the electrophilic site increases. Stabilization of the partial bond being formed seems to occur by polarizability effects only up to a certain point, and then apparently with respect to either the electrophile or the nucleophile, but not to both.

Experimentally, the question of whether the electrophile or nucleophile will supply the desired amount of polarizable character often can be answered by the types of products obtained. For example, neopentyl tosylate is attacked by methoxide ion at the more polarizable sulfur atom (Eq. 8.4) and by the more polarizable thiophenoxide ion at the less polarizable carbon site (Eq. 8.5).

$$(CH_3)_3CCH_2OSO_2C_7H_7 + {}^-OCH_3 \longrightarrow$$
$$(CH_3)_3CCH_2O^- + CH_3OSO_2C_7H_7 \quad (8.4)$$

$$(CH_3)_3CCH_2OSO_2C_7H_7 + {}^-SC_6H_5 \longrightarrow \quad (8.5)$$
$$(CH_3)_3CCH_2SC_6H_5 + {}^-O_3SC_7H_7$$

The question, then, is what model process may best be chosen to evaluate polarizability (Chapter 4-1). Edwards[1,3] chose oxidative dimerization of the nucleophile relative to water. The *Edwards equation*,[1,3,4] also called the oxibase scale,[5] is written as

$$\log \frac{k}{k_0} = \alpha E_N + \beta H \quad (8.6)$$

---

4. Reference J, pp. 451–452.
5. R. E. Davis, R. Nehring, W. J. Blume, and C. R. Chuang, *J. Amer. Chem. Soc.* **91**, 91 (1969), and references therein.

where α and β are substrate constants, H measures the basicity of the nucleophile relative to water ($H = pK_a + 1.74$), and E_N is the standard electrode potential of the nucleophile X^- for the equilibrium

$$2X^- \rightleftharpoons X_2 + 2e^-$$

relative to a similar equilibrium for water,

$$2H_2O \rightleftharpoons H_4O_2^{2+} + 2e^- \qquad (E_0 = -2.60)$$

such that $E_N^{X^-} = E_0^{X^-} + 2.60$. This use of an oxidation half cell for the second model process is realistic since the nucleophile is formally oxidized when it contributes electrons in the nucleophilic substitution process.

The Edwards equation (Eq. 8.6) exhibits a linear relationship with the Swain-Scott equation (Eq. 8.3) with β approximately zero, indicating that the oxidation half cell by itself provides a satisfactory model for many aliphatic nucleophilic substitutions. Variations in substrate behavior are incorporated into the Edwards equation since the use of two substrate parameters, α and β, makes allowance for the variable polarizability influences. One of the most serious limitations on Eq. 8.6 is that many $E_0^{X^-}$ and pK_a are not known, and some are not readily accessible because of the intervention of other factors, such as other reactions in the oxidation half cell. More seriously, in several cases negative β values were encountered, suggesting the possibility that an enhanced basicity might lead to a reduced nucleophilicity, which is an unacceptable state of affairs. Edwards assumed that this type of result indicated that the model processes were not sufficiently distinct, and attempted to further separate the processes.

The key to finding the model process to be used in conjunction with basicity is to find some adequate experimental measure of polarizability. Since polarizability averaged in a spherical sense is proportional to the molar refractivity at infinite wavelength, R_∞, a polarizability value[1,3] for a nucleophile, P_N, could be defined relative to water by log $(R_\infty^N/R_\infty^{H_2O})$ if these values could be determined. If

$$E_N = 3.60 P_N + 0.0624 H$$

substitution into Eq. 8.6 would give

$$\log \frac{k}{k_0} = AP_N + BH \qquad (8.7)$$

where $A = 3.60\alpha$ and $B = 0.0624\alpha + \beta$. Since B is always found to be positive, the problem connected with negative β values is solved and the model processes in this second Edwards equation[1,3] appear reasonably distinct. The major objection to Eq. 8.7 is that P_N does not refer to any physically real model process. It is defined by an assumed proportionality requiring the extrapolation of physical results to an imaginary infinite wavelength. On a different plane, the question is whether or not a model process must be based on physical reality. Pragmatically speaking, values of P_N may be determined with sufficient validity for rates to be predicted with some confidence using Eq. 8.7. Yet the second Edwards equation surely is no longer a true linear free-energy correlation. This dilemma between purity and pragmatism is a common one in science. The best model must always be sought, yet the many approximate models proposed to organize data and direct experimentation must be utilized for progress to occur.

Several generalizations may be derived from the Edwards basicity–polarizability approach to nucleophilic reactivity.[1] Basicity becomes more important as the charge on the substrate increases. Polarizable nucleophiles tend to possess empty orbitals having reasonably low energies. Polarizability of the nucleophile is therefore most significant when the electrophilic site includes many outer-orbital electrons which might otherwise interfere with nucleophilic attack. Nucleophile basicity appears to be dominant for attack on hydrogen (obviously), carbonyl carbon, boron, and tetrahedral sulfur and phosphorus, while nucleophile polarizability exerts the major influence when the electrophilic site is tetrahedral carbon, bivalent oxygen, platinum(II), and halogen atoms. Both nucleophile basicity and nucleophile polarizability play major roles in attack on bivalent sulfur or aromatic carbon. Significant exceptions exist to these generalizations, yet they provide a useful frame of reference until a comprehensive system can be established.

8-3 The Winstein Equations

One of the weaknesses in the Swain-Scott and Edwards approaches (Chapters 8-1 and 8-2) is the dependence on aqueous solvent systems. When a wider range of solvents is employed (Chapter 8-5) and when the attacking nucleophile is the solvent itself, significant dispersal of points results in any attempted correlation using these techniques. Different equations must therefore be sought to accommodate solvent effects and to evaluate solvent nucleophilicities.

The most well-known solvolysis correlation equation (1948) is the Winstein-Grunwald equation,[6]

$$\log \frac{k}{k_0} = mY \qquad (8.8)$$

---

6. Reference B, pp. 417–423; Reference C, pp. 293–298; Reference D, pp. 299–303; Reference E, Section 7-1; Reference F, Chapters 2.6 and 2.7; Reference G, pp. 299–300; Reference I, Chapter 8; C. Reichardt, *Angew. Chem. Intern. Ed. Engl.* **4**, 29 (1965); and A. Streitwieser, *Solvolytic Displacement Reactions*, McGraw-Hill, New York, 1962, pp. 43–49 and 63–66.

TABLE 8.3 *Solvent Ionizing Power Constants*[3,6]

Solvent[a]	Y
H_2O	3.49
40% EtOH	2.15
80% EtOH	0.00
EtOH	−1.97
69.5% MeOH	1.02
85.1% MeOH	0.09
MeOH	−1.05
HCO_2H	2.08
CH_3CO_2H	−1.63
t-BuOH	−3.26

[a] Percentages are volume–volume with the cosolvent being water.

in which m is a substrate parameter measuring the substrate sensitivity to changes in the "ionizing power" of the medium Y. The standard system was chosen as t-butyl chloride ($m = 1.00$) at 25°C in 80% aqueous ethanol (80% ethanol–20% water by volume; $Y = 0.00$) on the premise that nucleophilic solvent participation would be absent in $S_N 1$ reactions of tertiary substrates. Recent investigations utilizing bridgehead-substituted tertiary substrates[7] has supported this premise that nucleophilic solvent participation and rate-determining elimination reactions are absent in this standard butyl system. Typical Y values are indicated in Table 8.3, and m values for selected substrates are shown in Table 8.4.

TABLE 8.4 *Substrate Sensitivities to Solvent Ionizing Power*[3,6]

Compound[a]	m
1-Adamantyl bromide	1.20
t-BuCl	1.00
t-BuBr	0.94
α-Methallyl chloride	0.89
$(C_6H_5)_2CHCl$	0.76
Benzyl chloride (50°)	0.43
i-PrBr	0.43
Allyl chloride (45°)	0.40
EtBr (55°)	0.34
EtOTs (50°)	0.26
MeBr	0.22
MeOTs (75°)	0.23

[a] 25°C unless otherwise indicated.

---

7. D. J. Raber, R. C. Bingham, J. M. Harris, J. L. Fry, and P. v. R. Schleyer, *J. Amer. Chem. Soc.* **92**, 5977 (1970).

Plots of log k with Y for a given substrate in different mixed solvent systems generate a different straight line for each solvent system. This suggests multiple solvation mechanisms or at least two solvent properties contributing significantly to the solvent effects. The results shown in Table 8.4 indicate significant sensitivity of m to the amount of charge dispersal in the transition state, as well as leaving-group effects. The role of the solvent appears to include solvation of the leaving group, electrostatic stabilization of the central carbon atom in an S_N1 process, and hydrogen-bonding effects, which often are most apparent in deviations from linear correlations using Eq. 8.8 or in switches in m values for different leaving groups on changing solvent systems. Solvent effects on ion-pair formation and return are plausible but difficult to ascertain.

The trends in m values suggest their potential usefulness as probes for the amount of solvent nucleophilic participation. Those reactions which are S_N1 by other criteria exhibit m values near 1, while values for the S_N2 systems range from 0.25 to 0.35. Values of m between these extremes are characteristic of secondary substrates and other structural types which exist in the borderline area between pure S_N1 ("limiting") and pure S_N2. It appears that the larger the m value, the less the nucleophilic participation. While it is fairly reasonable that an S_N2 process, with a larger activated complex radius, might be less susceptible to solvent ionizing power, it is more reasonable that solvent ionizing power and solvent nucleophilicity are dual components in creating m values below some critical point around 0.8.

A model system potentially superior to t-butyl chloride for the evaluation of solvent ionizing power is p-methoxyneophyl tosylate (**I**), in which solvolysis occurs with exclusive phenyl participation[6] (Eq. 8.9). At 75°C,

log k_{ion} provides an excellent standard for solvent polarity, as suggested by Smith, Fainberg, and Winstein.[3,6] The log k_{ion} scale is equivalent to the Y-value scale for hydroxylic solvents and can be extended to nonhydroxylic systems, in which Y cannot be determined. The problem of a nucleophilic component in Eq. 8.9 might be expected to be less serious than in Eq. 8.8, yet the good correlation between $\log k_{ion}$ and Y leads to appreciable doubts as to whether this problem has been solved by changing to this new standard system.

8-4 Multiparameter Nucleophilicity Relationships

Winstein, Grunwald, and Jones[6] attempted to correlate solvent effects on solvolysis reactions by a four-parameter equation of the type shown in Eq. 8.10,[8]

$$\log \frac{k}{k_0} = lN + mY \tag{8.10}$$

where l and m are substrate parameters, Y is solvent ionizing power, and N measures solvent nucleophilicity. By assuming that water, ethanol, and methanol have comparable nucleophilicities, it would be possible to use results in these solvents to generate m and Y values. Comparison of relative reaction rates in solvents of high nucleophilicity and low nucleophilicity with assumed similar Y values (aqueous ethanol compared to acetic acid or formic acid) could lead to scales of l and N. In particular, compounds with m values near 1 (S_N1 processes) should be insensitive to solvent nucleophilicities, while substrates with fractional m values (S_N2 processes) should exhibit excellent sensitivity to solvent nucleophilicities.

A similar equation was proposed by Swain and Scott,[6]

$$\log \frac{k}{k_0} = sn + s'e \tag{8.11}$$

in which s and s' measure substrate sensitivity to solvent nucleophilic push n and solvent electrophilic pull e, respectively. Swain, Mosely, and Bown[6] employed Eq. 8.11 empirically after recasting it as Eq. 8.12 to avoid any possible confusion between the physical models in the former

$$\log \frac{k}{k_0} = c_1 d_1 + c_2 d_2 \tag{8.12}$$

equation and the arbitrary parameters used in Eq. 8.12. These parameters, chosen to provide the best fit to a large number of experimental rate constants, have been criticized[6] for lack of connection with physical reality. Nevertheless, Eq. 8.12 does have considerable predictive utility, provided substrate and solvent closely resemble those used in the original determination of the parameters. About twenty years after this equation was first

---
8. The original equation was expressed as a partial differential equation.

TABLE 8.5 *Scales of Solvent Nucleophilicities*

Solvent[a]	N (Peterson)[9]	N (Schleyer)[10]
H_2O		−0.26
EtOH	0.76	0.09
80% EtOH	0.00	0.00
50% EtOH		−0.20
MeOH		0.01
HCO_2H	−1.52	−2.05
CH_3CO_2H	−1.66	−2.05
CF_3CO_2H	−5.33	−5.55
56% w–w Acetone		−0.20
50% w–w Dioxane		−0.41

[a] Percentages are volume–volume unless indicated by w–w for weight–weight. The cosolvent is water.

suggested, Peterson and Waller[9] derived a set of equations converting the Swain-Mosely-Bown parameters for Eq. 8.12 into the more meaningful Eq. 8.10 assuming values for *l* of 0.0 for *t*-butyl chloride and 1.0 for methyl bromide, and equal *N* values for acetic acid and formic acid. This last assumption—equal nucleophilicity of acetic acid and formic acid—has been confirmed by Peterson and Waller,[9] who have proposed a scale of solvent nucleophilicities (Table 8.5) based on measurements of the logarithms of the rate constants for the reactions of carboxylic acid solvents with tetramethylenehalonium ions in liquid sulfur dioxide at −65°C (Eq. 8.13). The generality of this scale remains to be investigated.

$$RCO_2H + \overset{+}{\underset{Cl}{\square}} \longrightarrow R-C\overset{\overset{+}{O}\frown}{\underset{OH}{\diagdown}}\underset{Cl}{\diagdown} \qquad (8.13)$$

Two other approaches to the questions of solvolysis correlations and solvent nucleophilicities as in Eq. 8.10 have been proposed by Schleyer and coworkers.[10] The first approach, Eq. 8.14, uses two reference substrates,

$$\log \frac{k}{k_0} = (1 - Q)\log \frac{k^A}{k_0^A} + Q\log \frac{k^B}{k_0^B} \qquad (8.14)$$

---
9. P. E. Peterson and F. J. Waller, *J. Amer. Chem. Soc.* **94**, 991 (1972).
10. T. W. Bentley, F. L. Schadt, and P. v. R. Schleyer, *J. Amer. Chem. Soc.* **94**, 992 (1972).

substrate A, with high sensitivity to nucleophilicity, and substrate B, with low sensitivity to nucleophilicity. The reference solvent is 80% ethanol, and Q is an adjustable substrate-blending parameter reflecting substrate sensitivity to nucleophilicity relative to the standard substrates. Because of leaving-group and symbiotic effects (see Chapter 8-7), different standard substrates should be used whenever necessary. For tosylate correlations, Schleyer[10] proposes methyl tosylate as reference A and 2-adamantyl tosylate as reference B, having established in earlier work that the latter is essentially free of solvent nucleophilicity effects and, therefore, is a good model for S_N1 behavior. Values of Q (Table 8.6) fall into a pattern that

TABLE 8.6 Sensitivity to Solvent Nucleophilicity[10]

Tosylate[a]	Q	Tosylate[a]	Q
Methyl (50°)	0.00	Cyclopentyl	0.67
Ethyl (50°)	0.16	Cyclohexyl	0.75
Benzyl	0.22	2-Adamantyl	1.00
Isopropyl	0.56		

[a] At 25°C unless otherwise indicated.

lower values correspond to greater steric accessibility to nucleophilic attack. The assumption of a continuum of solvolytic behavior from methyl to 2-adamantyl appears reasonable from the results even though detailed mechanistic information is lacking and ion-pair effects are ignored.

A solvent-nucleophilicity scale (Table 8.5) was developed by Schleyer[10] using the idea that m values (Chapter 8-3) faithfully measuring substrate sensitivity to solvent ionizing power could be obtained using acetic and formic acid. Assigning $l = 1.00$ to methyl tosylate, which has the lowest m value to date (0.30), Eq. 8.10 becomes

$$N = \log \frac{k^{CH_3OTs}}{k_0^{CH_3OTs}} - 0.30 Y \tag{8.15}$$

in which 80% ethanol is the standard solvent. For tosylates used in Eq. 8.14, Eq. 8.15 defines values for l and m of

$$l \cong 1 - Q \quad \text{and} \quad m \cong 0.3 + 0.7Q \tag{8.16}$$

if the logarithmic term involving reference substrate B is equated with Y. Schleyer[10] has emphasized that the relationships expressed in Eq. 8.16 are limited and less than satisfying since the Winstein-Grunwald Y values (Table 8.3) do not relate perfectly to the 2-adamantyl series.

8-5 Solvation in Nucleophilic Processes

While solvent effects on solvolysis reactions can be analyzed and predicted on the basis of Eq. 8.10 and its analogs, the general area of solvent effects on nucleophilic processes of all types is far more complicated[1,11] (also see Chapter 9). Solvation is related to basicity and polarizability, and is therefore implicitly included in the Edwards equations (Chapter 8-2). For anionic nucleophiles, the usual order of nucleophilic strength is $I > Br > Cl > F$ and $S > O$ (p. 186), yet this order really holds primarily for polar protic solvents,[12] such as water and alcohols. In dipolar aprotic solvents, such as dimethylformamide (DMF), dimethyl sulfoxide (DMSO), acetonitrile, and acetone, which contain no hydrogens that can participate in hydrogen-bonding interactions, the order of nucleophilic strength is often reversed[12] and the S_N2 reaction rate greatly accelerated, particularly for attack by the smaller anions.

Several explanations have been advanced for both the changes in nucleophilic order and the dramatic rate differences observed when dipolar protic and aprotic solvents are compared. The explanations most commonly advanced are desolvation of the anionic nucleophile in the aprotic solvent relative to the protic solvent, thereby increasing the energy of the reactants; increased solvation of the transition state in the aprotic system relative to the protic system, thereby lowering the energy of the transition state; and a combination of these two effects. Since Parker[13] has established that these solvent effects are primarily reflected in enthalpies of activation, most approaches have been enthalpic in nature.

The first explanation—desolvation of anionic nucleophiles in aprotic media relative to protic media—has been partially supported by several groups studying enthalpies of solution.[13,14] Small anions are more strongly solvated in the protic media than are large anions, presumably because hydrogen bonding is greatest for the smaller and more densely charged anions. On transfer to the dipolar aprotic solvents, the degree of solvation for all anions is decreased (or at least is not increased), as is the difference in solvation between the smaller ions and the larger ones. The larger ions do not appear to be solvated to a significantly greater extent in protic media than in aprotic media.[13] Nevertheless, the smaller anions are still more solvated in the dipolar aprotic solvents than are the larger anions.[14] Therefore, anion-solvation effects probably contribute to

---

11. Reference I, Chapter 8; Reference J, Section 29; and M. H. Abraham, *J. Chem. Soc., Perkin II*, 1343 (1972).
12. M. S. Puar, *J. Chem. Educ.* **47**, 473 (1970).
13. A. J. Parker, *Chem. Rev.* **69**, 1 (1969), and B. G. Cox and A. J. Parker, *J. Amer. Chem. Soc.* **95**, 402 (1973), and references therein.
14. R. Fuchs and coworkers, *J. Amer. Chem. Soc.* **90**, 6698 (1968), and **91**, 5797 (1969); G. Choux and R. L. Benoit, ibid., **91**, 6221 (1969).

increased S_N2 reaction rates in dipolar aprotic media, yet do not in themselves account for the reversal in nucleophilic order.

Haberfield[15] has emphasized the importance of transition-state solvation. His approach has been to obtain enthalpies of activation for a given reaction in two solvents and the enthalpies of transfer for the reactants from one of these solvents to the other, thereby arriving at differences in transition-state enthalpies for the given reaction in these two solvents. For an uncharged nucleophile, as in the Menshutkin reaction (pyridine attacking benzyl halides), the slight rate enhancement in dipolar aprotic solvents relative to protic systems is completely the result of transition-state stabilization. For charged nucleophiles in S_N2 or S_NAr reactions, the rate enhancement is more dramatic and the solvent effects are more complex. Haberfield[15] asserts that hydrogen bonding to nucleophile and leaving group is less important than dipole–dipole transition state–solvent interactions, yet in most instances both nucleophile and transition-state effects are enthalpically significant. Transition-state solvation effects are more dominant when the nucleophile is a weak base, thereby agreeing with Parker's conclusion[13] that larger anionic nucleophiles, the weaker bases, are more or less equivalently solvated in protic and dipolar aprotic solvents.

Parker[13] has attacked the problem of solvent effects in bimolecular reactions using "solvent activity coefficients." These solvent activity coefficients, $^0\gamma_i{}^s$, are for transfer of an ideal species i from a reference solvent 0 to another solvent s at 25°C. The rate constant for a bimolecular reaction (e.g., S_N2) in solvent s is related to the rate constant in reference solvent 0 by

$$\log \frac{k^s}{k^0} = \log {}^0\gamma_{Y^-}{}^s + \log {}^0\gamma_{RX}{}^s - \log {}^0\gamma_{\ddagger}{}^s \qquad (8.17)$$

Since it is impossible to determine solvent activity coefficients for single ions, Parker[13] has employed a number of different independent extrathermodynamic assumptions to evaluate the terms for the ionic species in Eq. 8.17. Since these various extrathermodynamic approaches lead to similar conclusions, some confidence can be placed in the resulting single-ion activity coefficients. Parker obtains values for the various terms in Eq. 8.17 which lead him to conclude that $\log {}^0\gamma_{Y^-}{}^s$ dominates and that $\log ({}^0\gamma_{RX}{}^s/{}^0\gamma_{\ddagger}{}^s)$ is roughly constant for reactions of related substrates with different anionic nucleophiles. Equation 8.17 may therefore be recast as

$$\log \frac{k^s}{k^0} = \log {}^0\gamma_{Y^-}{}^s + C \qquad (8.18)$$

---

15. P. Haberfield, et al., *Chem. Commun.* 194 (1968), *J. Amer. Chem. Soc.* **91**, 787 (1969), and *J. Org. Chem.* **36**, 1792 (1971).

in which the C values will be constants characteristic of the substrate type with respect to the nature of the electrophilic atom,[13] the hydrogen-bonding ability of the leaving group, and the "tightness" or "looseness" (or electronic localization or delocalization) in the transition state. This linear free-energy approach might produce useful information about transition-state structure (Chapter 7-8) and charge distribution,[13] but its use thus far has not received much investigation.

The conclusions which have been reached with respect to solvation in nucleophilic processes are those which, in retrospect, might have been expected. Solvations of ionic species (see Chapter 7-8, note 52b, for significant evidence) and charged transition states are both of major importance, as they must be in transition-state theory. Solvation of uncharged species is of minor importance since the major solute–solvent interactions are ion–dipole and dipole–dipole in character.[11] Quantitative conclusions have been offered, yet controversy remains.[13,15] Solvation effects in solvents with low dielectric constants differ significantly from those in either dipolar protic or dipolar aprotic media because of the intervention of ion-pairing and aggregation phenomena. They have not been extensively studied, to some extent because of solubility-related difficulties. The detailed nature of any and all solvent effects remains a problem (see Chapter 9 for further discussion), with hydrogen bonding, polarizability effects, electrostatic interactions, and solvent structure, including ionic effects on structure making and structure breaking, all at best at a superficial level of analysis and comprehension.

8-6 Ambident Anions; The Alpha Effect

Two types of nucleophiles consistently produce deviations from almost any attempted correlation (Chapter 8-1 through 8-4). One of these types consists of certain nucleophiles possessing a lone pair of electrons adjacent to the nucleophilic site; these will be discussed later in this section. The other is those nucleophiles which contain two unlike nucleophilic sites. These are called *ambident anions*.[1,16]

Ambident Anions. Since the nucleophilic sites in a given ambident anion are different, they react at different rates, are affected by solvent changes to different extents, and even possibly react with different electrophilic sites in the substrate molecule. For example, the more basic site in the anion would react preferentially with the more polarizable substrate electrophilic site, and vice versa (see Eq. 8.4 and 8.5 for examples involving nonambident nucleophiles). Typical ambident anions are NO_2^-, CN^-,

---

16. Reference C, pp. 298–301; Reference D, pp. 296–298; and Reference I, Section 8.19.

SCN⁻, and enolates, such as those produced by ionization of 2-naphthol, malonic ester, and cyclohexanone.

Differential nucleophilic effects are reflected in percentages of O- and C-alkylation or -acylation of various enolates,[17] which are a function of the enolate, the solvent, the substrate (with emphasis on the nature of the electrophilic site and the leaving group), and the metal counterion.[18] The lack of cation effects in alkylations of enolates in DMSO suggested the use of the Swain-Scott equation (Eq. 8.3) and partial rate factors (Chapter 5-2) to determine relative nucleophilicities.[18] The calculated n values paralleled enolate basicities[18] (as measured by ketone pK_a's), with deviations reflecting steric phenomena.

Wigfield[19] has suggested that O/C product ratios on alkylation of ethyl acetoacetate anion and related compounds can be explained on the basis of Hammond-postulate analyses of the transition states (Chapter 7-8). For reactant-like transition states, O-alkylation is predicted, since the oxygen would most likely be the most nucleophilic site from ground-state electron densities. For product-like transition states, carbon would be the most nucleophilic site, because the C-alkylated product is the more stable. Dipolar aprotic solvents should increase the O/C alkylation ratio over that for hydroxylic solvents, since the less solvated enolate in the aprotic media should react in a more exothermic manner, thereby involving a more reactant-like transition state. The O/C alkylation ratio should similarly increase as the cation is varied in group I from Li^+ to Cs^+, since the more covalent lithium enolate should be the least reactive and involve the most product-like transition state. Since increased S_N2 reactivity of the alkyl halide corresponds to a decreased O/C ratio, the halide with the greatest S_N2 reactivity should correlate with the most product-like transition state (and the least exothermic reaction). Wigfield[19] presents spectral evidence for ground-state solvent–alkyl halide interactions in dipolar aprotic solvents consistent in order of strength with what is demanded by such a Hammond-postulate analysis. In other words, the alkyl halide with the greatest S_N2 reactivity is the most stabilized by interactions with the solvent in the ground state. This leads to a prediction using the Hammond postulate that it is this halide that will react via the least exothermic process and the most product-like transition state, giving the lowest O/C ratio under the given set of reaction conditions.

The Zook[18] and Wigfield[19] approaches appear somewhat complementary. Whether or not cation effects exist in dipolar aprotic media for

---

17. H. O. House, *Modern Synthetic Reactions*, 2nd ed., W. A. Benjamin, Menlo Park, Calif., 1972, Chapters 9 and 11.
18. H. D. Zook and J. A. Miller, *J. Org. Chem.* **36**, 1112 (1971).
19. D. C. Wigfield, *Can. J. Chem.* **48**, 2120 (1970).

such carbon–oxygen ambident anions seems somewhat confused from their results.

The Alpha Effect. The alpha effect[1,20] is a somewhat more complex phenomenon. Certain nucleophiles which possess a lone pair of electrons adjacent to the nucleophilic site sometimes react much faster than might be expected from their basicities and polarizabilities. Examples of these "supernucleophiles"[20,21] are hydrazines, hydroxylamines, hydroxamic acids, hypohalite anions, and the anions of hydroperoxides and oximes.

Of the several proposals advanced to explain this phenomenon, none has been effectively refuted. Edwards and Pearson[20] have suggested that the transition state is stabilized by delocalization of the partial positive charge (or decreased negative charge) generated on the nucleophilic site in the transition state. This stabilization would be analogous to that in the carbonium ion formed in an S_N1 reaction of an α-haloether (Eq. 8.19).

$$R-\ddot{O}-CH_2X \longrightarrow R-\ddot{O}-CH_2^+ \longleftrightarrow R-\overset{+}{O}=CH_2 \quad (8.19)$$

Ibne-Rasa and Edwards[1] have been proponents of an explanation based on ground-state destabilization resulting from repulsions involving the alpha lone pair and the nucleophilic electrons. Formation of a covalent bond would relieve these p_π–p_π repulsions by generating less severe p_π–σ interactions and reducing the electron density at the nucleophilic site.[1,22] An alternate way of stating this explanation is within a molecular orbital framework.[21,22] Overlap of electron pairs on adjacent atoms in the nucleophile would lead to orbital splitting and increase the energy of the highest occupied molecular orbital (HOMO), possibly, according to Ingold,[20] even destabilizing it sufficiently to make it antibonding or giving rise to inhomogeneous polarizability. The higher the energy of the HOMO, the more the stabilization which could be produced on removal of the orbital splitting.

If the nucleophile conformation is such as to minimize the p_π–p_π repulsions, as is probably the case in hydrazines and hydroxylamines, or if one of the lone pairs is part of a conjugated system, as in oximate anions, the enhanced nucleophilicity must arise for a different reason. Hudson[22] proposes the intervention of intra- or intermolecular catalysis in these

---

20. Reference J, pp. 452–453; J. O. Edwards and R. G. Pearson, *J. Amer. Chem. Soc.* **84**, 16 (1962). See Chapter 7-8, note 52b, for additional data which must be considered in any ultimate understanding of this effect.
21. G. Klopman, K. Tsuda, J. B. Louis, and R. E. Davis, *Tetrahedron* **26**, 4549 (1970).
22. J. D. Aubort and R. F. Hudson, *Chem. Commun.* 937 (1970).

instances, although Dixon and Bruice[23] feel this is not a reasonable possibility because of the absence of significant entropy changes when supernucleophiles are compared to other nucleophiles.

Another possibility is that factors which stabilize the product stabilize the transition state as well to a significant extent.[23] Substrate factors[1,20] and the necessity of considerable bond formation,[21,23] as evidenced from Brønsted β factors (Chapter 7-5) and a combination[21] of theoretical considerations and Edwards-equation analysis (see Chapter 8-8), point to product stability as a contributing factor. This type of product-stability explanation is closely related to the original Edwards and Pearson transition-state stabilization argument.

8-7 Hard and Soft Acids and Bases

As discussed in Chapter 8-1, a nucleophilic reaction has many of the characteristics of a generalized acid–base reaction.[24] An electrophilic site and a nucleophilic site combine in such a way that an electron pair is provided by the nucleophilic site and shared with the electrophilic site just as a Lewis base provides an electron pair to be shared with a Lewis acid. Pearson[25] has recognized the generalities involved in the multifarious types of acid–base reactions and proposed the *principle of hard and soft acids and bases* (HSAB principle) as a phenomenological way of correlating and predicting behavior in such systems. While there must be underlying theoretical bases for these generalities (Chapter 8-8), the HSAB principle itself is not theoretical, but rather is correlative and based on experimental observations. In fact, the major usefulness of this principle as such is in its ability to provide qualitative results, although some efforts have been directed toward quantification.

The basic definitions in the HSAB approach were originally recognized in the area of inorganic chemistry and illustrated the reliance (and, often, overreliance) of the inorganic chemist on the criteria of size and charge. *Soft bases* are highly polarizable, easily oxidized, and of low electronegativity and possess empty low-lying orbitals. *Hard bases* are opposite in these properties. *Soft acids* are large species with low positive charges and filled outer orbitals, being highly polarizable with low electronegativities. The opposite properties are associated with *hard acids*. The

---

23. J. E. Dixon and T. C. Bruice, *J. Amer. Chem. Soc.* **93**, 3248 (1971), and **94**, 2052 (1972).
24. Other common examples of generalized acid–base reactions are formation of salts, coordination complexes, and charge-transfer complexes; electrophilic reactions; and solute–solvent interactions.
25. R. G. Pearson, *Science* **151**, 172 (1966), *Chem. Brit.* **3**, 103 (1967), *J. Chem. Educ.* **45**, 581, 643 (1968); R. G. Pearson and J. Songstad, *J. Amer. Chem. Soc.* **89**, 1827 (1967).

HSAB principle states that extra stabilization exists in the interaction of a hard acid and a hard base or of a soft acid and a soft base compared to the interaction of a hard acid and a soft base or of a soft acid and a hard base. The softness or hardness of a base is to some extent a function of the acid involved in the reaction since, as discussed in Chapter 7-6, there is no universal scale of relative acidity and basicity. Hard and soft are not the same as strong and weak. An acid is characterized by at least two properties, which, in the HSAB approach, are its strength and its hardness or softness. The proton is the simplest hard acid, and the methylmercury cation is often used as the prototype of a soft acid.[25] For the generalized acid–base reaction

$$A + :B \rightleftharpoons A:B$$

log K assumes the following value

$$\log K = S_A S_B + \sigma_A \sigma_B \tag{8.20}$$

where S_A and S_B are strength factors for the acid and base and σ_A and σ_B are softness factors for the acid and base, the latter being negative for a hard species and positive for a soft species. The product of the sigmas will therefore be positive for either a hard–hard or a soft–soft interaction and negative for either type of hard–soft interaction.

The four-parameter approach of Eq. 8.20 is by no means unique. Drago[26] has developed an empirical equation for correlating and predicting the enthalpy of formation of acid–base adducts in the gas phase or in poorly solvating media (Eq. 8.21). Use of the enthalpy of formation as a

$$-\Delta H = E_A E_B + C_A C_B \tag{8.21}$$

measure of the interactions is justified by the enthalpy invariance in the gas phase, carbon tetrachloride, and hexane, even though the free energy is not constant because of solvent entropy influences.[26] The E and C parameters are empirical, yet are interpreted in terms of the electrostatic and covalent (or charge-transfer) interactions generated in the formation of the adduct. The scales were developed with iodine as the reference acid ($E_A = 1.00$, $C_A = 1.00$), and therefore have no meaning as absolute values. The E and C parameters, which are not related to any ground-state properties, are consistent with the HSAB approach, and provide evidence for steric effects whenever the enthalpy measured calorimetrically differs significantly from that calculated using the empirical parameters (e.g., as in Chapter 11-6).

The major success of the Drago approach[26] is its ability to correlate reversals in base donor strengths toward different acid acceptors. Relative softness or hardness can be considered as the C/E ratio, the larger ratio

---

26. R. S. Drago, *Chem. Brit.* **3**, 516 (1967); R. S. Drago, G. C. Vogel, and T. E. Needham, *J. Amer. Chem. Soc.* **93**, 6014 (1971).

corresponding to the softer acid or base, but this ratio does not completely agree with the qualitative scales evolved by the HSAB approach.[25] The source of this discrepancy probably is that the individual magnitudes of C and E are lost in the ratio. Drago concludes[26] that the C/E ratio for bases indicates whether hardness or softness is most important in that base's interactions, but does not permit prediction of the interaction strength toward a soft or hard acid even on a relative basis. In addition, Drago claims[26] that both Eq. 8.20 and 8.21 are not simply quantitative expressions of HSAB, but rather extend the scope of HSAB and make it more complete.

Another four-parameter equation related to HSAB is the Edwards equation, Eq. 8.6. Pearson[25] noticed that β varies just as S_A (Eq. 8.20)

$$\log \frac{k}{k_0} = \alpha E_N + \beta H \tag{8.6}$$

would be expected to behave, suggesting that H is a way of evaluating S_B. If αE_N is then equated with $\sigma_A \sigma_B$, the Edwards equation is identical with the HSAB equation (Eq. 8.20). Interesting as this relationship may be, the desirability of this correlation may be questioned because of the difficulties with the Edwards equation itself (Chapter 8-2). Nevertheless, the conclusions are useful. Acids with large α and β are soft, while those with small α and large β are hard. Therefore, soft acids should react more favorably with those bases which are highly polarizable, good reducing agents, and poorer bases in the thermodynamic sense (Eq. 8.1); hard acids should react more favorably with those bases which are the strongest in the usual thermodynamic sense, with polarizability criteria being of minor significance. That β is large for both hard and soft acids points out the importance of both the magnitude of α and β and their ratio.[26]

One of the important corollaries of the HSAB principle is the *symbiotic principle*,[25] which states that extra stabilization is associated with the clustering of several soft bases or several hard bases about a given acidic site. Examples of symbiosis are the softness of BH_3 relative to the hardness of BF_3, the extra stability of the more highly branched hydrocarbons, and the thermodynamic instability of hemiketals relative to disproportionation to the ketones and ketals. Symbiotic effects are also apparent in nucleophilic displacements in terms of substituent effects at the electrophilic site and the complex interrelationships between nucleophile, leaving group, and reaction rate.[25]

The solubility rule that "like things dissolve each other" falls within the HSAB framework.[25] Hard solutes dissolve best in hard solvents and soft solutes in soft solvents. Water is a hard solvent as both an acid and a base, indicating the necessity of careful extrapolation of aqueous behavior to softer nonaqueous systems. Lack of reactivity in aqueous media may be more closely related to solvent than to substrate (see Chapter 8-5 for illustrations), although both must be important and, according to HSAB concepts, are usually interrelated (see below).

One of the difficulties in the application of HSAB to organic systems is the ambiguity inherent in the continual necessity of mentally dissecting organic molecules into acid and base fragments.[25] For example, ethyl acetate may be viewed as a Lewis base being protonated, as a Lewis acid undergoing nucleophilic attack at the carbonyl carbon, or as a complex of an acidic acylium ion and a basic ethoxide ion; methane can be dissected into a carbonium ion and a hydride or into a carbanion and a proton, ignoring homolytic representations. These multiplicities are required because of the varied possibilities in reaction types, with the ultimate choice dependent primarily on an understanding of reaction mechanisms. For example, nucleophilic attack on methyl chloride can be accommodated best by viewing the alkyl halide as a carbonium ion acid complexed with a halide base, since the reaction is then a substitution of one base for another.

The HSAB approach can be used to rationalize thermodynamic stability.[25] The key to any question of relative stability—which is, after all, the main question answered by the HSAB principle—is the choice of the acid–base reaction to be used as the basis for comparison. Fluoride ion is a hard base and iodide is a soft base, so in competitive equilibria involving the proton, a hard acid, and the methyl cation, a borderline or soft acid, the favored compounds would be methyl iodide (soft–soft) and hydrogen fluoride (hard–hard). If hydroxide ion, a hard base, were substituted for iodide, prediction would become difficult. Replacing the hydrogens in the methyl cation with methyl groups leads to progressively harder acids. This trend is reasonable only if carbon is more electronegative than hydrogen— if alkyl groups are electron-withdrawing relative to hydrogen (see Chapter 4-4). Isomerizations of alcohols to more highly substituted alcohols (e.g., 1-propanol to 2-propanol) may be considered as reactions leading to systems in which the harder carbonium ion (acid) is bonded to the hard hydroxide ion (base). Acylium ions are harder than alkyl ions, a logical situation because the electronegative oxygen tends to create a more positive carbon.

The HSAB principle can be applied to kinetic phenomena[25] with the corollary that softness is more important in kinetic processes than in thermodynamic ones. In other words, borderline or soft acids will be more reactive to soft bases than product stability would suggest. In general, soft nucleophiles react best with soft electrophiles and hard with hard in consonance with generalizations determined by other methods (p. 188), as required by the empirical nature of the HSAB approach.

The S_N2 reaction (Eq. 8.22) provides a useful framework for analysis of the HSAB approach.[1,25]

$$Y: + SX + \text{solvent} \longrightarrow [Y \cdots S \cdots X \cdots \text{solvent}]$$
$$\downarrow \qquad (8.22)$$
$$YS + X: + \text{solvent}$$

Assume that S, the electrophilic site, is soft. The most efficient reaction would then occur (a) if X, the leaving group, were hard, giving a relatively less stable starting material, (b) if Y, the nucleophile, were soft, producing a relatively more stable product, and (c) if the solvent were hard, thereby most effectively solvating the leaving group and least effectively solvating the soft nucleophile. Metal catalysis of a process with the above-mentioned hard and soft characteristics would be most efficient if the metal ion were hard,[27] since the metal replaces the solvent in terms of interactions with the leaving group. In the transition state the electrophilic site has less positive (or greater negative) character than in the ground state and is therefore softer, leading to increased sensitivity to softness in these kinetic processes compared to equilibrium processes.

If the electrophilic site is an alkyl carbon, which is a borderline soft acid, and if the solvent is aqueous or alcoholic, and therefore hard, the softer nucleophiles, such as I^- and R_2S, would react faster. Switching to a dipolar aprotic solvent, which is softer, produces a reversal in reactivity order in such a way that Cl^- reacts faster than I^- (Chapter 8-5). This reversal occurs only because alkyl cations are borderline soft acids. Were they softer, the halide reactivity sequence favoring iodide would be maintained in dipolar aprotic solvents, although the reactivity differences in such a series would be reduced. If the electrophilic site is an acyl cation, which is hard, reaction with harder nucleophiles, such as Cl^- and R_2O, would be favored. In the terminology of the Edwards equation (Chapter 8-2), in the alkyl halides nucleophile polarizability is dominant, while in the harder acyl halides nucleophile basicity is dominant.

Relative substrate sensitivity to nucleophiles may also be predicted in more detail.[25] Since alkyl halides are most sensitive to nucleophile polarizability, sensitivity to the nature of the nucleophile in an S_N2 reaction should decrease in the series $CH_3X > C_2H_5X > i\text{-}C_3H_7X > t\text{-}C_4H_9X$ since the more branched systems are harder and therefore less sensitive to polarizability factors. For the same reason, differences in reaction rates in this series should be greater for the softer nucleophiles than for the harder ones. Both of these trends agree with the experimental results.

---

27. B. Saville, *Angew. Chem. Intern. Ed. Engl.* **6**, 928 (1967).

If the nucleophile is an ambident anion (Chapter 8-6), the position of nucleophilic attack can be predicted from HSAB considerations.[25] The harder nucleophilic site would react preferentially with hard electrophiles and the softer nucleophilic site with soft electrophiles. For example, enolate ions react with methyl iodide to produce C-alkylated product (soft–soft) and with methoxymethyl chloride, CH_3OCH_2Cl, to give the O-alkylated product (hard–hard). The same type of approach may be used for nucleophilic attack on a molecule with several electrophilic sites.[25] In particular, the competition between substitution (S_N2) and elimination (E2) is amenable to analysis. Hard ethoxide will react with an alkyl halide to give predominant elimination because of the hard–hard interaction between the ethoxide and a proton. On the other hand, soft malonate ion will lead to predominant substitution with an alkyl halide because of the soft–soft interaction with the alkyl electrophilic site.

8-8 Charge-controlled and Frontier-controlled Reactions

Several theories may be advanced to explain the principle of hard and soft acids and bases discussed in the preceding section. The oldest of these is implicit in Drago's equation[26] (Eq. 8.21). Hard acids have the highest positive charges and smallest sizes, and therefore interact most efficiently by ionic forces. The ionic interactions are most favorable with hard bases, which have large negative charges and small sizes. Soft acids, which have low or zero charge and large sizes, interact most efficiently by covalent bonding, which is best when the bond is formed between atoms of similar size and electronegativity, such as with soft bases. The favorable nature of soft–soft interactions is reinforced by London forces,[25] which are largest when the interacting groups are most polarizable.

A more elegant approach is the quantum mechanical perturbation approach developed by Klopman.[28] In this general treatment of all chemical reactivity, free-energy changes are evaluated in terms of a charge–charge interaction term representing electrostatic interactions and a covalent interaction term evaluating covalent bond formation. In Eq. 8.23,

$$\Delta G^{\ddagger} = \frac{-q_r q_s e^2}{R\varepsilon} + 2 \sum_{\substack{m \\ occ}} \sum_{\substack{n \\ unocc}} \frac{(c_r^m)^2 (c_s^n)^2 \beta^2}{E_m^* - E_n^*} \qquad (8.23)$$

q_r is the charge on the nucleophilic site, q_s is the charge on the electrophilic site, R is the distance between these sites, ε is the dielectric constant, c_r^m is the coefficient of atomic orbital r in molecular orbital m of energy E_m^*

---

28. G. Klopman, *J. Amer. Chem. Soc.* **90**, 223 (1968).

in the nucleophile donor, c_s^n is defined accordingly for the electrophile acceptor, and β is the resonance integral of the developing bond (Chapter 1).

The key concept is that the approach of two orbitals, the HOMO of the donor base and the LUMO of the acceptor acid, produces a mutual perturbation of these orbitals and a change in energy. When the difference in energy between these frontier orbitals m and n (Chapter 3) is large, the second term in Eq. 8.23 is small and the reaction, being dominated by ionic forces, is said to be *charge-controlled*.[28] When $E_m^* - E_n^*$ is small, strong electron transfer from donor to acceptor can occur, the interactions are primarily covalent, and the reaction is *frontier-controlled*.[28] A charge-controlled reaction is favored by donors which are difficult to ionize or polarize, by acceptors which have low tendency to accept electrons, by strongly solvated reactants, and by small values of β, indicating low tendency toward covalent-bond formation. All of these properties are consistent with small reactants of low polarizability, which correspond to the hard acids and bases. A frontier-controlled reaction is favored by reagents which are highly polarizable with low solvation energies and little effective overlap of the interacting orbitals, corresponding to the soft categories of acids and bases.

In addition, Klopman's theoretical approach[28] predicts that hard–hard interactions will be endothermic in water because of solvation effects and occur primarily because of entropy terms, while soft–soft interactions will be exothermic because of the offsetting stabilization provided by the formation of a covalent bond. The frontier orbital energies themselves may be obtained from ionization potentials, electron affinities, hydration energies, and ion sizes, and provide a quantitative basis[25,28] for evaluating hardness and softness without any prior knowledge of the chemical properties of the acids and bases. Solvents with low dielectric constants tend to promote charge-controlled reactions, while polar solvents favor frontier-controlled processes.

Klopman's approach[28] correlates well with HSAB concepts,[25] nucleophilic reactivities (Chapter 8-1), solvation effects (Chapter 8-5), and ambient reactivity (Chapter 8-6). In the latter instance, distribution of charge in an ambient anion often is not localized in one frontier molecular orbital, thereby requiring a more complicated approach involving consideration of total electronic charge densities instead of frontier electronic charge densities. Nevertheless, appropriate approximations lead to reasonably successful predictions.

The same type of approach has been applied to an analysis of the alpha effect[21] (Chapter 8-6). The results[21] are that orbital splitting raises the energy of the HOMO, thereby increasing the frontier-controlling properties of the reaction and requiring considerable covalent-bond formation in the rate-determining step. In Edwards-equation terminology (Chapters

8-2 and 8-7), the alpha effect requires that α/β be large, corresponding to frontier control, and that β be large, corresponding to considerable bond formation.

A similar frontier orbital perturbation approach has been used by Epiotis[29] to analyze donor–acceptor relationships in cycloaddition reactions (Chapter 3), leading to results differing sharply from the concerted–nonconcerted reactivity classification implicit in the Woodward-Hoffmann approach. Epiotis suggests that the Woodward-Hoffmann rules (Chapter 3) are applicable only to part of the reactivity spectrum possible in thermal and photochemical cycloadditions. In particular, Epiotis emphasizes that nonstereospecificity does not imply nonconcertedness. Several groups[30] have subsequently argued that orbital-symmetry-forbidden processes may be concerted (Chapter 3-5), thereby supporting Epiotis' suggestions.

---

29. N. D. Epiotis, *J. Amer. Chem. Soc.* **94**, 1924, 1935, 1941, 1946 (1972).
30. J. A. Berson, *Accounts Chem. Res.* **5**, 406 (1972); J. A. Berson and L. Salem, *J. Amer. Chem. Soc.* **94**, 8917 (1972); J. E. Baldwin, A. H. Andrist, and R. K. Pinschmidt, Jr., *Accounts Chem. Res.* **5**, 402 (1972); W. Schmidt, *Tetrahedron Lett.* 581 (1972).

Chapter **9**

Medium Effects

Significant emphasis has been placed in preceding sections (Chapters 5-4, 5-5, 7-4, 7-6, 7-7, and 8-3 through 8-8) on the influences of the medium on reaction parameters. Without repeating the arguments advanced for the various types of salt and solvent effects or reinvestigating the approaches proposed for evaluation and quantification of these effects, it should be readily apparent that a diversity of medium effects are possible and that the effects themselves are not well understood. It should also be apparent that most of these previous treatments were limited in scope and failed to come to grips with the general questions of solvent classifications and comprehensive solvent scales, much less attempt to analyze the detailed interactions producing the various phenomena. Salt-induced medium effects[1] will not be considered in this chapter, and the extremely difficult questions about detailed solvent interactions[2] and solvent structure[3] will be treated in an extremely superficial manner.

9-1 General Treatment

Any chemist working with solutions is constantly confronted with the need for a proper choice of solvent. Historically, most investigations were

---
1. P. Salomaa, A. Kankaanperä, and M. Lahti, *J. Amer. Chem. Soc.* **93**, 2084 (1971), and references therein.
2. For example, see M. D. Johnston, Jr., F. P. Gasparro, and I. D. Kuntz, Jr., *J. Amer. Chem. Soc.* **91**, 5715 (1969), for a discussion of several types of solvent effects on nmr spectra.
3. Reference F, Sections 2.0, 2.1, and 2.5.

directed toward correlation of solvent effects with solvent "polarity," an ambiguous concept most often expressed in terms of the dielectric constant,[4-6] the dipole moment, or the refractive index of the solvent. While these macroscopic properties[7] of solvents are often useful in the designation of general solvent classes, such as dipolar and nonpolar classes, no single macroscopic physical parameter could possibly account for the multitude of solute–solvent interactions on the molecular level. These microscopic interactions are typically Coulombic, charge-transfer, London, and hydrogen-bonding in nature.

The first satisfactory qualitative theory of solvent effects was the 1935 Hughes-Ingold electrostatic model,[4-6,8] which treated the effects of solvents on rates of aliphatic substitutions and eliminations in aqueous media in terms of transition-state theory (Chapter 5-3) and charge types (Chapters 5-4 and 8-5). Unfortunately, these ideas, emphasizing electrostatic stabilization of reactants or transition states, are useful only for the limited range of solvents in which specific nonelectrostatic solute–solvent or solvent–solvent interactions are not significant. No later theory has really been more successful, nor might any be expected to be since the specific interactions are so poorly understood. The entire question of the nature of the liquid state[3] and of solutions is both extremely difficult[2] and insufficiently studied.

Since no single macroscopic physical characteristic of a solvent is useful and since the detailed interactions are poorly understood, the problem of solvent parameters has been attacked empirically. Attempts have been made to find solvent-dependent standard chemical or physical processes which could be used to create an empirical solvent scale general enough to be useful. Naturally, the most useful model processes should be those best understood on a molecular basis. These empirical scales of solvent parameters have been either reactivity-based or spectral in origin.

9-2 Reactivity-based Solvent Scales

The most dramatic solvent effects on rate or equilibrium processes[4,5] should be those in systems in which the most charge is generated or destroyed. The previously discussed Winstein-Grunwald Y values (Chapter 8-3), the log k_{ion} scale (Chapter 8-3), and the related more complex multi-

---

4. C. Reichardt, *Angew. Chem. Intern. Ed. Engl.* **4**, 29 (1965).
5. M. R. J. Dack, *Chem. Brit.* **6**, 347 (1970).
6. Reference B, pp. 376–388; Reference F, Sections 2.2–2.4; Reference H, Chapter 7; and Reference I, Chapter 8.
7. For a table of several of these properties, see Reference F, Table 2.1.
8. Reference B, pp. 379–383; Reference J, Sections 29 and 40b.

parameter relationships (Chapter 8-4) are typical examples of this idea, with emphasis on nucleophilic substitution reactions.

Gielen and Nasielski[4] have proposed a solvent scale analogous to the Winstein-Grunwald equation (Eq. 8.8) for electrophilic substitutions (Eq. 9.1). The standard reaction (Eq. 9.2) is the reaction of bromine with

$$\log \frac{k}{k_0} = pX \qquad (9.1)$$

$$(CH_3)_4Sn + Br_2 \longrightarrow CH_3Br + (CH_3)_3Sn^+ + Br^- \qquad (9.2)$$

tetramethyltin ($p = 1.00$) in glacial acetic acid ($X = 0.00$). The solvent polarity parameters, X, were obtained for only four solvents (CH_3OH, DMF, CCl_4, and chlorobenzene) and exhibited no correlation with dielectric constant functions or Y values. Two reasonable explanations for this lack of correlation have been proposed. Just as nucleophilic substitution is assisted by solvent electrophilic "pull" (Chapter 8-4), electrophilic substitution should be assisted by solvent nucleophilic interactions with the leaving group, thereby suggesting the need for more parameters in any quantitative analysis. In addition, the reference process is believed to involve a polar transition state in polar solvents and a cyclic, less polar transition state in less polar solvents. This change in mechanism would prevent meaningful correlations with any but the most closely related systems. Somewhat surprisingly, however, the X values do correlate well with solvent effects on $J_{^{117}Sn - ^1H}$ values obtained in various nuclear magnetic resonance experiments.

Another empirical solvent parameter is that derived from an analysis of the Diels-Alder reaction (Chapter 3-4) by Berson, Hamlet, and Mueller.[4] The rate of this reaction is only slightly solvent-sensitive,[9] yet the ratio of *endo* product (N) to *exo* product (X) is markedly solvent-dependent. Since the Diels-Alder reaction is kinetically controlled, the ratio of the rate constants for formation of N relative to X can be determined from the product ratio,[10] leading to the solvent parameter Ω (Eq. 9.3). The standard reaction[4] is that of cyclopentadiene with methyl acrylate (Eq. 9.4), with Ω

$$\Omega = \log \frac{N}{X} = \log \frac{k_N}{k_X} \qquad (9.3)$$

ranging from 0.44 for triethylamine to 0.84 for methanol and limited by solubility factors to a small range of possible solvents. The physical basis for this scale is the greater dipole moment in the *endo* transition state compared to the *exo*. Increasing solvent polarity would improve *endo*

---

9. Reference B, pp. 376–379.
10. See Chapter 5-2 for a discussion of this approach.

solvation more than *exo*, lowering the activation energy for the *endo* process more than that for the *exo*, and increasing the N/X ratio. The Ω scale correlates well with Diels-Alder reactions involving other dienophiles, $\log k_{ion}$, dielectric constants (for the aprotic solvents only), and the Z and E_T scales to be discussed in the next section.

$$\text{cyclopentadiene} + H_3CO-\overset{O}{\overset{\|}{C}}-CH=CH_2 \xrightarrow{k_N, k_X} [\text{TS}]^{\ddagger} \rightarrow N, X \quad (9.4)$$

9-3 Solvatochromism Scales

Since the common spectroscopic techniques require a change in dipole moment on absorption of the incident electromagnetic radiation, it is reasonable that solvent effects resulting from differential solvation of the ground-state and excited-state dipoles would occur.[4,5] Changes in solvent cause changes in the position, intensity, and shape of both absorption and emission bands, a phenomenon called *solvatochromism*.[4] The solvent effect is primarily a function of the spectral technique. In spectra resulting from electronic transitions, the solvent effect is primarily dependent on the chromophore and the nature of the transition (Chapter 9-5), such as $n \rightarrow \sigma^*$, $n \rightarrow \pi^*$, $\pi \rightarrow \pi^*$, or charge-transfer, assuming the solvent does not somehow alter the nature of the chromophore itself.

Most quantitative evaluations of solvent effects on electronic transitions[4,5,11] have used intramolecularly ionic dyes, such as phenol blue

$$(CH_3)_2N-\bigcirc-N=\bigcirc=O$$

$$\updownarrow$$

$$(CH_3)_2\overset{+}{N}=\bigcirc=N-\bigcirc-O^-$$

I

(I), a merocyanine dye. These are systems in which an electron-donating group is linked by a conjugated system to an electron-accepting group. The electronic transition is usually[4] associated with a charge transfer between these two groups, producing an excited state with a dipole moment appreciably different from that in the ground state.

Two qualitatively different types of behavior are possible, depending on the direction of the change in dipole moment as a molecule undergoes electronic excitation (or emission). If the excited state is more polar than the ground state, as with phenol blue, the long-wavelength absorption band undergoes a bathochromic shift (to longer wavelengths and lower energies) as the solvent polarity increases. This type of behavior is termed *positive solvatochromism*.[4,11] If the ground state is more polar than the excited state, as with the pyridinium-N-phenol betaines (II), the opposite behavior, a hypsochromic shift, occurs, which is called *negative solvatochromism*.[4,11] In the vocabulary of resonance theory, the type of solvatochromism depends on whether the zwitterionic resonance contributor is more important in the ground state or in the excited state.

$$R-\underset{R}{\overset{R}{\bigcirc}}N=\underset{R'}{\overset{R'}{\bigcirc}}=O \longleftrightarrow R-\underset{R}{\overset{R}{\bigcirc}}\overset{+}{N}-\underset{R'}{\overset{R'}{\bigcirc}}-O^-$$

II

The first attempt to generate a solvent scale using a dye electronic transition was Kosower's analysis of the negative solvatochromism of the longest-wavelength charge-transfer band of 1-ethyl-4-carbomethoxypyridinium iodide[4,11] (Eq. 9.5). The positions of the absorption maximum in

---

11. Reference C, pp. 296–297; Reference F, Sections 2.6–2.8; Reference G, pp. 300 and 309–311; K. Dimroth, *Chem. & Eng. News* 99 (May 31, 1965).

$$\underset{\underset{Et}{|}}{\underset{N+}{\bigcirc}}\text{CO}_2\text{CH}_3 \quad I^- \xrightarrow{h\nu} \underset{\underset{Et}{|}}{\underset{\overset{\cdot}{N}}{\bigcirc}}\text{CO}_2\text{CH}_3 \quad I\cdot \longleftrightarrow \underset{\underset{Et}{|}}{\underset{\overset{\cdot}{N+}}{\bigcirc}}\text{CO}_2\text{CH}_3^{-} \quad I\cdot \longleftrightarrow \text{etc.} \quad (9.5)$$

various solvents lead to a scale of Z values (Table 9.1) corresponding to the molar transition energies, E_T (Eq. 9.6). The longer the wavelength of the

$$Z \equiv E_T = \frac{2.859 \times 10^5}{\lambda} \quad (\lambda \text{ in angstroms}) \quad (9.6)$$

absorption maximum, the less the energy required in the transition, and the lower the Z value. Since this system exhibits negative solvatochromism, the more polar solvents would stabilize the ground state more than the excited state, thereby requiring more energy for the electronic transition and giving a higher Z value.

Several limitations[4,11] are inherent in the Z-value approach. First of all, the Franck-Condon principle must be satisfied. In other words, the electronic transition must occur much more rapidly than the nuclei can move, leading to a nonequilibrium excited state in which the solvent arrangement around the solute is the same as it was in the ground state. Secondly, the standard pyridinium salt is not soluble in many nonpolar solvents ($\lambda > 452$ nm corresponding to $Z < 63.2$ kcal/mole). Use of secondary standards with greater solubility in these nonpolar solvents overcomes this limitation, provided that the results are extrapolated to zero ionic strength to remove ion-pair aggregation effects. Thirdly, in the most polar solvents ($\lambda < 331$ nm corresponding to $Z > 86.4$), the desired charge-transfer band becomes lost under the stronger $\pi \to \pi^*$ band. Indirect methods must then be used, the easiest of which is usually the careful extrapolation of results obtained in mixed solvent systems including the desired polar solvent.[11]

TABLE 9.1 *Typical Kosower Z Values*[11]

Solvent	Z	Solvent	Z
Water	94.6	2-Propanol	76.3
80% Ethanol	84.8	Acetonitrile	71.3
Methanol	83.6	t-Butyl alcohol	71.3
Ethanol	79.6	DMSO	71.1
Acetic acid	79.2	DMF	68.5
1-Propanol	78.3	Acetone	65.7
1-Butanol	77.7	Methylene chloride	64.2

While no correlation was observed between Z values and dielectric constants, excellent correlations were observed with Y, log k_{ion}, Ω, and a host of other solvent-sensitive absorptions, such as the longest-wavelength band in phenol blue (**I**), the charge-transfer band of tropylium iodide, the $n \to \pi^*$ of cyclohexanone, and both the $n \to \pi^*$ and $\pi \to \pi^*$ of mesityl oxide (Chapter 9-5).

While the Kosower Z-value scale is the most well-known and often-used scale, a more comprehensive solvent scale is that developed by Dimroth, Reichardt, and coworkers[4,5,11,12] based on the pyridinium phenol betaine (**II**, where R and R' are phenyl groups) and called the E_T scale. The major advantage of this scale is that the charge-transfer band is

TABLE 9.2 *Typical E_T Values*[4,11,12]

Solvent	E_T	Solvent	E_T
Water	63.1	Acetone	42.2
Methanol	55.5	Methylene chloride	41.1
80% Ethanol	53.6	Pyridine	40.2
Ethanol	51.9	Deuteriochloroform	39.0
1-Propanol	50.7	THF	37.4
1-Butanol	50.2	Benzene	34.5
2-Propanol	48.6	Toluene	33.9
Acetonitrile	46.0	CS_2	32.6
DMSO	45.0	CCl_4	32.5
t-Butyl alcohol	43.9	Cyclohexane	31.2

at longer wavelengths for **II** than for Kosower's dye, generating an extraordinarily large range for the solvatochromic behavior (from 810 nm, $E_T = 35.3$, for diphenyl ether; to 453 nm, $E_T = 63.1$, for water). Since much of this range is in the visible region, many estimates could even be performed visually. The E_T values (Table 9.2) are linear with Z over the entire range of overlapping solvent systems. Extrapolation of this scale to nonpolar solvents has been accomplished by using substituted dyes (such as **II**, R = p-tolyl).[4,12] The major limitations on the E_T scale are the assumption of the Franck-Condon principle and the unsuitability of the standard dye system, regardless of substitution pattern, in acidic solvents which can protonate the oxygen atom.

Several general phenomena are suggested by the Z scale, the E_T scale, or both. Addition of an electrolyte produces changes in the solvent param-

---

12. C. Reichardt, *Ann.* **752**, 64 (1971); K. Dimroth and C. Reichardt, *Ann.* **727**, 93 (1969); and references in both of these papers.

eter,[4,13] as does changing the temperature.[4] If a protic solvent and an aprotic solvent have the same dielectric constant, the protic solvent will appear more polar[4] (have higher Z and E_T). Steric hindrance is associated with decreasing polarity.[4] Steric effects cause protic solvents to be more similar to aprotic ones with similar dielectric constants, and lead to decreasing Z and E_T values in a given series (e.g., pyridine, $E_T = 40.2$; α-picoline, $E_T = 38.3$; and 2,6-lutidine, $E_T = 36.7$).

The conclusion is that there are many useful empirical solvent scales (including many not mentioned here), of which the solvatochromic ones are the most comprehensive. However, it is always necessary to assume that the interactions in the standard system used to develop a scale are similar to those in the system in which some question about solvent is being raised, and this is a drastic assumption. If anything, it is surprising that the existing solvent scales, varying widely in form and in the energies involved, agree so well in most cases with each other and with the many different types of systems to which they have been applied.

9-4 Multiple Correlations

In order to take into account two or more aspects of solvation, it is possible to use a multiparameter approach[5] of the general form

$$X = a + bK + cL + dM + \cdots$$

where X is the system being studied, K, L, and M are solvent parameters, and a, b, c, and d are constants chosen to best fit the experimental results. This approach is often useful in providing information about the type and magnitude of interactions with the solvent or in pinpointing the dominant solvent interaction in a given system.

The most recent and thorough study of multiparameter approaches to solvent effects on physical and chemical properties is that of Fowler, Katritzky, and Rutherford.[14] They conclude that multiparameter approaches can be significantly better than other empirical techniques, and that the most useful method is to use a function of E_T (Chapter 9-3) and either the dielectric constant or the refractive index.[14]

Caution should be emphasized in rate correlations if the reaction is multistep or involves an equilibrium prior to the rate-determining step. In either case, medium effects are a composite of the effect on each independent step, and correlations of any kind would be no more than fortuitous if the mechanism were correct.

---

13. Reference F, Section 2.9.
14. F. W. Fowler, A. R. Katritzky, and R. J. D. Rutherford, *J. Chem. Soc.*, B 460 (1971).

9-5 Solvent Effects on Electronic Transitions

Solvent effects on electronic transitions involving the common chromophores in organic chemistry are often predictable on the basis of the relative polarities of the ground and excited states, with hydrogen bonding most often accounting for aberrant behavior.

One of the most common chromophores absorbing in the readily accessible area of the ultraviolet region is the carbonyl group. From observation of the $n \to \pi^*$ absorption of acetone, it was found that increasingly polar protic solvents shifted the absorption maximum to shorter wavelengths[15] (279 nm in hexane and 265 nm in water), presumably because hydrogen bonding stabilizes the ground state more than the excited state. Similar trends were observed with aprotic solvents,[15] indicating that the above results are the result of dipolar effects as well as hydrogen bonding. Since the excitation takes an electron from a nonbonding oxygen orbital and moves it into an antibonding orbital involving both the carbon and oxygen, the excited state has a smaller dipole moment than the ground state, and would therefore be less stabilized as the solvent becomes more polar. This same behavior holds for all carbonyl compounds, including the $n \to \pi^*$ transition in conjugated substrates such as mesityl oxide[15,16] (**III**) (Table 9.3).

$$H_3C\diagdown\atop{H_3C\diagup}C=CH-\underset{\underset{\textstyle \textbf{III}}{}}{\overset{\overset{\textstyle O}{\|}}{C}}-CH_3$$

TABLE 9.3 *Solvent Effects in Mesityl Oxide* (**III**)

Solvent	$n \to \pi^{*\,a}$	$\pi \to \pi^{*\,a}$
Hexane	327 (97.5)	229.5 (12,600)
Ether	326 (96)	230 (12,600)
Ethanol	315 (78)	237 (12,600)
Methanol	312 (74)	238 (10,700)
Water	305 (60)	244.5 (10,000)

^{a} In nanometers, with ε values in parentheses.

SOURCE: G. Scheibe, *Ber.* **58**, 586 (1925); H. H. Jaffé and M. Orchin, *Theory and Applications of Ultraviolet Spectroscopy*, Wiley, New York, 1962, Table 10.5.

---

15. Reference F, Section 2.7.
16. H. H. Jaffé and M. Orchin, *Theory and Applications of Ultraviolet Spectroscopy*, Wiley, New York, 1962, Chapter 10; D. J. Pasto and C. R. Johnson, *Organic Structure Determination*, Prentice-Hall, Englewood Cliffs, N.J., 1969, Sections 3.5.1 and 3.5.2.

Solvent effects on $\pi \to \pi^*$ transitions are significantly different. Whether conjugated or not, carbon–carbon double bonds exhibit little or no solvent sensitivity.[16] The changes in dipole moment occurring on excitation are not of a type where solvent interactions can play a significant role. However, the $\pi \to \pi^*$ transitions of conjugated unsaturated carbonyl systems such as mesityl oxide (III) are markedly solvent-sensitive.[15,16] The carbonyl oxygen is electron-withdrawing, and more so in excited states than in ground states because of the higher energy and greater accessibility of an electron in a π^* orbital.[15] As shown in Table 9.3, the more polar solvents stabilize the excited $\pi \to \pi^*$ state more than the ground state, producing bathochromic shifts. Woodward[16,17] and the Fiesers[16,18] independently recognized this trend in the $\pi \to \pi^*$ band of steroidal conjugated carbonyl systems and proposed sets of correction factors to convert the wavelength of an absorption maximum to its value in ethanol as a standard solvent (Table 9.4).

TABLE 9.4 *Solvent Corrections to Ethanol for $\pi \to \pi^*$ Transitions of Conjugated Carbonyl Systems*[16]

From	Woodward[17,a]	Fiesers[18,a]
Methanol	−1	0
Chloroform	0	+1
Ether	+6	+7
Hexane	+7	+11
Water	—	−8

[a] Values in nanometers.

Solvent effects on benzenoid systems are extremely complex.[19] Using aniline as an example of a substituted system in which the excited state is more dipolar than the ground state, one observes a bathochromic trend with aprotic solvents (Table 9.5). Protic solvents, even if more polar, superimpose a hypsochromic shift (Table 9.5, last three entries). The lone pair of electrons on nitrogen is more accessible in the ground state than in the excited state, making hydrogen bonding a greater stabilizing influence in the ground state than in the excited state and producing a hypsochromic shift related to the strength of the solvent as a hydrogen-bond donor.[19]

---

17. R. B. Woodward, *J. Amer. Chem. Soc.* **63**, 1123 (1942).
18. L. F. Fieser and M. Fieser, *Steroids*, Reinhold, New York, 1959, pp. 15–21.
19. Reference B, pp. 187–190; H. H. Jaffé and M. Orchin, *Theory and Applications of Ultraviolet Spectroscopy*, Wiley, New York, 1962, Chapter 12.4.

TABLE 9.5 *Solvent Effects in Aniline*

Solvent	λ_{max} (nm)	Solvent	λ_{max} (nm)
Vapor	229.5	Dioxane	240.0
Heptane	234.0	Ethanol	235.0
Cyclohexane	234.5	Methanol	233.5
Acetonitrile	238.5	Water	229.5

SOURCE: W. M. Schubert and J. M. Craven, *J. Amer. Chem. Soc.* **82**, 1357 (1960); and sources listed in footnote 19.

9-6 Solvent Effects on Nuclear Magnetic Resonance Transitions

In the early 1960s, it became apparent that in the nuclear magnetic resonance spectra of dipolar compounds, aromatic solvents induce shifts in signals from positions observed in spectra obtained in "inert" solvents such as carbon tetrachloride and deuteriochloroform.[20] These shifts, called aromatic-solvent-induced shifts—ASIS's—, signified by Δ and expressed in parts per million, are defined using Eq. 9.7. They are found to be

$$\Delta = \delta_{CCl_4 \text{ or } CDCl_3} - \delta_{C_6H_6 \text{ or } C_6D_6} \tag{9.7}$$

as large as ± 1.5 ppm. The nature of the shift requires postulation of a specific solute–solvent interaction. A dipolar solute molecule is believed to polarize the aromatic solvent, inducing an opposed dipole in a molecule of solvent and forming a transient 1:1 collision complex involving a molecule of solute and a molecule of solvent. The specific interaction involves the aromatic π cloud, which has a large magnetic anisotropy, and an electron-deficient site in the solute molecule. The collision complex is believed to be nonplanar, with the aromatic ring interacting with the positive end of the solute dipole and positioned as far as possible from the negative end of the dipole. The lifetime, exact stoichiometry, and exact geometry of the collision complex are not known, yet the lifetime must be extremely short and the geometry more or less constant in each complex.[20] The solvent shifts are often additive when several polar groups are present in the solute, reflecting the location of the hydrogen nucleus with respect to each polar group and each associated aromatic solvent molecule. The shifts seem to be relatively insensitive to steric hindrance in the vicinity of

---

20. N. S. Bhacca and D. H. Williams, *Applications of NMR Spectroscopy in Organic Chemistry*, Holden-Day, San Francisco, 1964, Chapter 7; P. Laszlo and P. Stang, *Organic Spectroscopy*, Harper and Row, New York, 1971, pp. 208–209; J. Ronayne and D. H. Williams, *Chem. Commun.* 712 (1966); F. H. A. Rummens and R. H. Krystynak, *J. Amer. Chem. Soc.* **94**, 6914 (1972); K. Pihlaja and M. Ala-Tuori, *Acta Chem. Scand.* **26**, 1891, 1904 (1972).

the positive end of the dipole, indicating that the aromatic ring is not located within normal bonding distances of the polar group.

Aromatic-solvent-induced shifts have been most extensively investigated in steroids and other ring systems.[20] A typical shift is that of an axial 2-methyl group in a cyclohexanone ring 0.2–0.3 ppm upfield (in benzene relative to carbon tetrachloride) and that of the corresponding equatorial 2-methyl group 0.05–0.10 ppm downfield, thereby permitting relatively easy determination of configuration or conformation.

Somewhat related to the ASIS's are the lanthanide-induced shifts.[21] These phenomena involve addition of a lanthanide salt to form a complex, rather than a solvent effect.

---

21. J. K. M. Sanders, S. W. Hanson, and D. H. Williams, *J. Amer. Chem. Soc.* **94**, 5325 (1972); B. C. Mayo, *Chem. Soc. Rev.* **2**, 49 (1973); and references therein.

General References PART II

A. Liberles, A. *Introduction to Theoretical Organic Chemistry.* New York: Macmillan, 1968.
B. Wiberg, K. B. *Physical Organic Chemistry.* New York: Wiley, 1964.
C. March, J. *Advanced Organic Chemistry: Reactions, Mechanisms, and Structure.* New York: McGraw-Hill, 1968.
D. Gould, E. S. *Mechanism and Structure in Organic Chemistry.* New York: Holt, Rinehart and Winston, 1959.
E. Hine, J. *Physical Organic Chemistry*, 2nd ed. New York: McGraw-Hill, 1962.
F. Kosower, E. M. *An Introduction to Physical Organic Chemistry.* New York: Wiley, 1968.
G. Leffler, J. E., and E. Grunwald. *Rates and Equilibria of Organic Reactions.* New York: Wiley, 1963.
H. Frost, A. A., and R. G. Pearson. *Kinetics and Mechanism*, 2nd ed. New York: Wiley, 1961.
I. Hammett, L. P. *Physical Organic Chemistry: Reaction Rates, Equilibria, and Mechanisms*, 2nd ed. New York: McGraw-Hill, 1970.
J. Ingold, C. *Structure and Mechanism in Organic Chemistry*, 2nd ed. Ithaca, N.Y.: Cornell University Press, 1970.
K. Gilliom, R. D. *Introduction to Physical Organic Chemistry.* Reading, Mass.: Addison-Wesley, 1970.

PART III

Stereochemistry and Conformational Analysis

The importance of organic chemistry in the world in which we live is fundamentally dependent on the stereochemistry, or three-dimensional structure, of most of the key compounds. Chemical reactivity, physical behavior, and biochemical activity are all extremely sensitive functions of the detailed structural features of the chemical being used. Life itself, at all levels of development, is based on the ability of each portion of the living system to generate only one of many possible three-dimensional structures and to react with only certain designated topological arrangements. It therefore behooves all chemists to come into at least cursory contact with the principles and nomenclature of stereochemistry and some of their applications, all of which we will consider almost exclusively in the framework of organic compounds.
General references for Part III will be found on page 276.

Chapter 10

Stereochemistry

Compounds with identical molecular formulas but differing in the nature or sequence of bonding of atoms or in the arrangement of these atoms in space are called *isomers*.[1] *Stereoisomers* are those isomers differing only in the arrangement of their atoms in space. The arrangements may differ in two or in three dimensions, leading to different types of stereoisomeric systems. Stereoisomers have the same molecular formula and the same atomic sequence, and, in the most general sense, will be classified as either enantiomers or diastereomers (see below).

10-1 Chirality

Chirality[1,2,3] is the nonidentity of an object with its mirror image. The word itself denotes "handedness"; and the concept accordingly is illustrated by the nonidentity of a person's right hand and left hand, which form more or less nonsuperimposable mirror images. The term "chirality"[4] is synonymous with "dissymmetry"—chiral objects may have axes of symmetry. From the point of view of symmetry operations, chiral

---

1. References A and B; "IUPAC 1968 Tentative Rules for the Nomenclature of Organic Chemistry, Part E," *J. Org. Chem.* **35**, 2849 (1970); H. Hirschmann and K. R. Hanson, *J. Org. Chem.* **36**, 3293 (1971).
2. R. S. Cahn, C. K. Ingold, and V. Prelog, *Angew. Chem. Intern. Ed. Engl.* **5**, 385 (1966), and previous papers in series.
3. E. L. Eliel, *J. Chem. Educ.* **48**, 163 (1971); D. Arigoni and E. L. Eliel, *Topics in Stereochemistry* **4**, 127 (1969).
4. Ruch has discussed algebraic aspects of chirality in *Accounts Chem. Res.* **5**, 49 (1972).

objects do not have reflection symmetry; a combination of a rotation operation and a reflection operation does not convert the chiral object into itself. The principal chiral elements are chiral centers, which are equivalent to centers of "asymmetry" such as tetrahedral carbon atoms containing four different ligands (Chapters 10-2 and 10-3); chiral axes, such as in allenes and biphenyls (Chapters 10-3 and 10-5); and chiral planes, such as in paracyclophanes and *trans*-cyclooctene (Chapters 10-3 and 10-5).

Enantiomers are defined as stereoisomers which are mirror images of each other (or have opposite chirality). When equal amounts of enantiomers are present, the product is called *racemic*.[1] All other stereoisomers are called *diastereomers* (or diastereoisomers), including the special class often called "geometrical isomers" (Chapter 10-3). Diastereomers may be chiral or achiral.

Chirality is specified by the Cahn-Ingold-Prelog R-S system[1,2] (Chapter 10-2). Enantiomers have identical physical and chemical properties unless they are placed under diastereomeric conditions.[1] They will exhibit identical physical properties except under the influence of another chiral object, as in the cases of the direction of rotation of plane-polarized light or the nmr chemical shifts in chiral solvents or in the presence of chiral salts, and identical chemical properties except on reaction with a chiral substance (for an exception, see Chapter 10-4). Diastereomers will exhibit different chemical and physical properties (except for an occasional example of accidental equivalency in a given property of a given pair of diastereomers).

Chiral systems rotate plane-polarized light.[1] A system is termed *optically active* when this rotation of plane-polarized light is observed experimentally. The only limitation on a 1:1 correspondence between chirality and optical activity is the experimental and instrumental problem of observing extremely small rotations. If the plane-polarized light is rotated to the right (or clockwise), the rotation is called *dextrorotatory* and symbolized by d or a plus sign. Rotation to the left (or counterclockwise) is *levorotatory* rotation, symbolized by l or a negative sign. Enantiomers (under similar measurement conditions) will rotate plane-polarized light in opposite directions in identical magnitudes. Racemic systems will not rotate plane-polarized light, and are therefore often denoted as dl or $\pm$.

The process of separating a racemic system into its enantiomers is called *resolution*.[1] The most common resolution technique is reaction with a chiral reagent to produce diastereomers,[5] which can then be separated by physical methods. The individual diastereomers can be treated chemically to regenerate the desired enantiomers.

---

5. For examples, see P. H. Boyle, *Quart. Rev.* **25**, 323 (1971), and S. H. Wilen, *Topics in Stereochemistry* **6**, 107 (1971).

10-2 Configuration—Relative and Absolute

The most commonly encountered chiral element in organic chemistry is the chiral center[6] as exemplified by a tetrahedral carbon atom containing four different ligands. The usual two-dimensional representation of such a system (**I**) implies the three-dimensional structure (**II**) as a *Fischer projection*.[1] Ligands above or below the chiral carbon atom lie behind the plane of the paper (as symbolized by a dashed line), while those to the left and right of the carbon lie above the plane (as usually shown by a darkened line or triangle). Wavy lines are used to designate unknown stereochemistry or the presence of both stereoisomers at the given position. A Fischer projection can be rotated only end for end (180°); it can neither be rotated 90° nor be removed from the plane of the paper: Structures **I** and **III** are enantiomers.

$$\begin{array}{ccc} X & X & X \\ | & \vdots & | \\ W-C-Y & W-C-Y & Y-C-W \\ | & \vdots & | \\ Z & Z & Z \\ \textbf{I} & \textbf{II} & \textbf{III} \end{array}$$

The *configuration*[1] of a molecule is defined as the arrangement of its atoms in space without regard to rotations around single bonds.[7] Enantiomers and diastereomers will therefore always differ in configuration. The question remains as to which Fischer projection really represents a given enantiomer or diastereomer.

The first step in answering this question is to establish the *absolute configuration* of at least one enantiomer of all the myriad enantiomers known to chemists. This was first accomplished by Bijvoet in 1951 using x-ray diffraction methods on a salt of (+)-tartaric acid.

The second step is to use this absolute configuration as a standard and to relate the configurations of other enantiomeric systems to this known system.[1] This process of correlating configurations may be accomplished by (a) converting one enantiomer of one compound into one enantiomer of a second compound, or (b) converting one enantiomer of each of the above two compounds into an enantiomer of the same third compound, using reactions which either do not involve the chiral centers or affect the chiral centers in known stereochemical ways. If the reaction takes place at a position not involving bonding at the chiral center, the configuration

---

6. For planar and axial chirality, see note 1 and G. Krow, *Topics in Stereochemistry* **5**, 31 (1970).
7. For a discussion of the problems inherent in precise definition of configuration and conformation, see IUPAC Rules (note 1).

at the chiral center would be unchanged, and the reaction would be said to proceed with *retention* of configuration; reaction at the chiral center that does not result in change in the configuration also constitutes retention of configuration. Reaction at the chiral center to produce the opposite configuration, such as in an S_N2 process, proceeds with *inversion* of configuration. Loss of a unique configuration at the chiral center during reaction is called *racemization*. It is possible, then, to use a series of reactions of known stereochemistry to convert an enantiomer whose configuration we wish to know into an enantiomer whose absolute configuration has been established.[8]

Having established which Fischer projection should be used to represent a given enantiomer, it would be desirable to have some way in which to consistently and concisely describe a given absolute configuration. The method used is the *Cahn-Ingold-Prelog R-S system*,[1,2] which provides a symbol for each chiral element in a molecule. This method does *not* have any relation to the direction of rotation of plane-polarized light (d or l) or to any reaction sequence used to establish relative configuration; a reaction which converts an R enantiomer into an S enantiomer may proceed with retention of configuration.

The first step in use of the R-S system is factorization,[1,2] or recognition of the chiral elements which are present. The simplest case is a chiral center containing four unlike ligands attached to carbon (e.g., structure **I**). The sequence rules[1,2] are then used to arrange the ligands in an order of preference (with > used to denote "is preferred to"). The atoms directly attached to the chiral center are compared first. If relative priorities cannot be established using these atoms, those atoms bonded to the atoms of equal priority are compared, and so forth, working away from the chiral center in a bond-by-bond manner as required.

The sequence rules consist of several subrules,[1,2] which are applied in order. The first subrule is applicable only to chiral axes (as in allenes).[6] The second subrule gives preference to the atom of higher atomic number, while the third gives preference to the atom of higher mass number. This produces a typical priority order of

$$I > Br > Cl > O > N > C > D > H$$

If the ligancy at any position other than a hydrogen is less than four, phantom atoms of atomic number zero are added so as to create a ligancy of at least four. Double bonds are split into two bonds, with C=O treated

---

8. Notice that a sequence used to correlate configurations is dependent on the reactions used. A different reaction sequence might lead to the other enantiomer or to the opposite conclusion as to the relative configuration of a given enantiomer.

as $C{\overset{O}{\underset{(O)}{\diagdown}}}(C)$ and having lesser precedence than an actual $C{\overset{O}{\underset{O}{\diagdown}}}$ situation.

Triple bonds are split into three bonds, while aromatic rings are treated as Kekulé structures.

When the ligand sequence is established, the chirality rule[1,2] is applied. The molecule is viewed from the side farthest from the ligand of lowest priority. If the priority sequence of the remaining three ligands decreases clockwise, the configuration is called R (Latin *rectus* for "right"). If the priority sequence decreases counterclockwise, the configuration is S (Latin *sinister* for "left"). Converting one enantiomer into its mirror image changes each R to S and each S to R. A racemic element is therefore represented as RS (or SR).

An enantiomer of 2-chlorobutane (**IV**) furnishes a useful example. The molecule contains a chiral center (denoted by ∗), to which are directly

$$\underset{\textbf{IV}}{CH_3-CH_2-\overset{\overset{Cl}{|}}{\underset{\underset{H}{|}}{C^*}}-CH_3} \qquad \underset{\textbf{V}}{Cl-\overset{\overset{CH_2-CH_3}{|}}{\underset{\underset{CH_3}{|}}{C^*}}-H} \qquad \underset{\textbf{VI}}{Cl-\overset{\overset{CH_2-CH_3}{|}}{\underset{\underset{H}{|}}{C^*}}-CH_3}$$

attached a chlorine, two carbons and a hydrogen. The atomic number priority sequence rule assigns the highest priority to the chlorine and the lowest priority to the hydrogen, with the two carbon groups tied for the middle positions. Moving one atom farther from the chiral center along the carbon groups, the ethyl carbon is attached to one carbon and two hydrogens while the methyl carbon is bonded to three hydrogens. Again the atomic number sequence rule is sufficient to assign the higher priority: ethyl > methyl. The overall priority sequence therefore is:

$$Cl > ethyl > methyl > H$$

According to the Fischer projection representation, both the chlorine and the hydrogen are pointing away from someone looking at the page in the normal way, so it can be quickly discerned that the three groups of highest priority decrease in a counterclockwise direction. This enantiomer is therefore (2S)-2-chlorobutane.

Had enantiomer **V** been the molecule in question, the analysis would have been complicated by the group of lowest priority, the hydrogen, being situated above the plane of the paper, thereby requiring the observer to view the structure from a position behind the page. This problem, with its inherent visualization difficulties, can often be avoided for chiral centers by exchanging the positions of adjacent groups in a pairwise manner until the group of lowest priority is located at the bottom of the representation. Exchanging the positions of the hydrogen and the methyl would convert **V** to **VI**, which has the hydrogen at the bottom and would be assigned the R configuration. This resulting assignment must be corrected to take into account the number of pairwise exchanges. If the number of pairwise exchanges is an even number, the configuration assigned to the final representation is the same as that which should be assigned to the original representation. If an odd number of exchanges has been performed, the configuration has been reversed. Representation **VI** is R and has been produced by one exchange from **V**, so **V** must therefore be S and the same enantiomer as that represented by **IV**.

It must be reemphasized that the R-S system merely permits construction of an accurate model or representation of a given configuration at some chiral element. It provides communication without necessitating an elaborate descriptive presentation. It should never be assumed that an R configuration of one molecule will be converted to an R configuration of another molecule with retention of configuration. Rather, the two configurations and the reaction stereochemistry in a specified reaction are three dependent parameters; knowledge of two of these parameters produces the third.

10-3 Diastereomers and *meso* Systems

If a structure possesses two or more chiral elements, the Cahn-Ingold-Prelog system just described must be applied to each chiral element independently, with the configuration associated with the appropriate IUPAC numbering system.[1,9] For example, enantiomer **VII** would be designated as (2S,3R)-2,3-dichloropentane (Cl > C > H priority order).

$$\underset{\textbf{VII}}{\overset{\displaystyle\overset{\text{Cl}}{\underset{\text{H}}{|}}\overset{\text{Cl}}{\underset{\text{H}}{|}}}{\underset{5}{CH_3}-\underset{4}{CH_2}-\overset{}{\underset{}{C}}-\overset{}{\underset{}{C}}-\underset{1}{CH_3}}} \qquad \underset{\textbf{VIII}}{\overset{\displaystyle\overset{\text{Cl}}{\underset{\text{H}}{|}}\overset{\text{CH}_3}{\underset{\text{H}}{|}}}{CH_3-CH_2-\overset{}{\underset{}{C}}-\overset{}{\underset{}{C}}-Cl}}$$

Molecules which are enantiomers of each other will have the opposite configuration at each chiral element. Any other configurational relationship will relate diastereomers. For a system with two differently

---
9. See, for example, R. T. Morrison and R. N. Boyd, *Organic Chemistry*, 3rd ed., Allyn and Bacon, Boston, 1973, Chapter 4.

substituted chiral centers, such as the 2,3-dichloropentanes, diastereomers will have the same configuration at one center and the opposite configuration at the other. The (2R,3R)-2,3-dichloropentane (**VIII**) is therefore a diastereomer of the 2S,3R isomer (**VII**).

The total number of stereoisomers possible in a system containing n chiral centers (with no other chiral elements) is a maximum of 2^n. This total is less than maximum if a molecule is possible in which equal numbers of enantiomeric groups linked in an identical manner are present without any other chiral group being present. Such a *meso* compound is superimposable on its mirror image and is achiral even though it contains chiral elements. Examples are (2R,3S)-2,3-dichlorobutane and (2S,3R)-tartaric acid.

$$\begin{array}{cc}
{}^4CH_3 & {}^4CO_2H \\
| & | \\
H-{}^3C-Cl & H-{}^3C-OH \\
| & | \\
H-{}^2C-Cl & H-{}^2C-OH \\
| & | \\
{}^1CH_3 & {}^1CO_2H \\
\end{array}$$

(2R,3S)-2,3-Dichlorobutane   (2S,3R)-Tartaric acid

Since diastereomers include all stereoisomers which are not enantiomers, it is worthwhile to examine various types of double-bond stereoisomers and to see where they fall in this scheme. In any system containing a double bond in which the substituents at each end of the double bond are different[1] (structure **IX** where $a \neq b$ and $c \neq d$), restricted rotation about the double bond generates the possibility of diastereomers (e.g., **IX** and **X**). Interconversion of these diastereomers requires that the π bond

$$\begin{array}{cc}
a \diagdown \quad \diagup c & b \diagdown \quad \diagup c \\
X=Y & X=Y \\
b \diagup \quad \diagdown d & a \diagup \quad \diagdown d \\
\text{IX} & \text{X}
\end{array}$$

be broken, and therefore requires roughly 40–60 kcal/mole. If one substituent at each end is a hydrogen (e.g., **IX** with $b = d = H$) or if one substituent at each end is a lone pair of electrons (e.g., **IX** with $b = d =$ lone pair), the terms "*cis*" and "*trans*" may be used to differentiate the diastereomers.[1,3] Groups on the same side of the plane defined by the double bond are called *cis*, while those on opposite sides are called *trans*.

In other stereoisomers differing at a single double bond, ambiguity is generated by the *cis–trans* system.[1,3] For unambiguous specification of this kind of stereoisomerism, the *E-Z system*[1,3,10,11] has been developed with the Cahn-Ingold-Prelog sequence rules as building blocks. In implementing this system, the sequence rules are applied to the two ligands attached to one end of the double bond and then to the two ligands attached to the other end. The groups of higher priority are then compared with respect to the reference plane generated by the double bond. If they are on the same side of this plane, the configuration is called *Z* (German *zusammen* for "together"). If they are on opposite sides of this plane, the configuration is *E* (German *entgegen* for "opposite"). For example, compound **XI** is called (Z)-1-bromo-2-chloro-1-iodoethylene and compound **XII** is (E)-benzaldehyde oxime.

$$\underset{\textbf{XI}}{\underset{H}{\overset{Cl}{>}}C=C\underset{Br}{\overset{I}{<}}} \qquad \underset{\textbf{XII}}{\underset{H}{\overset{C_6H_5}{>}}C=\overset{\cdot\cdot}{N}\diagdown OH}$$

If a molecule has several double bonds, each is characterized within the E-Z system and identified with the number assigned each double bond in the IUPAC naming system.[12] For example, the following structure is (2Z,4E)-5-methyl-2,4-heptadiene.

$$\underset{H_3C}{\overset{H}{>}}C=C\underset{H}{\overset{H}{<}}C=C\underset{CH_2CH_3}{\overset{CH_3}{<}}$$

Cumulenes[1,6] present another type of double-bond stereoisomerism. Cumulenes with an odd number of double bonds (e.g., structure **XIII**) have

$$\underset{\textbf{XIII}}{\underset{c}{\overset{a}{>}}C\underset{\parallel}{\overset{\parallel}{C}}\underset{\parallel}{\overset{\parallel}{C}}\underset{d}{\overset{b}{<}}} \qquad \underset{\textbf{XIV}}{\underset{c}{\overset{a}{>}}C\underset{\parallel}{\overset{\parallel}{C}}\underset{d}{\overset{b}{<}}}$$

---

10. J. E. Blackwood, C. L. Gladys, K. L. Loening, A. E. Petrarca, and J. E. Rush, *J. Amer. Chem. Soc.* **90**, 509 (1968).
11. This system replaces the Cahn-Ingold-Prelog seq*cis* and seq*trans* as well as the more ambiguous terms "*syn*" and "*anti*" when applied to double-bond substituents.
12. See, for example, R. T. Morrison and R. N. Boyd, *Organic Chemistry*, 3rd ed., Allyn and Bacon, Boston, 1973, Chapter 5.

diastereoisomerism of the same type as simple alkenes.[13] However, cumulenes with an even number of double bonds (e.g., allene **XIV**) are capable of existing as enantiomers if $a \neq b$ and $c \neq d$. In this latter type of system, the ligands at one end of the system are in a plane perpendicular to that defined by the ligands at the other end of the cumulated system, thereby generating a chiral element in the presence of different ligands in each separate plane. Interconversion of enantiomeric or diastereomeric cumulene isomers usually requires around 25–40 kcal/mole, and is therefore easier than isomerization of simple alkenes.

10-4 Prochirality

If an achiral molecule has a pair of ligands which can be distinguished by reference to a chiral object, the molecule possesses a *prochiral* element.[1,3,14] Consider a tetrahedral carbon atom bonded to two identical groups (a) and to two other groups (b and c) which are different from a and from each other, giving a total of four ligands ($aabc$) of three different types. If one of the identical a ligands is replaced by a ligand d different from a, b, and c, the tetrahedral carbon atom now becomes a chiral center. The original carbon atom bonded to a, a, b, and c is therefore termed a prochiral center and the a ligands are called *stereoheterotopic*.

If substitution of d for one of the a ligands and separate substitution of d for the other a ligand leads to enantiomers, the a ligands are called *enantiotopic*[1,15] and can be distinguished only under chiral conditions, such as by relative or selective reactivity with a chiral reagent or by a chemical-shift difference in a nuclear magnetic resonance spectrum[15] obtained in a chiral solvent or in the presence of an added chiral solute. The methylene hydrogens in ethanol are enantiotopic, as demonstrated by substitution of deuterium for hydrogen to give (R)-**XV** and (S)-**XVI**.

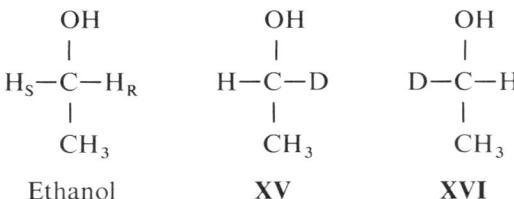

If replacement of d for each a separately leads to diastereomers, the a ligands are called *diastereotopic*[1,15] and can be distinguished physically

---
13. Consider an alkene as a cumulene with one double bond.
14. K. R. Hanson, *J. Amer. Chem. Soc.* **88**, 2731 (1966).
15. K. Mislow and M. Raban, *Topics in Stereochemistry* **1**, 1 (1967).

and chemically under normal achiral conditions. The methylene hydrogens in (R)-malic acid are diastereotopic, deuterium substitution producing the diastereomers **XVII** and **XVIII**.

$$\begin{array}{ccc}
\text{CO}_2\text{H} & \text{CO}_2\text{H} & \text{CO}_2\text{H} \\
| & | & | \\
\text{H}-\text{C}-\text{OH} & \text{H}-\text{C}-\text{OH} & \text{H}-\text{C}-\text{OH} \\
| & | & | \\
\text{H}_S-\text{C}-\text{H}_R & \text{D}-\text{C}-\text{H} & \text{H}-\text{C}-\text{D} \\
| & | & | \\
\text{CH}_3 & \text{CH}_3 & \text{CH}_3 \\
\text{(R)-Malic acid} & \textbf{XVII} & \textbf{XVIII}
\end{array}$$

Ligands at a given center which are identical even after a substitution process, and therefore are not stereoheterotopic, are called *homotopic* or *equivalent*. Homotopic ligands such as the methylene hydrogens in methylene chloride (CH_2Cl_2) and the methyl hydrogens in malic acid are indistinguishable under all conditions.

Hanson[1,14] has applied the Cahn-Ingold-Prelog system (Chapter 10-2) to prochiral elements. Assume that the ligand *d* used in substitution for one of two ligands *a* at a prochiral center has a higher priority in the R-S sequence rules than the remaining ligand *a*. If the resulting chiral center is R, the ligand at which the substitution has been performed is called pro-R and denoted by a subscript R; if the resulting chiral center is S, the ligand is pro-S and denoted by a subscript S (see structures of ethanol and malic acid, above).

Prochirality is extremely important in discussing the stereochemistry and stereospecificity of enzymatic reactions. Enzyme stereospecificity at a prochiral center is usually determined by use of stereospecifically labeled isotopic substrates.[3] For example, if enzymatic oxidation (yeast alcohol dehydrogenase or liver alcohol dehydrogenase) of (R)-ethanol-1-*d* (**XV**) in the presence of nicotinamide adenine dinucleotide (NAD^+) removes only the deuterium to produce unlabeled acetaldehyde and specifically deuterated (4R)-4-*d*-reduced nicotinamide adenine dinucleotide (NADD),[3,14] it follows that the pro-R hydrogen in unlabeled ethanol would be transferred and become the pro-R hydrogen at carbon 4 of the NADH. Another important biochemical example is the stereospecific enzymatic degradation of one of the enantiotopic CH_2CO_2H groups in citric acid[16] (see structure next page) in the citric acid cycle[3,15] (or Krebs cycle).

---

16. Notice also that the hydrogen atoms in a given CH_2CO_2H group in citric acid are diastereotopic, while one hydrogen atom in one CH_2CO_2H group is enantiotopic to one of the hydrogens and diastereotopic to the other hydrogen in the other CH_2CO_2H group.

$$\begin{array}{c} CO_2H \\ | \\ H-C-H \\ | \\ HO_2C-C-OH \\ | \\ H-C-H \\ | \\ CO_2H \end{array}$$

Citric acid

Unsymmetrically substituted double bonds also exhibit prochirality with stereoheterotopic faces.[3,14,15] For example, addition of hydrogen cyanide to benzaldehyde produces a racemic mixture of the enantiomeric addition products because the faces of benzaldehyde are enantiotopic. In the presence of a chiral influence, such as the enzyme emulsin, only one of the enantiomers is produced and the reaction is stereospecific. Grignard addition to a ketone with diastereotopic faces, such as (R)-α-phenethyl methyl ketone (**XIX**), produces diastereomeric carbinols (e.g., **XX** and **XXI**), whose relative proportions may be estimated using Cram's rule.[17]

$$\begin{array}{c} CH_3 \\ | \\ C=O \\ | \\ H-C-CH_3 \\ | \\ C_6H_5 \end{array} + RMgX \xrightarrow{H_2O} \begin{array}{c} CH_3 \\ | \\ HO-C-R \\ | \\ H-C-CH_3 \\ | \\ C_6H_5 \end{array} + \begin{array}{c} CH_3 \\ | \\ R-C-OH \\ | \\ H-C-CH_3 \\ | \\ C_6H_5 \end{array}$$

XIX   **XX**   **XXI**

Stereoheterotopic faces are named[3,14] by using the Cahn-Ingold-Prelog sequence rules (Chapter 10-2) in two dimensions. The prochiral group is viewed from the side of the face in question. A clockwise decreasing sequence is called a *re*-face and a counterclockwise sequence a *si*-face. Carbinol **XX** therefore results from attack on the *si*-face and **XXI** from attack on the *re*-face. In an addition to an unsymmetrical substituted alkene (or some other double-bond system containing two different substituents at each end of the double bond; e.g., **IX** with $a \neq b$ and $c \neq d$), both ends will be subject to the *re–si* system. Notice that *trans* addition to a carbon–carbon double bond may be *re–re*, *re–si*, *si–re*, or *si–si* depending on the substituents at each end of the double bond.

---

17. Reference A, Section 4-4; Reference D, pp. 90–91 and 651; D. R. Boyd and M. A. McKervey, *Quart. Rev.* **22**, 95 (1968); T. J. Leitereg and D. J. Cram, *J. Amer. Chem. Soc.* **90**, 4011, 4019 (1968); E. Ruch and I. Ugi, *Topics in Stereochemistry* **4**, 99 (1969).

10-5 Atropisomerism

Enantiomers or diastereomers which have axial or planar chirality and which are separable because of hindered rotation about a single bond are called *atropisomers*.[18] These torsional isomers (Chapter 11) are isolable because of the presence of structural elements which somehow provide some degree of rigidity to the molecule. The most common examples are biphenyls containing bulky *ortho* substituents (**XXII**) which provide large amounts of nonbonded interactions in the transition state for racemization by rotation about the bond connecting the two phenyl rings. The stability of these atropisomers is related to the bulk of the *ortho* groups.[18,19] Most appropriately substituted 2,2′,6,6′-biphenyl systems (**XXII** where none of the R's is hydrogen) are separable,[18] as are derivatives of 2,2′-diiodobiphenyl (**XXII** where $R_1 = R_3 = H$ and $R_2 = R_4 = I$); but derivatives of 2,2′-dimethylbiphenyl (**XXII** where $R_1 = R_3 = H$ and $R_2 = R_4 = CH_3$) are not. Other examples of atropisomers are compounds such as III, IV, and V in Chapter 7-3, and appropriately substituted *ansa* ("handle") compounds[18,20] with smaller rings, such as the [10]-paracyclophane **XXIII** and the [2.2]-paracyclophane **XXIV**. Interconversion of various atropisomers usually requires between 15 and 30 kcal/mole, with the lower limit set by the requirement in their definition that they be isolable at or near room temperature.

18. Reference A, Section 6-4; Reference B, pp. 78–81; and Reference C, p. 12.
19. F. H. Westheimer in *Steric Effects in Organic Chemistry*, ed. by M. S. Newman, Wiley, New York, 1956, Chapter 12.
20. B. H. Smith, *Bridged Aromatic Compounds*, Academic Press, New York, 1964.

Chapter 11

Conformational Analysis of Acyclic Systems

Conformational analysis is essentially concerned with the arrangement of the atoms of a molecule in space with respect to rotations about single bonds.[1] The structures generated by rotation are called conformational isomers, conformers, rotational isomers, or rotamers. While an infinite number of such isomers is possible, the number of conformational isomers that really occurs in each case is limited by the existence of energy maxima and minima resulting from the rotation-dependent interactions involving the various substituents and electron clouds. The nature of these interactions is still open to speculation,[2,3] yet the presence of energy minima and an energy barrier to rotation, the *torsional energy*, is consistently observed.

11-1 Ethane Systems

The presence of a torsional energy was first postulated in 1935 by Pitzer, who surmised that hindered rotation about a carbon–carbon bond

---
1. See the IUPAC rules, note 1 of Chapter 10, for a discussion of the problems involved in precisely defining *conformation* and *configuration*.
2. J. M. Lehn in *Conformational Analysis*, ed. by G. Chiurdoglu, Academic Press, New York, 1971, p. 129; L. C. Allen, *Ann. Rev. Phys. Chem.* **20**, 315 (1969); W. L. Jorgensen and L. C. Allen, *J. Amer. Chem. Soc.* **93**, 567 (1971); L. Radom, W. J. Hehre, and J. A. Pople, *J. Amer. Chem. Soc.* **94**, 2371 (1972); J. P. Lowe, *J. Amer. Chem. Soc.* **92**, 3799 (1970); L. J. Oosterhoff, *Pure Appl. Chem.* **25**, 563 (1971); and Reference C, Section 1-2.
3. J. P. Lowe, *Progr. Phys. Org. Chem.* **6**, 1 (1968).

was required to explain the unexpected value obtained for the entropy of ethane. The two most obvious conformational isomers for ethane are the *staggered* form, in which the H—C—C—H dihedral angles are 60°, and the *eclipsed* form, with 0° dihedral angles. Sawhorse diagrams and *Newman projections* of these conformers are shown in Fig. 11.1. Assuming that

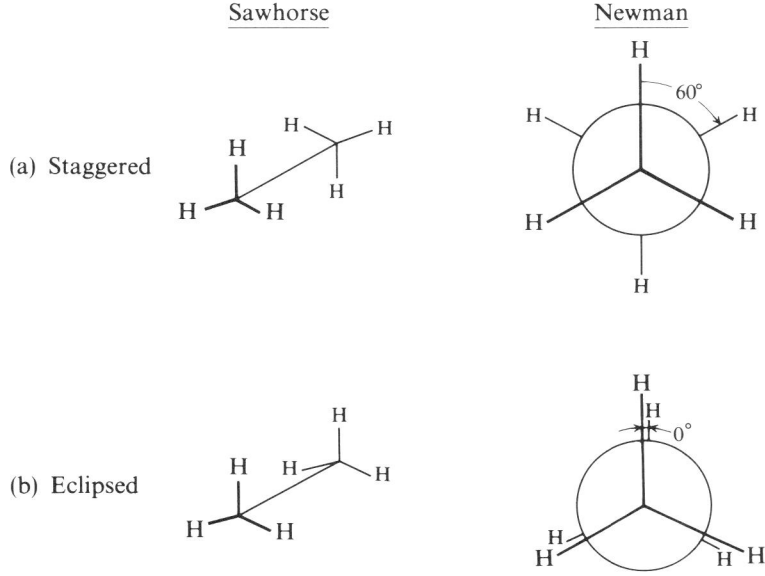

FIGURE 11.1 *The staggered* (a) *and eclipsed* (b) *conformations of ethane.*

the torsional energy can be described as a threefold cosine function (Eq. 11.1 with E_0 the barrier height and ω the dihedral angle), and assuming that the eclipsed form is the higher-energy structure and constitutes

$$E = \tfrac{1}{2}E_0(1 + \cos 3\omega) \qquad (11.1)$$

an energy maximum because of the nonbonded repulsive interactions between hydrogen in a 0° dihedral angle relationship, the change in energy as a function of the dihedral angle (or angle of rotation) may be diagrammed as in Fig. 11.2. Conformations intermediate between eclipsed and staggered forms and not representing energy maxima or minima are called *skewed* forms. The height of the barrier in ethane has been measured as 2.8 kcal/mole, suggesting that all of the infinite number of conformational isomers are present to some extent but that the staggered isomers greatly predominate.

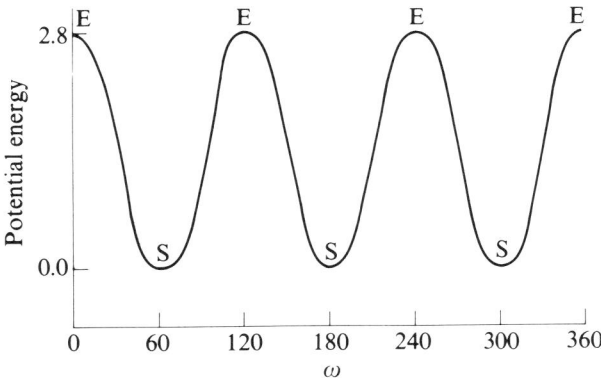

FIGURE 11.2 *The torsional energy in ethane* (E = *eclipsed conformation*, S = *staggered conformation*).

Replacing any single hydrogen in ethane with any other group produces an increase in the barrier to rotation[3] (Table 11.1). Apparently, the size of the group is a factor contributing to the torsional energy. However, the size factor must also be dependent on the distance between the eclipsing groups. For example, the barrier in ethyl iodide is smaller than that in any of the other monohaloethanes. Presumably, the length of the carbon–iodine bond is sufficient to compensate for the greater size of the iodine relative to the other halogens.

If one of the carbon atoms in ethane is replaced with some other atom which prefers a ligancy of less than four, the rotational barrier is decreased (Table 11.1). The series ethane–methylamine–methanol suggests that a lone-pair–hydrogen eclipsed interaction is approximately 0.9 kcal/mole less unfavorable than a hydrogen–hydrogen eclipsed interaction.[2]

TABLE 11.1 *Barriers to Rotation*[3]

Compound	Barrier (kcal/mole)
H_3C-CH_3	2.8
H_3C-CH_2F	3.3
H_3C-CH_2Cl	3.7
H_3C-CH_2Br	3.7
H_3C-CH_2I	3.2
$H_3C-CH_2CH_3$	3.4
H_3C-NH_2	1.9
H_3C-OH	1.1

11-2 n-Butane Systems

In 1,2-disubstituted ethanes, such as *n*-butane, there are four possible extreme conformations to consider (Fig. 11.3). Two of these conformations, the *anti* (or *trans*) and the *gauche*, are staggered rotamers, while the other two, the "unlike eclipsed" and the "like eclipsed", are eclipsed. Two *gauche* forms are possible, and these are enantiomers. Similarly, enantiomers of the "unlike eclipsed" structure would exist. Energetically, the staggered forms are more stable than the eclipsed, which are energy maxima, as might be expected from an analysis based on unfavorable eclipsed interactions (Chapter 11-1). Using similar reasoning, it can be argued that the *anti* conformer should be more stable than either of the *gauche* forms (which must be equal in energy) because the largest groups, the methyls, are the farthest apart in the *anti* conformation. This energy difference between staggered conformational isomers (or between eclipsed conformers) is called *van der Waals strain*.

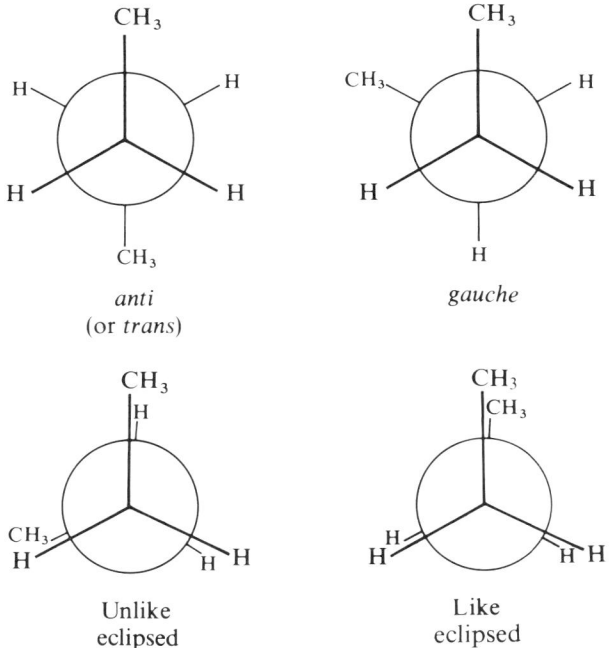

FIGURE 11.3 *Newman projections of the conformational isomers of n-butane.*

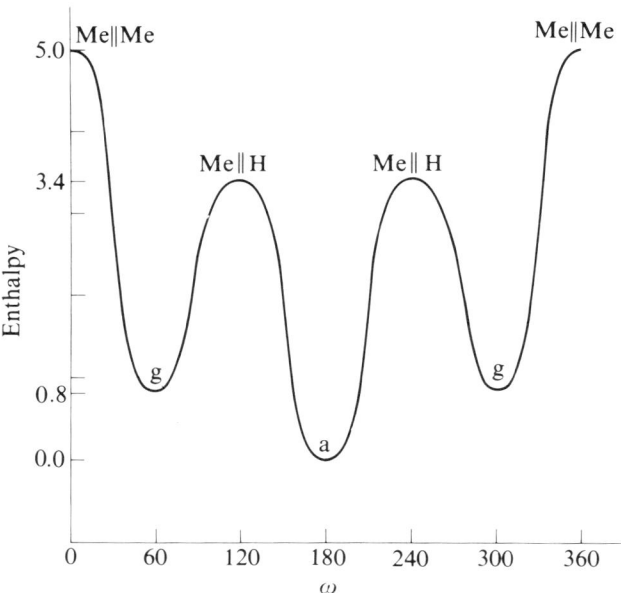

FIGURE 11.4 *Torsional enthalpy in n-butane* (a = *anti*, g = *gauche*, Me∥H = *unlike eclipsed*, Me∥Me = *like eclipsed*).

The enthalpy relationships[4] in *n*-butane are indicated in Fig. 11.4 and are obtained in a variety of ways. In this system, the more stable *anti* form is favored, but no single conformer can be isolated because of the low barriers to interconversion. The various physical properties will be either the average or the sum of the properties of the various conformers. This average or sum will be weighted according to the relative abundances of the various conformers on a mole fraction (or percentage) basis. In general, spectral properties will be of the summed type, since the spectrum results from the ability of a given instantaneous structure to absorb or emit incident radiation. Other physical properties will usually be averages of the behavior expected for each conformational isomer.

Since in most cases the exact dihedral angle is neither known nor required, Klyne and Prelog[5] have proposed a descriptive nomenclature for the relationship between two substituents on adjacent carbon atoms. One of the substituents is placed at the top of the front carbon atom in a Newman projection. The various possible dihedral angles are defined as a circle with respect to this reference point, and this circle is then divided

---

4. See Chapter 11-4 for a discussion of the entropy term.
5. W. Klyne and V. Prelog, *Experientia* **16**, 521 (1960); Reference C, p. 10.

into three pairs of zones: plus–minus, *syn–anti*, and clinal–periplanar (Fig. 11.5). The relative relationship of the two substituents is then described by the position of the second substituent in terms of these three pairs of zones. For example, the *anti* conformer of *n*-butane (Fig. 11.3) would be called $\pm$*anti*-periplanar, and the *gauche* form would be described as $-$*syn*-clinal.

This nomenclature is important since it is extremely unlikely that the dihedral angle between the methyl groups in the *gauche* form of *n*-butane will be exactly $60°$. The van der Waals repulsive forces between the methyls tend to increase this angle in an attempt to minimize the repulsion,[6] leading to a concomitant decrease in the dihedral angles between the hydrogens. The term "*gauche*" will therefore be used with reference to a conformational isomer at an energy minimum and in which the dihedral angle for the largest substituents on the two atoms about which rotation is being evaluated is between $30°$ and $90°$, whether these substituents are +*syn*-clinal or $-$*syn*-clinal.

All attempts to generalize the results from *n*-butane and state that *anti* isomers are more stable than *gauche* ones must be limited to nonpolar substituents. The presence of dipolar substituents requires consideration of the dipole interactions and introduces solvent dependence whenever the measurements are performed on solutions. Hydrogen bonding often leads to a *gauche* preference. Considerable effort has been expended in an attempt to better understand the so-called *gauche effect*,[2,7] which is a tendency of systems containing polar groups and/or electron pairs on adjacent atoms to favor that conformational isomer having these groups

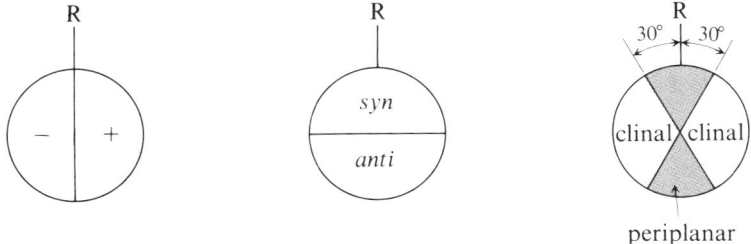

FIGURE 11.5 *The zones in the Klyne-Prelog system*[5] *for describing dihedral angle relationships* (R = *reference substituent*).

---

6. Reference C, Section 1-2; and N. L. Allinger, M. A. Miller, F. A. VanCatledge, and J. A. Hirsch, *J. Amer. Chem. Soc.* **89**, 4345 (1967).
7. Reference C, Section 1-3a; S. Wolfe, *Accounts Chem. Res.* **5**, 102 (1972); S. Wolfe, A. Rauk, L. M. Tel, and I. G. Csizmadia, *J. Chem. Soc.*, B 136 (1971); S. Wolfe, L. M. Tel, J. H. Liang, and I. G. Csizmadia, *J. Amer. Chem. Soc.* **94**, 1361 (1972).

gauche. Examples of such systems are hydrazines, peroxides, disulfides, halomethanols, 2-haloethyl acetates, dialkoxyethanes, and a whole host of related compounds[7] (Fig. 11.6).

An analysis of the results in such *gauche-effect* systems by Pople and coworkers[2] seems especially promising. This approach rationalizes molecular orbital calculations performed on these systems in terms of contributions from three principal effects. The first is the usual threefold torsional energy favoring staggered conformations (Chapter 11-1) and resulting from some form of bond–bond or group–group repulsion. The second effect is a twofold component in which the axis of a lone-pair orbital prefers to be coplanar with an electron-withdrawing polar group and/or perpendicular to a lone pair on the adjacent atom. The third effect, a onefold component, is caused by the preference of local dipoles at the ends of the rotation axis to be antiparallel to each other. This third component is illustrated by the usual tendency of two magnets connected by similar poles (e.g., positive poles) to have the other similar poles (the negative poles) as far from each other as possible, and represents the typical picture of dipolar interactions (Chapter 4-2).

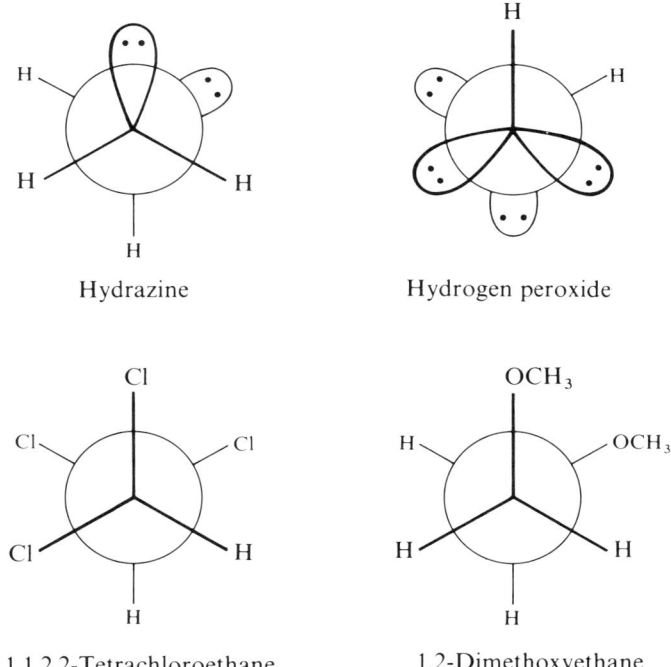

FIGURE 11.6 *Selected systems in which gauche conformations are preferred.*[7]

11-3 Unsaturated Systems

Rotational isomerism about sp^2–sp^3 carbon–carbon single bonds[8] poses new problems. In unsubstituted examples of such systems, two rotamers (**I** and **II**) are reasonable, one of which (**I**) involves a single bond eclipsing the double bond and the other of which (**II**) involves a single

H   X   H   X
H⋯⧸⧹⧸   H–⧸⧹⧸
H   Y   H   Y
I   **II**

bond eclipsing the other single bond at the sp^2 center. Since the key structural unit is the double bond, the former conformer is called the eclipsed isomer and the latter is called the *bisected* isomer.[9] Somewhat surprisingly, the eclipsed conformational isomer is favored enthalpically by 1–2 kcal/mole over the bisected one in all of the unsubstituted systems[8] (e.g., acetaldehyde,[2] acetyl halides, acetic acid, acetate esters, acetone, propene, and N-alkylacetaldimine).

If the unsaturated system contains a nonhydrogen substituent at the sp^3 center, four rotamers are possible (**III–VI**), two of which (**III** and **IV**)

R   X   H   X   H   X   H   X
H⋯   H⋯   R–   H–
H   Y   R   Y   H   Y   R   Y
III   **IV**   **V**   **VI**

are eclipsed and two of which (**V** and **VI**) are bisected. In propionaldehyde ($R = CH_3$, $X = O$, $Y = H$), the rotamer with the methyl group eclipsed (**III**) is favored by 0.8–1.0 kcal/mole in enthalpy over the other eclipsed rotamer (**IV**), which is itself about 1.0 kcal/mole more stable than the bisected rotamers (**V** and **VI**). On the other hand, in 1-butene[8] ($R = CH_3$, $X = CH_2$, $Y = H$), the conformer with a hydrogen eclipsing the double bond (**IV**) appears to be about 0.4 kcal/mole in enthalpy more stable than the other eclipsed conformer (**III**). Presumably, the nonbonded repulsions between the methyl and the CH_2 in conformer **III** of 1-butene are greater (or the attractions are less) than those involving the methyl and the oxygen in the same conformer of propionaldehyde.[8] Alternatively, through the *gauche* effect (Chapter 11-2), this may be an example of the stabilization of conformer **III** of the aldehyde, since rotamers of type **III** show increased relative stabilities over those of type **IV** in all of those systems where R or

---

8. Reference C, Section 1-3b; G. J. Karabatsos and D. J. Fenoglio, *Topics in Stereochemistry* **5**, 167 (1970); D. F. Ewing, *J. Chem. Soc., Perkin II* 701 (1972).
9. The double bond more or less bisects the angle defined by two of the bonds at the sp^3-hybridized center.

X has unshared or polarizable electrons.[8] The explanation based on nonbonded repulsion is supported by Riddell and Robinson,[10] since their work indicates that the presence of the double bond is the only significant factor and that the electronegativity of the X group plays no role.

The above discussion has pointed out the importance of the nature of R and X on the rotational isomerism and relative stabilities of **III** and **IV**. If at least one of these groups is not carbon or hydrogen, the rotamer stabilities are dependent on the dielectric constant of the medium because of the presence of dipolar interactions.

In addition, the rotamer distribution is a function of the sizes of R and Y. The conformer of type **III** is decreased in stability as the R group becomes larger. In the monosubstituted aldehydes, conformer **III** is favored by 0.8 kcal/mole when R is methyl, by 0.7 kcal/mole when R is ethyl, by 0.6 kcal/mole when R is *n*-propyl, and by 0.4 kcal/mole when R is isopropyl; and it is disfavored by 0.25 kcal/mole when R is *tert*-butyl.[8] Similar behavior is observed as the sp^3 center becomes disubstituted or trisubstituted with progressively larger groups, although rotamers of type **IV** where hydrogen or the smaller group eclipses the double bond become favored at an earlier stage of size increase. The same trends are observed in the 3-substituted propenes,[8] with the conformer of type **IV** more stable in all cases.

Few studies have been performed to investigate the effect of the size of Y on rotamer distribution, but increased size at this position would be expected[8] to destabilize all bisected forms and to destabilize those conformers of type **IV** which have an R of at least some critical size.

Rotational isomerism also exists about sp^2–sp^2 bonds. The principal planar conformers are the *s-trans* and the *s-cis*, although extensive amounts of nonplanar forms are sometimes observed. The planar forms would be

The *s-trans* conformer   The *s-cis* conformer

expected to predominate since they are stabilized by maximal π-electron delocalization[11] (Chapter 1-5). This prediction is supported by evidence[12] that *s-trans*-1,3-butadiene is favored over the *s-cis* by 2.3 kcal/mole and that the barrier to rotation via a species in which the double bonds are

---

10. F. G. Riddell and M. J. T. Robinson, *Tetrahedron* **27**, 4163 (1971).
11. There have been proposals that this delocalization is more efficient in the *s-trans* conformation.
12. Reference C, pp. 22–23; A. A. Bothner-By and D. Jung, *J. Amer. Chem. Soc.* **90**, 2342 (1968), and references therein.

orthogonal is about 5 kcal/mole. It has also been shown[12] that substituents in the 1- and 3- (or 2- and 4-) positions of the diene system destabilize the *s-trans* conformers because of steric repulsions, leading to large amounts of skewed nonplanar conformations.

The *s-trans* and *s-cis* forms are more comparable in energy in *trans*-3-penten-2-one,[12] suggesting that at least part of the energy difference in the butadienes is the result of nonbonded repulsive interactions between substituents at the ends of the diene system (positions 1 and 4) in the *s-cis* rotamer. Other interactions must also be significant, since infrared evidence[13] indicates that while the *s-trans* form is favored by about 0.5 kcal/mole enthalpy in methyl vinyl ketone and *trans*-3-penten-2-one, it is almost the exclusive conformer in the correspondingly substituted aldehydes.

The conclusion must be reached that the interactions in unsaturated systems are at best poorly understood. While further experimental results are required, it is conceivable that true understanding must await improved quantum mechanical techniques.

11-4 Conformational Entropy

Informative as it may be to know the enthalpy differences between conformational isomers, the actual proportions of these isomers present in a sample are dependent on the free-energy differences (Eq. 11.2). Therefore, if we want to know the isomer proportions, we must evaluate the entropy

$$\Delta G° = \Delta H° - T \Delta S° \tag{11.2}$$

terms.[14] In general, the entropy of a given structure is composed of vibration, rotation, and translation terms. For an isomerization, be it conformational or configurational, all terms will cancel to a first approximation except those related to entropy of mixing and entropy of "symmetry number".

The entropy of mixing is expressed in terms of N_i, the mole fraction of the *i*th component, as shown in Eq. 11.3,

$$S_{\text{mix}} = -R \sum_i N_i \ln N_i \tag{11.3}$$

where R is the gas constant.

Symmetry numbers enter the picture whenever an isomer has an element of rotation symmetry. The *symmetry number*, σ, is the number of indistinguishable spatial arrangements a molecule can have as the result of intramolecular rotations. A symmetric molecule will have less rota-

---
13. A. J. Bowles, W. O. George, and W. F. Maddams, *J. Chem. Soc.*, B 810 (1969).
14. Reference A, pp. 214–215; Reference C, pp. 10–11 and Section 1-4.

tional entropy than an asymmetric molecule, so the symmetry-number entropy is really part of the rotational entropy which is not canceled in comparing the isomers. The symmetry-number entropy is expressed as

$$S_\sigma = -R \ln \sigma \qquad (11.4)$$

with typical values of σ being 2 for *n*-propane, 2 for *cis*-2-butene, and 3 for isobutane.

Having considered these entropy terms, it is useful to evaluate a typical conformational equilibrium, such as the composition of *n*-butane at 25°. In the conformational equilibrium

$$anti \xrightleftharpoons{K} gauche$$

the *anti* form is enthalpically favored by 0.8 kcal/mole ($\Delta H° = +0.8$ kcal/mole; Chapter 11-2). Since there is only one *anti* conformation, $S_{mix}^{anti} = 0$. However, the *gauche* form is a *dl* pair (p. 237) which, being inseparable enantiomers, must be present in equal amounts: $N_d = N_l = 0.5$. The entropy of mixing of the *gauche* system would therefore be calculated (Eq. 11.3) as:

$$\begin{aligned}S_{mix}^{gauche} &= -R(0.5 \ln 0.5 + 0.5 \ln 0.5) \\ &= -R \ln 0.5 \\ &= +R \ln 2 \\ &\cong 1.4 \text{ eu}\end{aligned}$$

Both the *anti* form and the *gauche* form have symmetry numbers of 2, so there would be no entropy difference from this term. The free-energy difference for this equilibrium can then be evaluated using Eq. 11.2:

$$\begin{aligned}\Delta G° &= 800 \text{ cal/mole} - 298°\text{K } (1.4 \text{ eu}) \\ &= 400 \text{ cal/mole}\end{aligned}$$

Since $\Delta G° = -RT \ln K$, the equilibrium constant is found to be approximately 1/2. Since $K = [gauche]/[anti]$, there will be about two molecules of *anti* conformer per molecule of *gauche* conformer (or the mixture will be about 33% *gauche* and 67% *anti*).

11-5 Analysis of Configurational Equilibria

Conformation often exerts a considerable influence on configurational equilibria and on reactivity, the latter being affected both in reaction rates and in stereochemistry of products. Evaluation of a configurational equilibrium requires prior evaluation of the overall enthalpy and entropy of each configurational isomer, which itself requires some information about the relative stabilities of the individual conformational isomers.[14]

Since these relative stabilities are evaluated primarily in terms of *gauche* interactions, it must be assumed that all free-energy differences between the configurational isomers result from these interactions only.

A useful example of an analysis of a configurational equilibrium is the chemical equilibration of *meso-* and *dl-*2,3-dibromobutane (Eq. 11.5).

$$\begin{array}{ccc}
\text{CH}_3 & & \text{CH}_3 \\
| & & | \\
\text{H}-\text{C}-\text{Br} & \rightleftharpoons & \text{H}-\text{C}-\text{Br} \\
| & & | \\
\text{H}-\text{C}-\text{Br} & & \text{Br}-\text{C}-\text{H} \\
| & & | \\
\text{CH}_3 & & \text{CH}_3 \\
meso\text{-Dibromobutane} & & dl\text{-Dibromobutane}
\end{array}$$
(11.5)

The *meso* compound and each of the enantiomers consist of three conformational isomers, structures **VII–IX** for the *meso* and **X–XII** for one of the enantiomers.

[Newman projections VII, VIII, IX, X, XI, XII]

Assuming that methyl–methyl *gauche* interactions are 0.8 kcal/mole less favorable than the *anti*, that the bromine–bromine value is 0.7 kcal/mole, and that the methyl–bromine value is 0.2 kcal/mole, the relative enthalpies of these six conformations can be calculated first:

$$\begin{aligned}
H_{\text{VII}} &= 2(0.2) & &= 0.4 \\
H_{\text{VIII}} &= 0.8 + 0.7 + 0.2 &&= 1.7 \\
H_{\text{IX}} &= 0.8 + 0.7 + 0.2 &&= 1.7 \\
H_{\text{X}} &= 0.8 + 0.7 &&= 1.5 \\
H_{\text{XI}} &= 0.8 + 2(0.2) &&= 1.2 \\
H_{\text{XII}} &= 0.7 + 2(0.2) &&= 1.1
\end{aligned}$$

The mole fraction of each conformational isomer in the *meso* compound at 25°C is determined by first calculating the mole-fraction ratios using Eq. 11.6,

$$\frac{N_j}{N_k} = e^{-(H_j - H_k)/RT} \tag{11.6}$$

producing $N_{VIII}/N_{VII} = N_{IX}/N_{VII} = 0.11$. Since the total of the mole fractions must equal 1, normalization to this basis by $N_i / \sum_i N_i$ gives $N_{VII} = 0.82$, $N_{VIII} = 0.09$, and $N_{IX} = 0.09$. The enthalpy of the *meso* isomer is therefore:

$$H_{meso} = \sum_i N_i H_i$$
$$= 0.82(0.4) + 0.09(1.7) + 0.09(1.7)$$
$$= 0.63 \text{ kcal/mole}$$

A similar procedure for one of the enantiomers leads to $N_X = 0.22$, $N_{XI} = 0.36$, $N_{XII} = 0.42$, and $H_{enan} = 1.22$ kcal/mole.

In order to evaluate the free energy of each of the configurational isomers, the entropies of mixing and of symmetry number must be determined (Chapter 11-4). The entropies of mixing may be obtained using Eq. 11.3, giving $S_{mix}^{meso} = 1.18$ eu and $S_{mix}^{enan} = 2.12$ eu. Since structures X–XII represent only one of the enantiomers, a value of $R \ln 2$ (or 1.38 eu) must be added to the entropy of mixing of the enantiomer to convert it to the required entropy of mixing of the racemate: $S_{mix}^{dl} = 3.50$ eu. Symmetry numbers are identical in the configurational isomers and cancel.

The free energies of each configurational isomer resulting from group interactions can now be evaluated at 25°C. For the *meso* compound in Eq. 11.5, $G_{meso} = 630 - 298(1.18) = 278$ cal/mole, while the *dl* compound gives $G_{dl} = 1220 - 298(3.50) = 177$ cal/mole. The *dl* isomer would therefore be expected to be approximately 100 cal/mole more stable than the *meso* system. As mentioned earlier (p. 239), the result is applicable only to certain solvents since the bromine–bromine *gauche* interaction and the methyl–bromine *gauche* interaction are both solvent-dependent. It must also be understood that this calculation depends on the assumption that the diastereomers have the same enthalpies and entropies other than those resulting from *gauche* interactions.

11-6 F-strain and B-strain

The conformational analysis of acyclic systems revolves around the concepts of torsional strain and van der Waals strain. Both of these strains probably contain a component dependent on some type of repulsive force, be it between nonbonded atoms or between electron clouds.

Two other types of strain phenomena have been analyzed, primarily by H. C. Brown, and considered to have general validity. The first of these is *F-strain*,[15,16] where the "F" stands for face, front, or frontal. When two molecules combine to form a chemical bond, primarily in the sense of a Lewis acid combining with a Lewis base, steric repulsions can occur between the substituents in the two molecules. For example, consider the combination of a boron compound with an amine to form the addition compound:

$$R_3N: + BR'_3 \xrightleftharpoons{K} R_3N:BR'_3$$

For a given Lewis acid, the equilibrium constant would be expected to increase as the base strength of the amine increased (see Chapters 4-4 and 8-7). With trimethylboron and the methylamines, the electron-donating base-strengthening effects of methyl groups (Chapter 4-4) should lead to an order of equilibrium constants of $NH_3 < CH_3NH_2 < (CH_3)_2NH < (CH_3)_3N$. Experimentally, this order was observed except that the equilibrium constant was found to decrease when going from the dimethylamine system to the trimethylamine. Evaluation of the ethylamines exposed this reversal in order at an earlier stage: $NH_3 < CH_3CH_2NH_2 > (CH_3CH_2)_2NH > (CH_3CH_2)_3N$. An even more drastic result was observed in the *tert*-butyl series: $NH_3 > (CH_3)_3CNH_2$. Identical trends were obtained when the boron compounds were varied and trimethylamine was used as the base. All of these results are consistent only if steric hindrance intervenes in bond formation in the transition state whenever some value for the total steric bulk on the faces of the molecules being joined is exceeded.

The idea that F-strain is steric in nature has been put to use in the study of steric kinetic isotope effects (Chapter 7-3). Relative equilibrium constants for 2,6-dimethylpyridine and the corresponding compound having completely deuterated methyl groups with trimethylboron and the perdeuterated analog[17] provide some of the most dramatic evidence for the smaller effective steric bulk of deuterium compared to hydrogen.

Brown also attempted to explain the trends encountered in the basicities of the alkylamines.[15] The proton is such a small acid that F-strain would not be a viable concept for explaining these systems. The strain postulated in its place was called *B-strain*[15,18] for "back" strain. Brown

---

15. H. C. Brown, *J. Chem. Soc.* 1248 (1956), and *J. Chem. Educ.* **36**, 424 (1959); V. Gold, *Progr. Stereochem.* **3**, 169 (1962).
16. Reference D, p. 229; J. Hine, *Physical Organic Chemistry*, 2nd ed., McGraw-Hill, New York, 1962, pp. 64–65.
17. H. C. Brown and G. J. McDonald, *J. Amer. Chem. Soc.* **88**, 2514 (1966); H. C. Brown, M. E. Azzaro, J. G. Koelling, and G. J. McDonald, *J. Amer. Chem. Soc.* **88**, 2520 (1966).
18. H. C. Brown, *Science* **103**, 385 (1946).

suggested that steric congestion between the three groups attached to the nitrogen of the amine would be increased on the formation of a fourth bond to any species, however small. Later work[15-17] refuted this idea by establishing that the bond angles in the amines and in the ammonia salts are almost identical, thereby ruling out any increased steric congestion. The trends in amine basicities, it was proposed, are the result of differential solvation effects,[19] an interpretation which is supported by basicities determined in the gas phase (Chapter 4-4).

However, another type of system for which B-strain was proposed[18] turns out to require this type of steric-congestion concept. Strain in a bulky alkyl halide provides steric assistance to S_N1 reactivity (as measured by increased reaction rates).[15,20] In the rate-determining ionization process, the hybridization is changed from sp^3 to sp^2 and the bond angles are increased from tetrahedral to trigonal (from 109.5° to 120°). Repulsions between the substituents in a tertiary system would be relieved by this increase in the bond angles. The destabilization resulting from these repulsions in the alkyl halide (or other leaving group) would not be present in the carbonium ion intermediate, and the reaction would be accelerated. The presence of B-strain may be noted by changing the substitution pattern until the reaction rate shows a sudden increase greater than the increase associated with the electronic effects of the substituents (Chapter 4). Anchimeric assistance resulting from neighboring-group participation must be eliminated as an explanation for this rate increase by demonstrating lack of rearrangements, and seemingly has been.[21] Typical examples requiring the postulation of B-strain are shown in Table 11.2.

TABLE 11.2 *B-strain in Solvolysis Rates*[20,21]

Compound	Relative Rate
t-BuCl	1.0
$(CH_3)_2$(neopentyl)CCl	21
CH_3(neopentyl)$_2$CCl	580
t-BuONs	1.0
(*t*-Bu)$_3$CONs	13,000
(*t*-Bu)$_2$(neopentyl)CONs	19,000
(*t*-Bu)(neopentyl)$_2$CONs	68,000

---

19. Reference D, pp. 229–230; notes 26–28 in Chapter 4.
20. Reference D, pp. 127, 233–234, 282; A. Streitwieser, Jr., *Solvolytic Displacement Reactions*, McGraw-Hill, New York, 1962, p. 93.
21. V. J. Shiner and G. F. Meier, *J. Org. Chem.* **31**, 137 (1966); P. D. Bartlett and T. T. Tidwell, *J. Amer. Chem. Soc.* **90**, 4421 (1968).

Chapter 12

Conformational Analysis of Cyclic Systems

In addition to the various factors influencing conformational preferences in acyclic systems (Chapter 11), angle strain must be considered in cyclic molecules. The most frequently encountered nonaromatic ring system is that of cyclohexane, which is often present in various guises in many diverse families of natural products. For this reason, the cyclohexane system has been extensively studied,[1] is very well understood, and, therefore, provides a useful starting point for the conformational analysis of other cyclic systems.

12-1 Cyclohexane

If the cyclohexane ring were planar, the bond angles would be 120° and each hydrogen would be eclipsed with a vicinal hydrogen on each side. Since tetrahedral bond angles are strain-free and staggered conformations are more favorable than eclipsed ones, it is to be expected that the cyclohexane ring will pucker and exist in some nonplanar form. Angle strain is minimized in either a rigid form with D_{3d} symmetry called the *chair* form (Fig. 12.1), or various kinds of flexible forms. The chair conformation has

---

[1]. Reference A, Chapter 8; Reference B, pp. 76–78; Reference C, Chapter 2; Reference D, pp. 104–108; R. T. Morrison and R. N. Boyd, *Organic Chemistry*, 3rd ed., Allyn and Bacon, Boston, 1973, Chapter 9.

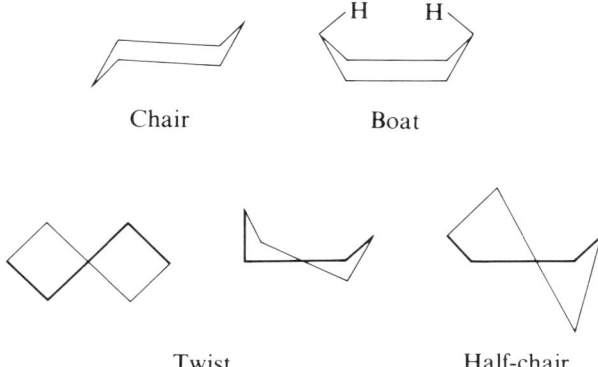

FIGURE 12.1 *Conformations of cyclohexane.*

roughly tetrahedral angles (111.1°) and almost perfectly staggered bonds, suggesting that this conformation will be extremely stable since angle strain and torsional strain are minimal. Electron diffraction studies[1,2] have verified that the chair conformation is the overwhelmingly predominant one for cyclohexane.

The bonds to hydrogen in chair cyclohexane may be divided into two sets.[1,2] At each carbon atom one of the bonds to hydrogen more or less lies in a plane defined by four of the carbons or parallel to this plane. These bonds (Fig. 12.2) are called *equatorial*. The other

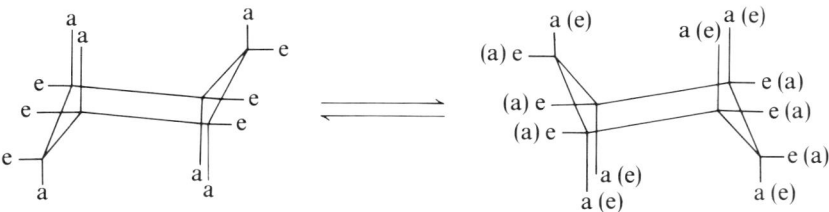

FIGURE 12.2 *Ring-flipping in cyclohexane. The equatorial or axial nature of each hydrogen is indicated by " e " or " a ", with the symbols in parentheses in the structure on the right indicating the nature of that hydrogen in the structure on the left prior to the ring-flipping.*

---

2. The original 1943 article by O. Hassel is now available in English translation in *Topics in Stereochemistry* **6**, 11 (1971); slight geometrical modifications have been proposed from a recent reexamination by H. R. Buys and H. J. Geise, *Tetrahedron Lett.* 2991 (1970).

set of six bonds, which are perpendicular to this plane, are called *axial* (Fig. 12.2). The diequatorial and equatorial–axial pairs of hydrogens on adjacent carbons and the diaxial hydrogens in a 1,3 relationship with respect to their carbon atoms (i.e., one carbon atom intervenes) are the same distance apart as staggered hydrogens in ethane, a distance usually considered to be very favorable and to involve attractive van der Waals forces.

The axial and equatorial hydrogens on a given carbon atom in cyclohexane are interconverted by ring-flipping[1] (Fig. 12.2). This process of chair inversion proceeds rapidly at room temperature (10^4–10^5 per second), and the proton magnetic resonance spectrum contains only one line, thereby implying the equivalence of all of the hydrogens. The ring-flipping process requires that the cyclohexane ring be distorted into a *half-chair* conformation (Fig. 12.1), which is an enthalpy maximum (10.8 kcal/mole above the chair, Fig. 12.3) where four of the carbons are coplanar, and beyond which are the flexible conformations,[1] where the axial and equatorial positions may be quickly interchanged. The D_2 *twist* forms (or skewed boat forms) (Fig. 12.1) represent the enthalpy minima for the flexible series (Fig. 12.3). In the twist conformation, which is 5.4 kcal/mole less stable than the chair, some angle strain and torsional strain are present. However, the total strain in the twist is less than that in the *boat* (Fig. 12.1), which is the C_{2v} conformation through which twist forms are interconverted. In the boat form, angle strain is fairly unimportant, but four pairs of hydrogens are eclipsed and a repulsive "flagpole interaction" exists between the hydrogens at the points of the boat pointing toward the center. The boat is therefore an energy maximum about 1.2 kcal/mole less stable than the twist conformation.

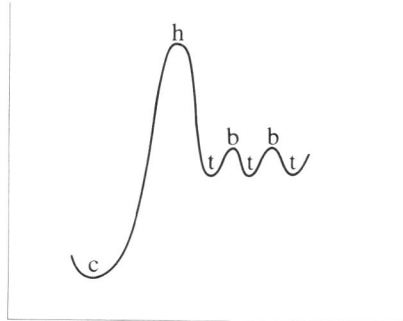

FIGURE 12.3 *Relative enthalpies of cyclohexane conformers* (c = *chair*, h = *half-chair*, t = *twist*, b = *boat*).

If the proton magnetic resonance spectrum of cyclohexane is obtained[3] at successively lower temperatures, the signal broadens and then (at approximately $-65°C$) begins to split into two signals; each becomes somewhat sharper as the temperature is further lowered. These observations are consistent with the slowing of ring-flipping at the lower temperatures below the speed of the act of measurement, so that two distinct diastereotopic families of hydrogens are observed, one equatorial and the other axial. The free-energy barrier to ring inversion of 10.3 kcal/mole is obtained from the coalescence temperature, the temperature at which the rate of an exchange phenomenon is comparable to the frequency separation of the components in the absence of exchange.

12-2 Substituted Cyclohexanes

In a monosubstituted cyclohexane,[1] two diastereomeric chair forms are possible, one with the substituent equatorial and the other with the substituent axial. These isomers can be interconverted by ring-flipping through a twist form, and their relative stability is a function of the substituent. Almost every substituent prefers to be equatorial,[1,4] presumably because of repulsive 1,3-diaxial interactions whenever the substituent is axial. Consider the ring-flipping equilibrium shown in Eq. 12.1. The axial isomer (II) contains two interactions of the *gauche*-

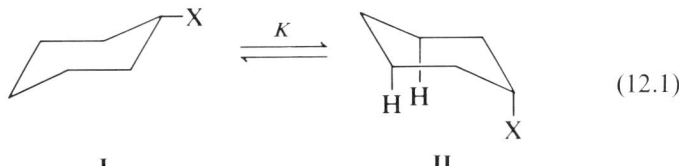

$$\text{I} \qquad\qquad\qquad \text{II} \qquad\qquad (12.1)$$

butane type (Chapter 11-2) which are not present in the equatorial isomer (I), and is therefore less stable than the equatorial form. If X is a methyl group, the equatorial isomer might be expected to be favored by 2×0.8, or 1.6, kcal/mole. The experimental value is approximately 1.7 kcal/mole.[4] Table 12.1 lists estimated best values for the free-energy differences for the equilibrium of Eq. 12.1.

---

3. In order to avoid extensive broadening from spin–spin coupling effects, a cyclohexane containing eleven deuteriums was employed by F. A. L. Anet and A. J. Bourn, *J. Amer. Chem. Soc.* **89**, 760 (1967), and other groups, whose work is referenced therein.

4. For a table of free-energy preferences, see J. A. Hirsch, *Topics in Stereochemistry* **1**, 199 (1967). For a newer consistent set of experimental results, see F. R. Jensen, C. H. Bushweller, and B. H. Beck, *J. Amer. Chem. Soc.* **91**, 344 (1969), and F. R. Jensen and C. H. Bushweller, *Advan. Alicycl. Chem.* **3**, 139 (1971).

TABLE 12.1 *Approximate Free-energy Differences in Monosubstituted Cyclohexanes*[4] (Eq. 12.1)

Substituent	$\Delta G_x°$ (kcal/mole)
Methyl	1.7
Ethyl	1.75
Isopropyl	2.1
Fluoro	0.15–0.27
Chloro	0.4–0.5
Bromo	0.4–0.5
Hydroxy	0.5,[a] 0.87–0.93[b]
Amino	1.2,[a] 1.6[b]
Carboethoxy	1.0–1.2

[a] Nonpolar solvents. [b] Polar solvents.

Many substituents are profoundly sensitive to concentration and solvent effects. In general, the preference of alkyl groups for the equatorial position is dependent on size, while the preference of polar groups does not appear to be overly sensitive to size. This may be another manifestation of the barrier-height trends in substituted ethanes[5] (Chapter 11-1). The ring carbon–polar group bond length increases as the polar group becomes larger, so a distance dependence (as expected for a non-bonded repulsive force) counteracts the size effect.[1] On the other hand, the ring carbon–alkyl group carbon-bond length should be fairly constant, leaving the size of the group as the dominant variable.[1] The key factors in these conformational preferences are therefore steric bulk and bond lengths, as well as the shape of the substituent and sometimes its polarizability.

The energy required to convert an axial isomer (**II**) into an equatorial one (**I**) by ring-flipping is approximately 10 kcal/mole, just as for the unsubstituted cyclohexane. At room temperature, sufficient energy is present for the relative amounts of the isomers to be governed only by their relative stabilities (Table 12.1). However, by working at lower temperatures, Jensen and Bushweller[6] have been able to freeze out the ring inversion process and to isolate pure equatorial isomers of chlorocyclohexane (**I**, X = Cl) and trideuteriomethoxycyclohexane (**I**, X = OCD_3) as solids and in solution.

Equatorial–axial free-energy differences have been determined by a variety of methods,[4] the most important classes of which are equilibration,

---
5. H. R. Buys and E. Havinga, *Tetrahedron Lett.* 3759 (1968); J. E. Anderson and H. Pearson, *Chem. Commun.* 871 (1971).
6. F. R. Jensen and C. H. Bushweller, *J. Amer. Chem. Soc.* **88**, 4279 (1966), and **91**, 3223 (1969).

kinetic methods, and physical methods (see Chapter 12-6). In all cases, either more highly substituted rings are utilized directly or as models, or low-temperature studies are performed. The use of 1,4-disubstituted systems with a *tert*-butyl holding group as one substituent was pioneered by Winstein and Holness in 1955.[1] The bulky *tert*-butyl group acts as a "conformational bias", having an almost overwhelming tendency (4–5 kcal/mole) to remain in an equatorial position.[1] For example, the base-catalyzed equilibration (Eq. 12.2) of the *cis*- and *trans*-4-*tert*-butyl-1-carboethoxycyclohexanes (III and IV, respectively) could be measured directly by a variety of techniques. Assuming that the tendency of the *tert*-butyl group to be equatorial prevents contributions from ring-flipped forms V and VI and that this group has no effect on the ring or on the position of the equilibrium, the position of the equilibrium could be considered to directly measure the equatorial preference of the carboethoxy group. Unfortunately, it has been shown that the *tert*-butyl group

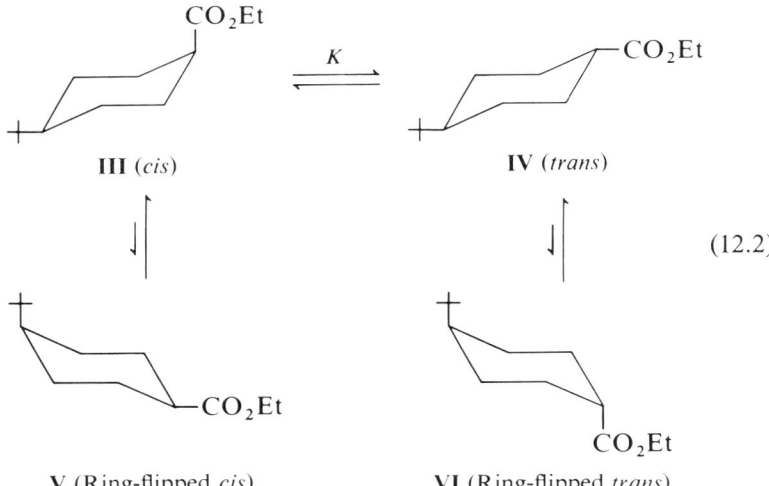

III (*cis*)   IV (*trans*)

(12.2)

V (Ring-flipped *cis*)   VI (Ring-flipped *trans*)

does influence the shape of the ring and the position of the equilibrium (as well as most of the physical and chemical properties of the carboethoxy substituent—or any other substituent) to at least some slight extent.[7] Nevertheless, in most instances the influence is almost negligible and the approximation acceptable if "ballpark" results are all that is desired. Sicher[1] has used the rigid *trans*-2-decalyl system (Eq. 12.3) as a model.

---

7. S. Wolfe and J. R. Campbell, *Chem. Commun.* 872 (1967); F. R. Jensen and B. H. Beck, *J. Amer. Chem. Soc.* **90**, 3251 (1968); R. Parthasarathy, J. Ohrt, H. B. Kagan, and J. C. Fiaud, *Tetrahedron* **28**, 1529 (1972); P. L. Johnson, et al., ibid., **28**, 2893, 2901 (1972); and references therein.

While this type of approach seems slightly better than the use of a *tert*-butyl group, it may be open to the same criticism.

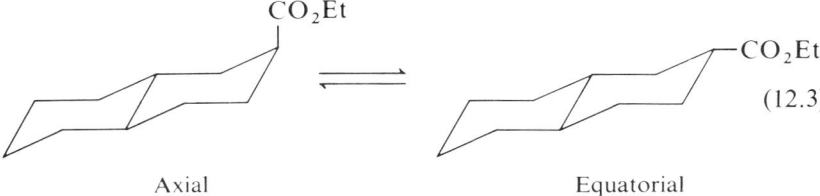

In di- or polysubstituted cyclohexanes, the rule is that the favored conformer will have the most groups equatorial and/or the largest groups equatorial. The free-energy differences of Table 12.1 may often be added as a first approximation, although additivity fails with geminal substituents and often with vicinal substituents.[1,4] If two or more polar substituents are present, dipolar interactions must also be considered.[8]

The 1,2-, 1,3-, and 1,4-disubstituted cyclohexanes each exist in *cis* and *trans* forms. If the substituents are identical, none of the *cis* isomers exhibits optical activity, nor does the *trans*-1,4 system. Analysis of the configurational equilibrium between each pair of *cis–trans* isomers of the dimethylcyclohexanes provides a useful format for analysis of both the configurational differences and the conformational situation in each isomer.

Looking first at the 1,4-dimethylcyclohexanes (Eq. 12.4), the *trans* isomer exists predominantly as a diequatorial conformation (**VII**). The

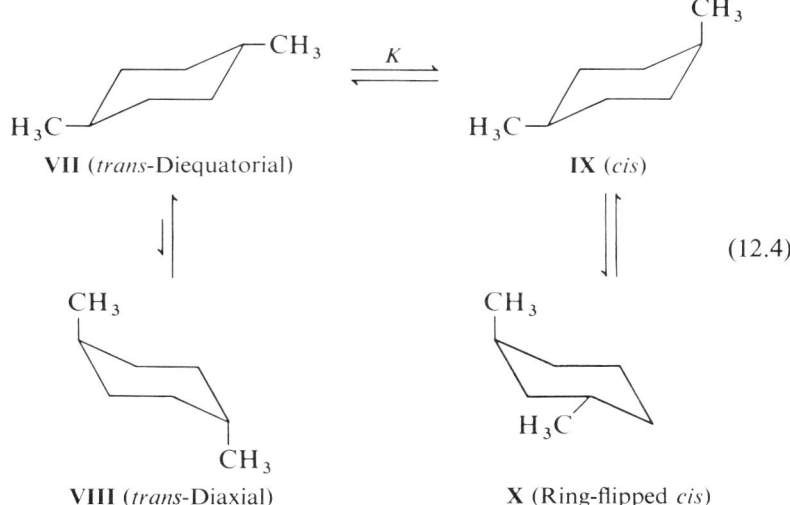

---

8. M. J. T. Robinson, *Pure Appl. Chem.* **25**, 635 (1971).

diaxial conformer (**VIII**), having four *gauche*-butane interactions (two axial methyl groups), is therefore approximately 3.4 kcal/mole higher in enthalpy than the diequatorial and so may be ignored. The *cis* isomer exists as two conformations (**IX** and **X**) which are superimposable and which contain two *gauche*-butane interactions (1.7 kcal/mole enthalpy). Both the *trans*- and the *cis*-1,4-dimethylcyclohexanes are achiral. The *trans* isomer has a symmetry number of 2 (Chapter 11-4) for an entropy of symmetry of $-R \ln 2$ (Eq. 11.4), as well as a small entropy of mixing from the diequatorial–diaxial conformational equilibrium which can be ignored as a first approximation. The free-energy difference for the equilibrium of Eq. 12.4 may then be calculated using Eq. 11.2 as:

$$\Delta G° = \Delta H° - T \Delta S°$$
$$= 1.7 \text{ kcal/mole} - 298(1.4)/1000 \text{ kcal/mole}$$
$$\cong 1.3 \text{ kcal/mole}$$

In the 1,3-dimethylcyclohexanes (Eq. 12.5), the *cis* system is the more stable. The most important conformation is diequatorial (**XI**), the diaxial (**XII**) being destabilized by two *gauche*-butane interactions and an ex-

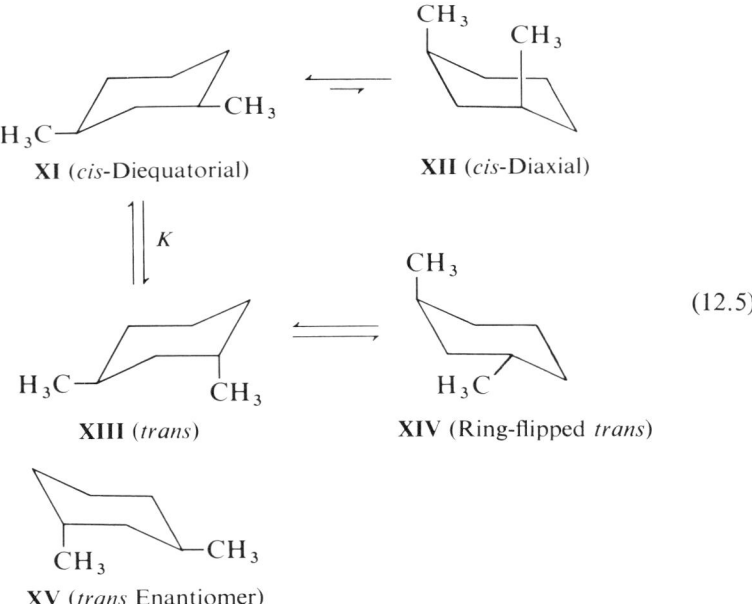

(12.5)

tremely unfavorable methyl–methyl diaxial interaction for a total of about 5.5 kcal/mole. The *cis* isomer is achiral. In the *trans* isomer, one methyl is equatorial and the other axial, so this isomer is 1.7 kcal/mole less stable enthalpically than the *cis* compound. Ring-flipping converts a *trans* conformer (**XIII**) into itself (**XIV**); and, since a given *trans*-1,3-dimethylcyclohexane conformer is chiral, the *trans* system is a potentially resolvable *dl* mixture (**XIII, XV**) for which a term for entropy of mixing of $R \ln 2$ must be taken into account. The free-energy difference in the configurational equilibrium of Eq. 12.5 can be calculated as:

$$\Delta G° = 1.7 \text{ kcal/mole} - 298(1.4)/1000 \text{ kcal/mole}$$
$$\cong 1.3 \text{ kcal/mole}$$

The 1,2-dimethylcyclohexanes (Eq. 12.6) remain to be considered. The *trans* isomer possesses a diequatorial form (**XVI**) and a diaxial form (**XVII**), the former of which has one *gauche*-butane interaction (between the methyl groups themselves), and the latter of which has four. The diequa-

$$\underset{\textbf{XVI }(\textit{trans}\text{-Diequatorial})}{\text{[structure]}} \overset{K}{\rightleftharpoons} \underset{cis}{\text{[structure]}}$$

$$\updownarrow \qquad\qquad \updownarrow \qquad\qquad (12.6)$$

$$\underset{\textbf{XVII }(\textit{trans}\text{-Diaxial})}{\text{[structure]}} \qquad \underset{\text{Ring-flipped }cis}{\text{[structure]}}$$

torial form is favored by 2.55 kcal/mole, yet the enthalpy difference is not sufficient for the entropy of mixing of these ring-flipped conformers to be ignored. Use of Eq. 11.3 permits evaluation of this S_{mix} term as approximately 0.1 entropy units. The *trans* diequatorial–diaxial pair are not superimposable on their mirror images, so *trans*-1,2-dimethylcyclohexane exists as a *dl* pair ($S_{mix} = R \ln 2$). This isomer also has a symmetry number of 2 ($S_\sigma = -R \ln 2$), so only the entropy of mixing of 0.1 eu need be considered. The *cis* compound has two conformers. Each conformer contains one methyl group equatorial and the other methyl group axial and has three *gauche*-butane interactions (one of which is between the methyls themselves). These conformers are enantiomers of one another: Ring-flipping converts one of the conformers into its mirror image. The free-energy difference for the equilibrium of Eq. 12.6 would therefore be calculated as:

$$\Delta G° = 1.7 \text{ kcal/mole} - 298(1.4 - 0.1)/1000 \text{ kcal/mole}$$
$$\cong 1.3 \text{ kcal/mole}$$

A recent set of experimental results[9] for the dimethylcyclohexane equilibria[1] is shown in Table 12.2. The most notable discrepancy between

TABLE 12.2 *Dimethylcyclohexane Equilibria*[9]

Series	$\Delta H_{540°K}$ (kcal/mole)	$\Delta S_{540°K}$ (entropy units)	$\Delta G_{298°K}$ (kcal/mole)
1,2-	1.72	0.79	1.46
1,3-	1.81	1.16	1.47
1,4-	1.78	1.14	1.43

the calculated and experimental values involves the decreased enthalpy and entropy observed in the 1,2-dimethylcyclohexane system, which effectively cancel to give a free-energy difference similar to those for the other equilibria. This result is ascribed to "cogwheeling", especially in the *cis* isomer.[1,9] The methyl groups in this compound are believed to interfere with each other's free rotation, thereby decreasing the rotational entropy. Other evidence supports the presence of this cogwheeling. While cogwheeling might also be expected to be present to at least some extent in the *trans*-diequatorial conformer, it has not been observed there.

12-3 Conformation and Reactivity in Cyclohexanes

The relationship between conformation and reactivity, with primary emphasis on cyclohexane systems, was first discussed by Barton in 1950.[10] This relationship hinges on steric factors, stereoelectronic factors, positions of ring inversion equilibria, and dipolar factors, to name only those phenomena generally accepted[11] as being significant even though the detailed nature of the relationship involving a given factor may be unresolved.

When the reaction occurs at an exocyclic atom, such as in a saponification or esterification, steric factors predominate. The axial position is more hindered and reaction occurs more readily at the equatorial position.

---
9. N. L. Allinger, W. Szkrybalo, and F. A. VanCatledge, *J. Org. Chem.* **33**, 784 (1968).
10. D. H. R. Barton, *Experientia* **6**, 316 (1950)—reproduced as *Topics in Stereochemistry* **6**, 1 (1971).
11. Reference A, Section 8-5; Reference C, Section 2-5.

In most saponifications and esterifications, the tetrahedral intermediate requires more space than the ground state, and the axial position reaction is said to be subject to steric hindrance. As a result, saponification of the ethyl 4-*tert*-butylcyclohexanecarboxylates at 25°C in 70% ethanolic sodium hydroxide[11] occurs twenty times faster for the *trans* compound (equatorial CO_2Et) than for the *cis* compound. Similarly, *trans*-N,N-dimethyl-4-*tert*-butylcyclohexylamine reacts with methyl iodide about fifty times faster than the *cis* compound.[12]

When the reaction occurs at a ring atom, the situation is considerably more complicated,[11] with both steric hindrance and steric assistance possible and stereoelectronic factors often playing the predominant role. Consider the chromic acid oxidation of cyclohexanols as a first example.[11,13] Axial alcohols are almost always oxidized faster than their equatorial isomers (*cis*-4-*tert*-butylcyclohexanol reacts 3.2 times faster than the *trans* compound), suggesting steric assistance in the axial systems. Additional support for the steric assistance argument is that additional axial substituents accelerate the reaction regardless of whether the hydroxyl is equatorial or axial, but to a much greater extent if it is axial. For example, *cis*-3,3,5-trimethylcyclohexanol (**XVIII**) is oxidized almost twice as fast as *trans*-4-*tert*-butylcyclohexanol, in both of which the hydroxyl is equatorial, while *trans*-3,3,5-trimethylcyclohexanol (**XIX**) reacts twenty

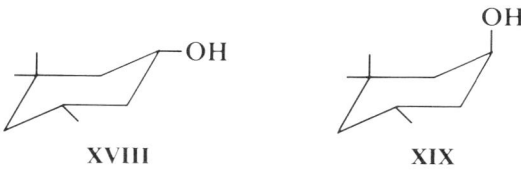

XVIII   XIX

times faster than the axial *cis*-4-*tert*-butylcyclohexanol. The major difference in energies must be in the ground states, since other evidence[11,13] supports the idea that the transition states for the oxidation of *cis–trans* isomers are similar in energy. The reaction-rate differences (and steric assistances), therefore, result from ground-state compressions in the various alcohols themselves.[14]

Another example of a reaction occurring at a ring position is solvolysis of a halide or sulfonate. While these systems are often interpreted within the S_N1 framework[11]—in which steric assistance is a reasonable assumption in proceeding from the sp^3-hybridized reactant to the sp^2-hybridized

---

12. N. L. Allinger and J. C. Graham, *J. Org. Chem.* **36**, 1688 (1971).
13. E. L. Eliel, S. H. Schroeter, T. J. Brett, F. J. Biros, and J.-C. Richer, *J. Amer. Chem. Soc.* **88**, 3327 (1966).
14. These arguments must be tempered by the knowledge that alcohol oxidation mechanisms often involve rate-determining cleavage of the carbon–hydrogen bond after formation of an ester of the alcohol. The relative rates may therefore be a composite of several steps and not be explicable in a straightforward manner.

carbonium ion, especially with axially oriented leaving groups—the S_N1 character of a given reaction is often open to question.[15] In particular, leaving-group effects, nucleophilic steric effects, kinetic isotope effects, and salt, solvent, and temperature effects all point to second-order components, ion-pair phenomena, hydride shifts, and nonchair conformations in various cyclohexyl reaction systems.[15] The most recent investigations[15] provide undeniable evidence for the absence of carbonium ion intermediates even in the best ionizing solvents and for predominant inversion of configuration.

Stereoelectronic effects[11] are those factors concerned with the conformational requirements of the groups involved in the reaction with respect to the electron orientations in the transition state. Whereas acyclic systems can often rotate to satisfy stereoelectronic eclipsed transition-state demands, cyclic systems such as cyclohexanes are often too rigid to adjust. The frequent result is that a reaction fails to proceed or occurs slowly by a different, less favorable pathway. On the other hand, cyclohexane compounds provide an almost ideal framework for studying processes requiring a staggered arrangement in the transition state, since rotation is essentially prohibited and $anti \rightleftharpoons gauche$ equilibria can occur only by ring-flipping.

Stereoelectronic effects are most influential in addition and elimination reactions and in rearrangements.[11] For example, the best understood bimolecular elimination process[16] (E2) requires *anti* coplanar orientation of the groups being eliminated.[17] A *trans*-2β,3α-dibromosteroid (**XX**), which has both bromines axial, loses 91% of its bromine on treatment with potassium iodide at 40°C for fourteen days, while the *trans*-diequatorial 2α,3β isomer (**XXI**) loses only 1%.[11] Addition reactions behave similarly,

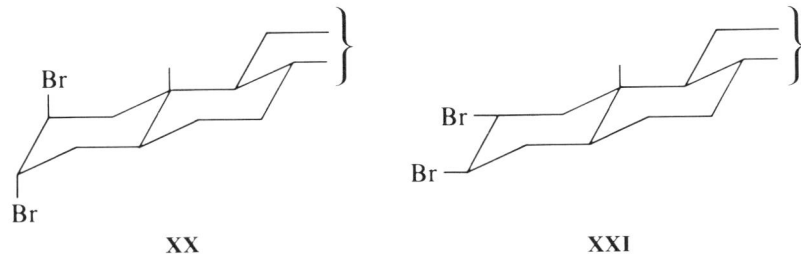

XX   XXI

---

15. N. C. G. Campbell, D. M. Muir, R. R. Hill, J. H. Parish, R. M. Southam, and M. C. Whiting, *J. Chem. Soc.*, B 355 (1968); D. S. Noyce, et al., *J. Org. Chem.* **34**, 463, 1247, 1252 (1969); J. B. Lambert, G. J. Putz, and C. E. Mixan, *J. Amer. Chem. Soc.* **94**, 5132 (1972); J. E. Nordlander and T. J. McCrary, Jr., ibid., **94**, 5133 (1972).
16. R. T. Morrison and R. N. Boyd, *Organic Chemistry*, 3rd ed., Allyn and Bacon, Boston, 1973, Chapters 7.11 and 14.22.
17. For a discussion of *syn* components, see J. Sicher, *Pure Appl. Chem.* **25**, 655 (1971), and *Angew. Chem. Intern. Ed. Engl.* **11**, 200 (1972). For a theoretical calculation of *syn* and *anti* preferences, see J. P. Lowe, *J. Amer. Chem. Soc.* **94**, 3718 (1972).

with *trans*-diaxial addition predominating;[18] e.g., the $2\beta,3\alpha$-dibromosteroid **XX** is the major product in the bromination of a Δ^2 steroid.

A striking example of stereoelectronic factors is the stereospecificity observed in the formation and cleavage of epoxides[11] (Furst-Plattner rule for the opening and closing of epoxides). Hydrobromination of a 2β-epoxysteroid leads to the 2β-hydroxy-3α-bromo product (Eq. 12.7), while similar treatment of the 2α isomer produces the 2β-bromo-3α-hydroxysteroid (Eq. 12.8). In each of these reactions, the *trans*-diaxial product has

been formed even though the *trans*-diequatorial isomer would be, at worst, of comparable stability.

It is unwarranted to conclude that stereoelectronic effects always require *trans*-diaxial orientations in additions and eliminations. Pyrolyses of xanthates and acetates are only two of many known processes which are favored when the orientations are *cis*.[11,18] It is fairly safe to suggest that *cis* processes usually involve cyclic transition states or aggregation phenomena such as ion-pair formation.

12-4 Unsaturated and Heterocyclic Six-membered Rings

Conversion of one or two of the ring carbons of cyclohexane into trigonal carbon atoms,[19,20] or replacement of the ring carbons by noncarbon elements,[19,21] introduces new conformational problems.

The parent systems containing a single sp^2-hybridized carbon atom are cyclohexanone (**XXII**, where $X = O$) and methylenecyclohexane (**XXII**,

---
18. R. C. Fahey, *Topics in Stereochemistry* **3**, 237 (1968).
19. J. B. Lambert, *Accounts Chem. Res.* **4**, 87 (1971).
20. Reference A, Section 8-7; Reference C, Section 2-6.
21. F. G. Riddell, *Quart Rev.* **21**, 364 (1967); E. L. Eliel, *Accounts Chem. Res.* **3**, 1 (1970), and *Pure Appl. Chem.* **25**, 509 (1971); Reference C, Section 4-6; C. Romers, C. Altona, H. R. Buys, and E. Havinga, *Topics in Stereochemistry* **4**, 39 (1969).

where $X = CH_2$). Both of these compounds prefer slightly flattened

⬡=X

XXII

chair conformations,[19] with the flexible forms less unfavorable ($\Delta H \sim 3$ kcal/mole[22]) and the barriers to inversion smaller ($\Delta G^{\ddagger} \cong$ 5–8 kcal/mole[23]) than in cyclohexane itself. When methyl substituents are present on the ring, the interactions are somewhat different than in cyclohexane. The single nonring substituent on the sp^2 carbon (the oxygen or vinyl carbon) occupies a position more or less halfway between the normal equatorial and axial arrangements. The preference of an alkyl group in the 3 position for the equatorial orientation is therefore less than in cyclohexane by roughly 0.5 kcal/mole, since one of the *gauche*-butane interactions destabilizing the axial orientation is considerably diminished in magnitude.[22,24]

Presence of a polar substituent at the 2 position introduces dipolar interactions in the cyclohexanone series which tend to destabilize the equatorial orientation of the substituent.[25] In fact, in most solvents chlorine and bromine prefer to be axial. These 2-halocyclohexanones have been studied by a variety of methods. When the halogen is equatorial, the carbonyl stretching frequency is increased by 15–30 cm^{-1}, while an axial halogen produces a shift of only ± 10 cm^{-1}.[25] Dipole moments are also often used.[25] Model systems, such as the 4-*tert*-butyl analogs, are measured first, and, assuming that the conformation-fixing group introduces no distortions[7] (Chapter 12-2), the results can be applied to the conformational analyses of systems undergoing ring inversions. Ultraviolet spectra and chiroptical techniques (Chapter 12-7) give similar results.[25] In all cases, the percent of axial substituent is greatest in the least polar solvent, as expected from an analysis of the dipolar orientations.

X=⬡=X

XXIII

---

22. N. L. Allinger, H. M. Blatter, L. A. Freiberg, and F. M. Karkowski, *J. Amer. Chem. Soc.* **88**, 2999 (1966).
23. F. R. Jensen and B. H. Beck, *J. Amer. Chem. Soc.* **90**, 1066 (1968)—recalculated in N. L. Allinger, J. A. Hirsch, M. A. Miller, and I. J. Tyminski, ibid., **91**, 337 (1969); J. T. Gerig, ibid., **90**, 1065 (1968); J. T. Gerig and R. A. Rimerman, ibid., **92**, 1219 (1970); M. St.-Jacques, et al., *Can. J. Chem.* **48**, 2386, 3039 (1970), and *Chem. Commun.* 1097 (1970).
24. W. D. Cotterill and M. J. T. Robinson, *Tetrahedron* **20**, 765, 777 (1964); Reference C, Section 2-6.
25. Reference C, pp. 114–115 and Section 7-3.

Cyclohexane-1,4-dione (**XXIII**, where X = O) is a source of controversy, the questions[19,26] being whether the preferred conformation is chair or twist in various phases and, if twist, whether the twist is symmetrical or not. The same questions are being debated for 1,4-dimethylenecyclohexane (**XXIII**, where X = CH_2) and its derivatives,[19,27] while cyclohexane-1,4-dioxime (**XXIII**, where X = NOH) is apparently primarily a twist boat in the solid state and in solution.[28] It appears reasonably certain that all of the systems of type **XXIII** exist as twist boats in the crystalline state. The debates swirl primarily around the conformational questions in solution and in the gas phase.

When two sp^2-hybridized carbons are adjacent, as in cyclohexene, the ring assumes a half-chair conformation.[29] The bonds at the allylic positions are imperfectly staggered and are called pseudoequatorial (e')

e'⟍   a'
   >=<
a'⁄   e'

and pseudoaxial (a') positions. Because of the half-chair geometry, reactions at the allylic carbon occur primarily at the pseudoaxial position, which overlaps more effectively with the π cloud. Cyclohexene is dissymmetric and is converted into its enantiomer by ring inversion (as for cis-1,2-dimethylcyclohexane, Chapter 12-2), a process requiring approximately 5.4 kcal/mole free energy.[30] Some few studies[4,30] have been performed on conformational preferences in 4-substituted cyclohexenes. Fluoro, chloro, and bromo substituents exhibit slight preferences for the equatorial position (14, 200, and 77 cal/mole at $-157°C$, respectively), while the iodo group prefers to be axial (by 16 ± 7 cal/mole) in perdeuteriovinyl chloride solvent. Halogens at the 3 position appear to prefer the pseudoaxial orientation[30] by 500–800 cal/mole.

Heterocyclic six-membered rings[19,21] prefer chair conformations[31] unless many sulfur atoms are present as part of the ring.[32] The barriers

---

26. Reference C, p. 474; P. Dowd, T. Dyke, and W. Klemperer, *J. Amer. Chem. Soc.* **92**, 6327 (1970), and references therein.
27. M. St.-Jacques and M. Bernard, *Can. J. Chem.* **47**, 2911 (1969); N. L. Allinger, J. A. Hirsch, M. A. Miller, and I. J. Tyminski, *J. Amer. Chem. Soc.* **90**, 5773 (1968).
28. P. Groth, *Acta Chem. Scand.* **22**, 128 (1968); H. Saito and K. Nukada, *J. Mol. Spectry.* **8**, 355 (1965).
29. Reference C, Section 2-6; L. H. Scharpen, J. E. Wollrab, and D. P. Ames, *J. Chem. Phys.* **49**, 2360 (1968); J. F. Chiang and S. H. Bauer, *J. Amer. Chem. Soc.* **91**, 1898 (1969); H. J. Geise and H. R. Buys, *Rec. Trav. Chim.* **89**, 1147 (1970).
30. F. R. Jensen and C. H. Bushweller, *J. Amer. Chem. Soc.* **91**, 5774 (1969), and references therein.
31. For example, R. A. Spragg, *J. Chem. Soc.*, B 1128 (1968); R. K. Harris and R. A. Spragg, ibid., 684 (1968); W. B. Smith and B. A. Shoulders, *J. Phys. Chem.* **69**, 579 (1965); V. M. Rao and R. Kewley, *Can. J. Chem.* **47**, 1289 (1969).
32. C. H. Bushweller, *J. Amer. Chem. Soc.* **89**, 5978 (1967), and **90**, 2450 (1968).

to ring inversion and the energy differences between chair and flexible forms parallel those for cyclohexane (p. 251) when one or two heteroatoms are present.[21,31,33] Substituent conformational equilibria and configurational equilibria are markedly different,[21] primarily because of the differences in bond lengths and bond angles in the ring, the presence of dipole effects involving the heterocyclic ring atoms, *gauche* effects (Chapter 11-2), and diminished nonbonded repulsive interactions involving axial substituents (heteroatoms possess lone pairs of electrons rather than hydrogens). Most striking are axial preferences for substituents on the heteroatoms[34] (such as in thiane-1-oxide) or at position 3 or 5 relative to a heteroatom.[21]

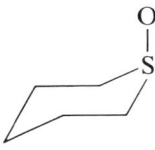

Thiane-1-oxide

12-5 Rings Other than Six-membered

In monocyclic systems of other than six ring atoms, the ring atoms usually are not equivalent,[35] so that substituents may be located at more than two types of positions. Since some torsional energy is present in each possible conformation, interactions involving substituents can sometimes produce conformational changes in the ring itself. For those rings having more than six and fewer than twelve to fourteen ring atoms, several conformations of more or less equal energy content are possible, resulting in considerable flexibility and subtle conformational influences on both physical and chemical phenomena.[35,36]

Cyclic systems are usually classified as *small* (cyclopropane and cyclo-

---

33. For example, the $\Delta G^{\ddagger}$ for ring inversion of tetrahydropyran is 10.2 kcal/mole [G. Gatti, A. L. Segre, and C. Morandi, *J. Chem. Soc.*, B 1203 (1967)] and that for 1,4-dioxan at $-97°C$ is 9.7 kcal/mole [F. R. Jensen and R. A. Neese, *J. Amer. Chem. Soc.* **93**, 6329 (1971)].
34. R. O. Hutchins and B. E. Maryanoff, *J. Amer. Chem. Soc.* **94**, 3266 (1972); and J. B. Lambert, C. E. Mixan, and D. S. Bailey, ibid., **94**, 208 (1972), *J. Org. Chem.* **37**, 377 (1972), and references therein.
35. Reference C, Chapter 4.
36. J. D. Dunitz, *Pure Appl. Chem.* **25**, 495 (1971), and *Perspectives Structural Chem.* **2**, 1 (1968); J. Dale, *Angew. Chem. Intern. Ed. Engl.* **5**, 1000 (1966).

butane), *common* (cyclopentane and cyclohexane),[37] *medium* (cycloheptane[37] through cycloundecane), and *large* (cyclododecane and larger). Other than cyclopropane, none of the rings is planar. From the point of view of the thermodynamic parameters for forming a ring from an acyclic system,[35] the common rings are favored, possessing the least angle and torsional strain involving the ring constituents and reasonable entropy terms for bringing together the acyclic reaction centers. The small rings are favored entropically but overwhelmingly disfavored enthalpically because of the requisite angle and torsional strains. Medium and large rings are entropically disfavored, with intermolecular reactions the predominant mode for the acyclic reactants.

Cyclopropane must be planar, while cyclobutane may be planar or puckered, depending on the substitution pattern and the phase.[35,38] In most cases, the puckered conformation is favored, the relief in torsional strain associated with the puckering being large enough to overcome the slightly increased bond angle strain. When the ring is puckered, pseudoaxial and pseudoequatorial positions are present. Studies of the *cis–trans* equilibrations of the 1,3-dihalocyclobutanes have been performed.[39] The *cis* isomers predominate, which is not surprising since these isomers would be diequatorial in a puckered conformation and presumably preferred.

Cyclopentane could be planar without significant angle strain.[35] However, a planar conformation would involve eclipsing at every position. The result is puckering to an envelope conformation with four carbons coplanar (**XXIV**) or a half-chair conformation with three carbons coplanar (**XXV**). The puckering rotates around the ring by a facile *pseudorotation* from an envelope to a half-chair to a different envelope so that all

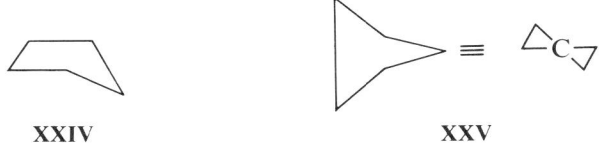

XXIV　　　　　　　　　**XXV**

---

37. The traditional classification system considers cycloheptane a common ring. It is the opinion of the author, however, that the physical and chemical properties of cycloheptane systems more closely parallel those of the medium rings. From the conformational point of view, four types of conformers must be evaluated as possible potential-energy minima, with considerable flexibility inherent in each type. This situation requires a type of conformational analysis different from that required for the pseudorotating cyclopentane system or for the more rigid (predominantly chair) cyclohexane system. The cycloheptane system is therefore very similar to those rings traditionally considered to be medium rings.
38. J. S. Wright and L. Salem, *Chem. Commun.* 1370 (1969); E. Adman and T. N. Margulis, *J. Amer. Chem. Soc.* **90**, 4517 (1968).
39. K. B. Wiberg and G. M. Lampman, *J. Amer. Chem. Soc.* **88**, 4429 (1966).

of the ring atoms are equivalent. Substituents restrict the pseudorotation and impose nonequivalence on the ring atoms. The puckering in cyclopentane has been detected[35] by electron diffraction and entropy measurements, and similar phenomena are observed in cyclopentanone,[40] tetrahydrofuran,[21,41] and tetrahydrothiophene.[21,42]

Those rings with sizes from seven to eleven exhibit considerable flexibility[35,36] and many conformations roughly equal in energy. Pseudorotation is extremely complex, so Hendrickson[43] has devised a method of mapping conformational interconversion paths. Saunders[44] has suggested that even-membered rings with six to twenty-four carbons will prefer structures related to a diamond lattice, unless none is available which does not contain sizeable nonbonded repulsive interactions. This suggestion is supported reasonably well by most of the available data. For example, cyclooctane appears to prefer a boat-chair conformation (**XXVI**) unless there is extensive replacement of the ring atoms with noncarbon species or extensive substitution for the hydrogens,[36,45] in which case crown conformations (**XXVII**) predominate. Cyclononane appears to prefer twist boat-chair forms,[46] which for reasons of symmetry is probably a typical type of result for the odd-membered rings.

XXVI   **XXVII**

Regardless of the ring size, extreme care must be taken in the interpretation of any chemical or physical property determined in rings larger than six-membered. Surprises and subtle effects abound.[47]

---

40. H. Kim and W. D. Gwinn, *J. Chem. Phys.* **51**, 1815 (1969); H. J. Geise and F. C. Mijlhoff, *Rec. Trav. Chim.* **90**, 577 (1971).
41. A. Almenningen, H. M. Seip, and T. Willadsen, *Acta Chem. Scand.* **23**, 2748 (1969); H. J. Geise, W. J. Adams, and L. S. Bartell, *Tetrahedron* **25**, 3045 (1969).
42. Z. Nahlovski, B. Nahlovsky, and H. M. Seip, *Acta Chem. Scand.* **23**, 3534 (1969).
43. J. B. Hendrickson, *J. Amer. Chem. Soc.* **89**, 7047 (1967). For "molecular mechanics" calculations (see below) of the conformations of cycloalkanes and methyl-substituted derivatives by this same worker, see *J. Amer. Chem. Soc.* **89**, 7036, 7044 (1967), and **86**, 4854 (1964).
44. M. Saunders, *Tetrahedron* **23**, 2105 (1967).
45. R. Srinivasan and T. Srikrishnan, *Tetrahedron* **27**, 1009 (1971).
46. F. A. L. Anet and J. J. Wagner, *J. Amer. Chem. Soc.* **93**, 5266 (1971); S. Dahl and P. Groth, *Acta Chem. Scand.* **25**, 1114 (1971).
47. See, for example, J. A. Hirsch and F. J. Cross, *J. Org. Chem.* **36**, 955 (1971), and references therein.

12-6 Methods of Conformational Analysis

The structures of molecules can be obtained by three types of methods.[48] The so-called total methods—x-ray diffraction, electron diffraction, and microwave spectrometry—provide detailed structural information on vapor-phase or crystalline samples, provided the molecule is amenable to analysis; e.g., a dipole moment must be present for a microwave determination. These total methods are all reasonably laborious and provide no information about the structures of compounds in solution, which is the most important phase to most organic chemists. A second type of method encompasses quantum chemical calculations[49] and molecular mechanics calculations.[43,50] While both of these techniques exhibit considerable promise, the former is limited by its inability to handle large systems with reasonable accuracy, while the latter is limited by the problems associated with the choice of a particular mechanical model to be used, especially with respect to the forces associated with the interactions of nonbonded atoms. Structure assignment therefore most often relies on the more easily obtainable partial structural methods, such as thermodynamic criteria and spectroscopic techniques, common examples of which are infrared, nuclear magnetic resonance, chiroptical (Chapter 12-7), and dipole moment approaches. Several of these partial methods will be discussed using cyclohexane systems as models (see also p. 262).

In most cyclohexyl systems, a higher stretching frequency in the infrared region is associated with an equatorial substituent than with the corresponding axial substituent. Conformationally biased systems of one type or another are used to establish the exact frequencies, which then may be applied to the determination of the conformational equilibria in unbiased systems. Examples from the biased compounds are given in Table 12.3.

TABLE 12.3 *Infrared Stretching Frequencies for Cyclohexyl Substituents*[48]

Substituent	Equatorial (cm^{-1})	Axial (cm^{-1})
D	2174	2146
OH	1062	955
Cl	742	688

---

48. Reference C, Chapter 3.
49. See Chapter 11, note 2.
50. J. E. Williams, P. J. Stang, and P. v. R. Schleyer, *Ann. Rev. Phys. Chem.* **19**, 531 (1968); N. L. Allinger, M. T. Tribble, M. A. Miller, and D. H. Wertz, *J. Amer. Chem. Soc.* **93**, 1637 (1971); and references therein.

Infrared spectroscopy may also be used to evaluate hydrogen bonding.[48] In very dilute solution in carbon tetrachloride, cyclohexanol exhibits a very sharp O—H stretch only at 3630 cm^{-1}; this is called the monomer band. As the solution is made more concentrated, intermolecular hydrogen bonding becomes apparent through a broad band at around 3400 cm^{-1} in addition to the monomer band. Intramolecular hydrogen bonding would not exhibit concentration-dependence in a nonpolar solvent. The difference in frequency, Δv, between the monomer band and the hydrogen-bonded band provides a measure of the strength of the intramolecular hydrogen bond. This Δv is 32 cm^{-1} for *trans*-cyclohexane-1,2-diol and 39 cm^{-1} for the *cis* compound, suggesting a stronger hydrogen bond in the *cis* system. The cyclohexane ring can

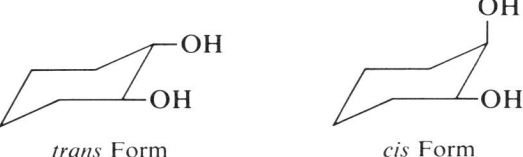

trans Form   *cis* Form

apparently bring *cis* groups into closer proximity than *trans*-diequatorial groups. A similar infrared analysis of the cyclopentane-1,2-diols shows a Δv of 61 cm^{-1} for the *cis* compound and no intramolecular hydrogen bond for the *trans*. The conclusion must be that a cyclopentane ring does not accept as a reasonable conformation any structure in which vicinal *trans* substituents are in close proximity.

One of the most useful partial methods is nuclear magnetic resonance spectroscopy,[48] which can often provide both structural and kinetic[51] information (see Chapter 12-1). In general, an axial hydrogen (or fluorine) absorbs at higher field than an equatorial one. This phenomenon may be approached in the framework of the McConnell approximation[52]

$$\Delta \sigma = \Delta \chi \frac{1 - 3 \cos^2 \theta}{3r^3} \qquad (12.9)$$

in which $\Delta \sigma$ is the shielding shift of a given nucleus caused by a given bond relative to some standard; $\Delta \chi$ is the diamagnetic anisotropy, which is a constant for a given type of bond and may be assumed to be zero for a carbon–hydrogen bond; r is the vectorial distance from the midpoint of the given bond to the given nucleus; and θ is the angle defined by the bond

---

51. Reviewed by G. Binsch, *Topics in Stereochemistry* **3**, 97 (1968); N. Booth, *Progr. NMR Spectry.* **5**, 149 (1969); H. Kessler, *Angew. Chem. Intern. Ed. Engl.* **9**, 219 (1970).
52. P. Laszlo and P. J. Stang, *Organic Spectroscopy*, Harper and Row, New York, 1971, Section 4.3.4.

axis and the vector r. As evident from this equation, the effect of bond anisotropy on the chemical shift falls off quickly with distance (and is most meaningful when $r = 4 - 5\text{Å}$)[52] and changes from deshielding at θ less than 55.7° to shielding at θ greater than 55.7°.

For example, consider the qualitative effects of the ring carbon–carbon bonds in a conformationally fixed cyclohexane (**XXVIII**) on the chemical shifts of an equatorial–axial pair of hydrogens. The bonds

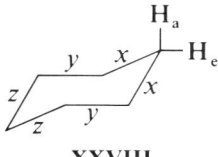

XXVIII

marked x will affect both hydrogens equally, while those designated z are too distant to be influential. The key ring bonds are those shown as y, for which θ to the equatorial hydrogen H_e is approximately 31° and θ to the axial H_a is approximately 67°. The equatorial hydrogen would therefore be deshielded by the y carbon–carbon bonds and the axial shielded. From a slightly different point of view, an *anti* arrangement (Chapter 11-2) leads to deshielding and a *gauche* to shielding.

Next, consider the additional effects if an axial methyl group were present on a vicinal carbon atom (**XXIX**). This methyl group is *anti* to H_a,

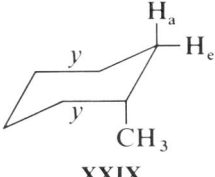

XXIX

and therefore would exert a deshielding influence on this nucleus in opposition to the shielding influence of the y ring bonds. Similarly, the methyl is *gauche* to H_e and exerts a shielding influence. While H_a would be unambiguously predicted to be at higher field than H_e in **XXVIII**, the chemical-shift difference between H_a and H_e would be less in **XXIX**, and H_e might even be the nucleus absorbing at the higher field. The relative relationship of a given axial–equatorial pair of nuclei in a noninverting cyclohexane ring is therefore substituent-dependent, although the natural bias from the ring atoms is for the axial nucleus to absorb at the higher field.

An additional complication is introduced whenever a substituent is within van der Waals distance of the nuclei being considered. In general, substituents at positions M in structure **XXX** are influential, while those at N are not.

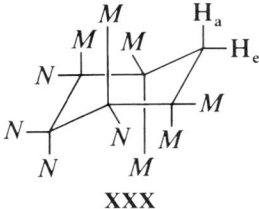

XXX

When cyclohexane ring-flipping is prevalent, the chemical shift for a given nucleus is the weighted average of its shift in the two contributing conformations. Assuming that the latter can be obtained from conformationally fixed systems (Chapter 12-2), Eq. 12.10 can be utilized,

$$v_{av} = N_a v_a + N_e v_e \tag{12.10}$$

where v_a and v_e are the values from the reference systems and N_a and N_e are the mole fractions of each conformation ($N_a + N_e = 1.0$). Care must be taken when hydrogen nuclei are used for the measurement in that the usual nuclei are those geminal to the substituent affecting the ring-flipping equilibrium; therefore, an axial hydrogen corresponds to an equatorial substituent and the chemical shifts and mole fractions must be defined in a consistent manner.

Vicinal spin–spin coupling constants ($^3J_{vic}$) are a function of the dihedral angle ω (Chapter 11), a useful modification of the more exact Karplus equation being:

$$\begin{aligned} ^3J_{vic} &= 10 \cos^2 \omega & 0° \leqslant \omega \leqslant 90° \\ ^3J_{vic} &= 16 \cos^2 \omega & 90° \leqslant \omega \leqslant 180° \end{aligned} \tag{12.11}$$

Vicinal coupling constants are greater for *anti* hydrogens than for *gauche* hydrogens. In substituted cyclohexanes, the $^3J_{vic}$ for *trans*-diequatorial hydrogens are 2–3 Hz, for *cis* hydrogens 4–5 Hz, and for *trans*-diaxial hydrogens 10–12 Hz. These coupling constants, or simply band widths, may be used to decide if a given cyclohexyl hydrogen is axial or equatorial if it is coupled to at least one vicinal pair of hydrogens. The key is the much greater magnitude of the axial–axial coupling constant.

12-7 Chiroptical Methods

Two of the most often used techniques for the determination of relative or absolute stereochemistry or conformation are the chiroptical methods: optical rotatory dispersion (ORD)[53,54] and circular dichroism (CD).[53,55] Optical rotatory dispersion involves the measurement of the rotation of plane-polarized light by a chiral compound as a function of the wavelength, while circular dichroism involves the differential absorption of this plane-polarized light as a function of the wavelength. Figure 12.4 illustrates the four major types of ORD curves, two of which are *plain* curves associated with *Cotton effects* at wavelengths shorter than the range being scanned, and two of which are Cotton-effect curves themselves. Cotton-effect curves are the more useful, producing information by empirical comparisons with curves of compounds of known structure, configuration, and conformation. Enantiomers give opposite Cotton-effect curves.

The relationship between an ultraviolet absorption band associated with a positive Cotton effect in the ORD and CD is shown in Fig. 12.5. Notice that the CD resembles the UV but has a sign associated with it. The position of the maximum for the ultraviolet absorption corresponds fairly

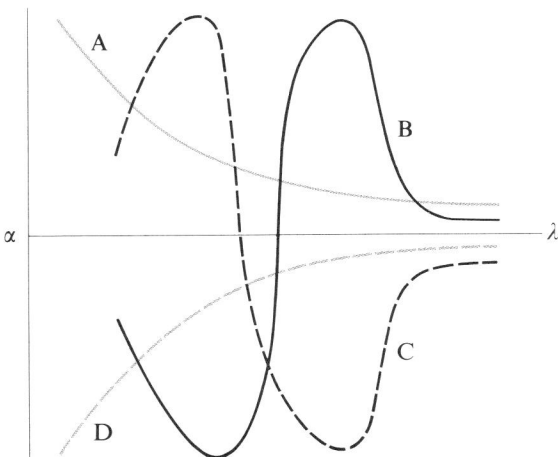

FIGURE 12.4 *Typical optical rotatory dispersion curves:* A, *a plain positive curve;* B, *a positive Cotton-effect curve;* C, *a negative Cotton-effect curve; and* D, *a plain negative curve.*

---

53. Reference D, pp. 109–116; P. Crabbé, *Optical Rotatory Dispersion and Circular Dichroism in Organic Chemistry*, Holden-Day, San Francisco, 1965, and *Topics in Stereochemistry* **1**, 93 (1967); G. Snatzke, *Angew. Chem. Intern. Ed. Engl.* **7**, 14 (1968).
54. Reference A, Chapter 14; Reference C, Chapter 3-6; C. Djerassi, *Optical Rotatory Dispersion*, McGraw-Hill, New York, 1960.
55. L. Velluz, M. Legrand, and M. Grosjean, *Optical Circular Dichroism*, Academic Press, New York, 1965.

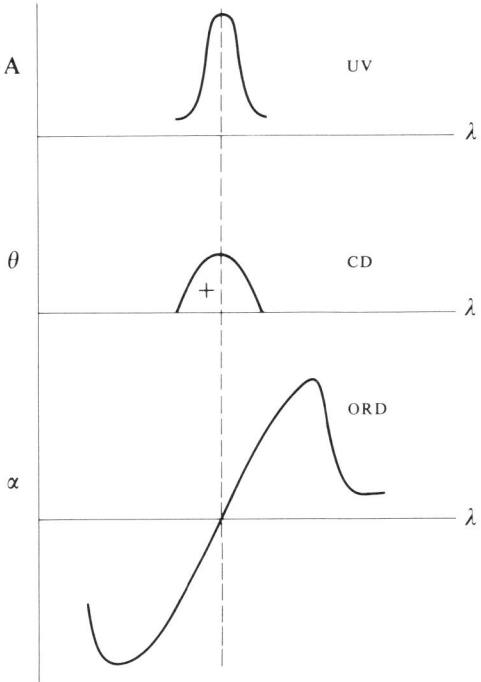

FIGURE 12.5 *A comparison of the ultraviolet, circular dichroism, and optical rotatory dispersion behavior associated with a positive Cotton effect.*

well to the position of the circular dichroism maximum and to the wavelength where the ORD curve crosses the zero rotation line.[56]

The advantage of CD lies in its simplicity, while the more complicated ORD curve often compensates by providing more information. For example, ORD is the method of choice for determining the position of a keto group in a normal steroid framework. The methods of ORD and CD are often nicely complementary.

The types of chromophores which can be investigated by the chiroptical methods fall into two major categories. The first category includes inherently dissymmetric chromophores, such as biphenyls, dienes, and unsaturated ketones. In these systems, the chromophore itself is chiral (Chapter 10-1). The second category is dissymmetrically perturbed

---

56. Deviations from these relationships usually result either from a plain curve background on which the Cotton-effect curve is superimposed, or from the presence of several overlapping chromophores.

symmetric chromophores. Symmetric chromophores, such as carbonyls, nitro groups, and aromatic rings, are associated with chiroptical behavior whenever they are present in chiral molecules. An element of chirality, then, induces dissymmetry in the normally symmetric chromophoric system.

The systems studied to the greatest extent have been those involving cyclohexanone rings.[53-55] The experimental results led to the postulation of the *octant rule*.[53-55] Cyclohexanone is considered to be divided into eight octants using three planes (Fig. 12.6). One of these planes (the Z plane) passes through the carbonyl carbon (C_1) and oxygen and C_4 and includes the substituents at C_4. A second plane (the Y plane) includes C_1, C_2, and C_6, with the equatorial substituents at C_2 and C_6 almost in this plane. The third plane (the X plane) is perpendicular to the other two and cuts through the carbonyl bond. The substituents on the cyclohexanone are therefore located in most instances in the four rear octants (positive X coordinates). In order to qualitatively predict the sign of the Cotton

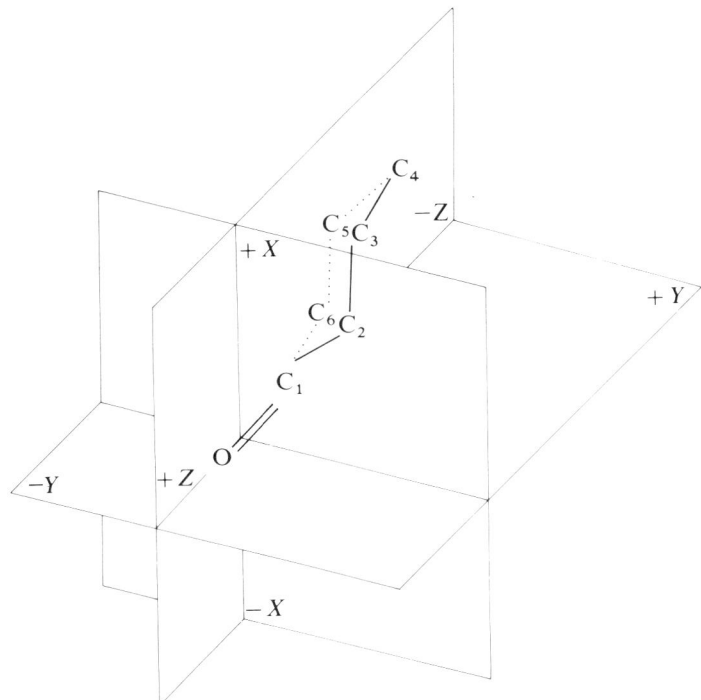

FIGURE 12.6 *The orientation of cyclohexanone in the octant rule.*

```
         a(+)  a(0) e(0)  a(-)
           |      \  /      |
   (+)e—C₅────── C₄ ──────C₃—e(-)
           |               |
           |               |
    (0)e—C₆────── C₁ ──────C₂—e(0)
           |      ||       |
          a(-)    O       a(+)
```

FIGURE 12.7 *Schematic representation of octant-rule contributions in cyclohexanones.*

effect associated with the carbonyl chromophore, the coordinates of each substituent are multiplied in terms of their signs to obtain the effect exerted by that substituent. Axial substituents at carbon 2 and equatorial and axial substituents at carbon 5 provide positive contributions, while axial substituents at carbon 6 and equatorial and axial substituents at carbon 3 provide negative contributions (Fig. 12.7).

Consider (+)-*trans*-10-methyl-2-decalone (**XXXI**) as an example for the prediction of the sign of the Cotton effect. Because of the required orientation of the cyclohexanone ring (Fig. 12.6), the molecule must be turned over and viewed as in **XXXII**. Schematic representation as in

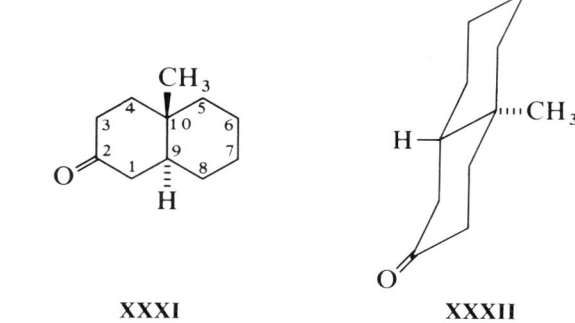

XXXI   XXXII

Fig. 12.7 gives Fig. 12.8, which shows that carbons 6, 7, and 8 exert a positive contribution. The observed Cotton effect is positive as predicted.

Attempts at quantification[53-55] by assigning empirical values to group contributions at the desired positions, and assuming additivity of these values, have been moderately successful. In general, the more polarizable group has the larger Cotton-effect contribution:

$$I > Br > Cl > alkyl, OH, OAc, NR_2 > H > F$$

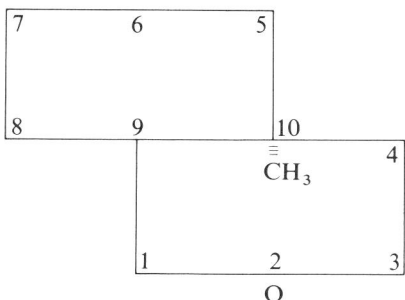

FIGURE 12.8 *Schematic representation of (+)-trans-10-methyl-2-decalone* (XXXII).

The general rule for the chiroptical method correlations is that the configuration must be known for the conformation to be determined and vice versa. In monocyclic systems, ring-flipping equilibria can be evaluated. For example, one of the enantiomers of *trans*-2-chloro-5-methylcyclohexanone has the absolute configuration indicated in Eq. 12.12:[54]

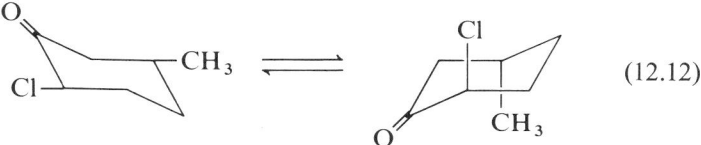
(12.12)

The Cotton-effect curve is very solvent-dependent, being negative in isooctane and positive in methanol. The dipole moment of the equatorial form is much greater than that of the axial, so the position of the equilibrium would be expected to be shifted toward the equatorial conformer as the solvent polarity was increased. Octant-rule considerations support this idea, since the equatorial conformation would be expected to exhibit a positive Cotton effect and the axial a negative one.[57] The ORD solvent effects therefore are as expected from dipole-moment predictions.

Anomalous results (qualitatively or quantitatively) usually indicate errors in configurational or conformational assignments, the presence of flattened chair or flexible conformations, or, rarely, behavior that is not in accordance with the octant rule.[53-55] The chiroptical methods are powerful tools in cyclohexanone systems, and their scope and predictive power is steadily being advanced into other structural areas.

---

57. The axial conformer must be turned over to satisfy the required octant-rule orientation.

General References PART III

A. Eliel, E. L. *Stereochemistry of Carbon Compounds.* New York: McGraw-Hill, 1962.
B. Mislow, K. *Introduction to Stereochemistry.* New York: W. A. Benjamin, 1965.
C. Eliel, E. L., N. L. Allinger, S. J. Angyal, and G. A. Morrison. *Conformational Analysis.* New York: Wiley, 1965.
D. March, J. *Advanced Organic Chemistry: Reactions, Mechanisms, and Structure.* New York: McGraw-Hill, 1968.

Author Index

Abraham, M. H., 194
Abrahamson, E. W., 65
Adams, W. J., 266
Adman, E., 265
Ala-Tuori, M., 217
Alberty, R. A., 122
Allen, L. C., 99, 234
Allinger, N. L., 233, 234, 238, 239, 241, 242, 243, 249, 258, 259, 261, 262, 263, 264, 267, 271, 276
Almenningen, A., 266
Altmann, S. L., 47
Altona, C., 261
Ames, D. P., 263
Anderson, G. L., 97, 105
Anderson, J. E., 253
Andrist, A. H., 84, 206
Anet, F. A. L., 252, 266
Angyal, S. J., 233, 234, 238, 239, 241, 242, 243, 249, 258, 261, 262, 263, 264, 267, 271, 276
Archie, W. C., Jr., 66
Arigoni, D., 222
Arnett, E. M., 104, 105, 167
Aubort, J. D., 198
Aue, D. H., 104
Azzaro, M. E., 159, 247

Badger, G. M., 48
Bailey, D. S., 264
Baird, N. C., 57
Baker, F. W., 97
Baker, J. W., 105
Baldwin, J. E., 82, 84, 206
Bartell, L. S., 266
Bartlett, P. D., 248
Barton, D. H. R., 258
Bauer, S. H., 263
Beauchamp, J. L., 104
Beck, B. H., 252, 254, 262
Bell, R. P., 153
Benoit, R. L., 194
Benson, S. W., 45
Bentley, T. W., 192
Berliner, E., 155
Bernard, M., 263
Bernstein, H. J., 48
Berson, J. A., 70, 84, 206
Berwin, H. J., 108
Bhacca, N. S., 217
Bingham, R. C., 156, 189
Binsch, G., 268
Biros, F. J., 259
Bitterwolf, T. E., 112
Blackwood, J. E., 229
Blair, L. K., 104
Blatter, H. M., 262
Blume, W. J., 186

Boekelheide, V., 158
Bohme, D. K., 104
Bolton, P. D., 109, 112, 167
Booth, H., 268
Borcic, S., 156
Bordwell, F. G., 153, 163
Bothner-By, A. A., 242
Bouis, P. A., 107
Bourn, A. J., 252
Bowden, K., 96, 98, 131, 135, 171
Bowers, M. T., 104
Bowles, A. J., 243
Bowman, N. S., 150
Boyd, D. R., 232
Boyd, R. N., 44, 53, 59, 95, 103, 106, 119, 176, 227, 229, 249, 260
Boyer, J. P. H., 170
Boyle, P. H., 223
Boyle, W. J., Jr., 153, 163
Branch, G. E. K., 92
Brauman, J. I., 66, 104
Breslow, R., 48, 57
Brett, J. J., 259
Briggs, A. G., 170
Briggs, J. P., 104
Brodsky, L., 78
Brown, H. C., 108, 110, 118, 129, 159, 178, 247
Brownlee, R. T. C., 99
Bruice, T. C., 153, 199
Buckley, A., 135
Buddenbaum, W. E., 156
Burkoth, T. L., 49
Bushweller, C. H., 252, 253, 263
Buys, H. R., 250, 253, 261, 263

Cahn, R. S., 222
Calvin, M., 92
Campbell, J. R., 254
Campbell, N. C. G., 260
Carboni, R. A., 96
Carter, R. E., 158
Castellano, S., 99
Cava, M. P., 57
Chapman, N. B., 135
Chapman, O. L., 59, 83
Charton, M., 134
Chen, H. L., 154, 162
Chiang, C. C., 51
Chiang, J. F., 263
Chiang, Y., 153, 162
Chiurdoglu, G., 234
Choux, G., 194
Chuang, C. R., 186
Clough, S., 78
Cocivera, M., 103
Cohen, A. D., 99
Cohen, I., 20
Collins, C. J., 150

277

Conkling, J. A., 156
Cooper, G H., 159
Corriu, R. J. P., 170
Cotterill, W. D., 262
Cotton, F. A., 20, 39, 48
Cottrell, T. L., 45
Coulson, C. A., 3, 19, 21, 22, 25, 26, 27, 43, 47, 86, 99
Cox, B. G., 153, 194
Crabbé, P., 271
Cram, D. J., 39, 44, 53, 59, 65, 99, 119, 232
Craven, J. M., 217
Criegee, R., 66
Cross, F. J., 266
Csizmadia, I. G., 239

Dack, M. R. J., 208
Dahl, S., 266
Dahlgren, L., 158
Dale, J., 264
Daniels, F., 122
Dauben, H. J., Jr., 49
Davis, J. C., Jr., 3
Davis, R. E., 186, 198
Del Bene, J., 20
de Llano, C. J., 58
Depuy, C. H., 59, 83
Dewar, M. J. S., 28, 34, 36, 37, 42, 43, 44, 47, 54, 56, 57, 58, 84, 86, 96, 107, 176
Dimroth, K., 211, 213
Dixon, J. E., 153, 199
Djerassi, C., 271
Doering, W. von E., 82
Doorakian, G. A., 66
Dougherty, R. C., 83
Dowd, P., 263
Dowd, W., 155
Drago, R. S., 200
Dunitz, J. D., 264
Dunlap, R. P., 159
Dyke, T., 263
Dzidic, I., 104

Eaborn, C., 108
Eaton, D. R., 84
Eckell, A., 79
Edward, J. T., 94
Edwards, J. O., 115, 183, 198
Ehrenson, S., 94
Eliason, R., 154
Eliel, E. L., 222, 232, 233, 234, 238, 239, 241, 242, 243, 249, 258, 259, 261, 262, 263, 264, 267, 271, 276
Epiotis, N. D., 84, 206
Ewing, D. F., 241
Exner, J. H., 153

Fahey, R. C., 261
Farrell, P. G., 94
Faulkner, D. J., 83
Favini, G., 82
Feldman, M., 57
Fenoglio, D. J., 241

Ferguson, L. N., 91
Fiaud, J. C., 254
Fieser, L. F., 53, 216
Fieser, M., 53, 216
Firestone, R. A., 79
Fisher, R. D., 155
Fleming, K. A., 112
Flythe, W. C., 57
Fowler, F. W., 214
Frater, G., 83
Freedman, H. H., 66
Freiberg, L. A., 262
Frost, A. A., 36, 114, 120, 122, 126, 148, 161, 165, 173, 208, 219
Fry, J. L., 189
Fuchs, R., 194
Fujimoto, H., 59
Fukui, K., 59, 63

Garratt, P. J., 49
Gasparro, F. P., 207
Gatti, G., 264
Geise, H. J., 250, 263, 266
Geissman, T. A., 44
George, W. O., 243
Gerig, J. T., 262
Gilchrist, T. L., 59
Gill, G. B., 59
Gilliom, R. D., 86, 219
Gladys, C. L., 229
Glass, M. A. W., 158
Goddard, W. A., III, 84
Godfrey, M., 137
Gold, V., 247
Golden, D. M., 66
Golden, R., 97
Goldstein, M. J., 44, 55
Gordon, A. J., 158
Gordon, M., 98
Gould, E. S., 86, 89, 95, 99, 103, 106, 109, 114, 128, 139, 140, 142, 148, 150, 161, 165, 173, 184, 188, 196, 219
Graeve, R., 158
Graham, J. C., 259
Graham, W. A. G., 131
Gramaccioni, P., 82
Grant, M. W., 107
Grashey, D., 79
Greene, E. F., 115
Grosjean, M., 271
Groth, P., 263, 266
Grunwald, E., 103, 109, 118, 120, 126, 161, 165, 173, 176, 178, 184, 188, 211, 219
Grutzner, J. B., 56
Gurka, D. F., 108
Gwinn, W. D., 266

Haberfield, P., 182, 195
Habich, A., 83
Halevi, E. A., 155
Hall, F. M., 112
Hall, R. E., 155

Hammett, L. P., 99, 109, 114, 120, 122, 125, 126, 128, 139, 142, 148, 150, 159, 160, 161, 165, 173, 188, 194, 196, 208, 219
Hammond, G. S., 39, 44, 59, 65, 99, 119
Hanna, M. W., 3
Hanna, S. B., 153
Hansch, C., 145
Hansen, H. J., 83
Hanson, K. R., 222, 230
Hanson, S. W., 218
Hanstein, W., 108
Harmony, M. D., 153
Harris, J. C., 182
Harris, J. M., 155, 189
Harris, R. K., 263
Hassel, O., 250
Hautala, J. A., 163
Havinga, E., 253, 261
Hehre, W. J., 99, 104, 234
Henderson, R. M., 97
Henderson, W. G., 104
Hendrickson, J. B., 39, 44, 53, 59, 65, 99, 119, 266
Hepler, L. G., 109
Hermann, R. B., 98, 104
Herndon, W. C., 30, 36
Hess, B. A., Jr., 58
Hill, R. K., 80
Hill, R. R., 260
Himoe, A., 108
Hine, J., 86, 99, 106, 109, 114, 128, 139, 140, 142, 143, 148, 150, 159, 161, 163, 165, 173, 184, 188, 219, 247
Hirsch, J. A., 39, 239, 252, 263, 266
Hirschmann, H., 222
Hoffmann, R., 44, 59, 65, 79, 86
Hogben, M. J., 131
Holtz, D., 104
Holtz, H. D., 94
Hopps, H. B., 158
House, H. O., 197
Hudson, R. F., 198
Huisgen, R., 79
Humski, R., 156
Hutchins, R. O., 264

Ibne-Rasa, K. M., 183
Ingold, C. K., 89, 139, 140, 142, 143, 186, 194, 198, 208, 219, 222
Irving, J. G., 131

Jackman, L. M., 48, 104
Jacobus, J., 83
Jaffé, H. H., 38, 39, 49, 109, 128, 215, 216
Jefferson, A., 83
Jencks, W. P., 143, 148, 150, 159, 161, 181
Jensen, F. R., 105, 252, 253, 254, 262, 263, 264
Jermini, C., 153
Jewett, J. G., 159
Job, J. L., 94
Johnson, C. D., 146, 167, 170
Johnson, C. R., 39, 215
Johnson, P. L., 254

Johnston, M. D., Jr., 207
Jones, A. J., 48
Jones, F. M., III, 104
Jones, J. P., 170
Jorgensen, W. L., 234
Jung, D., 242

Kagan, H. B., 254
Kamlet, M. J., 167
Kankaanpera, A., 207
Kaplan, M. S., 82
Karabatsos, G. J., 156, 241
Karkowski, F. M., 262
Katritzky, A. R., 96, 131, 167, 170, 214
Katz, A. M., 153
Katz, H., 59, 83
Kauzmann, W., 3
Kebarle, P., 104
Kelly, D. P., 104
Kerr, J. A., 45
Kessler, H., 268
Kewley, R., 263
Kim, H., 266
Kirby, J. A., 135
Klemperer, W., 263
Klopman, G., 198, 204
Klyne, W., 238
Kniepp, K. G., 153
Kobayashi, S., 178
Koelling, J. G., 159, 247
Koeppl, G. W., 159
Kosower, E. M., 59, 126, 160, 184, 188, 207, 208, 211, 214, 215, 219
Kostelnik, R., 99
Kouba, J., 153
Kreevoy, M. M., 154
Kresge, A. J., 153, 154, 157, 159, 162, 164, 176
Krow, G., 224
Krystynak, R. H., 217
Kuhn, D. A., 181
Kuntz, I. D., Jr., 207
Kurz, J. L., 182
Kurz, L. C., 182
Kwart, H., 104

Lahti, M., 207
Laidler, K. J., 114
Laidlow, W. G., 3
Laity, J. L., 49
Lamaty, G., 156
Lambert, J. B., 260, 261, 264
Lampman, G. M., 265
Lane, C. A., 107
Larsen, J. W., 105, 107
Laszlo, P., 26, 217, 268
Laurie, V. W., 104
Lee-Ruff, E., 104
Leffler, J. E., 109, 113, 118, 120, 126, 128, 161, 165, 173, 176, 178, 184, 188, 211, 219
Legrand, M., 271
Lehn, J. M., 234
Lehr, R. E., 59

Leitereg, T. J., 232
Leung, C., 96
Levine, I. N., 3
Lewis, E. S., 153
Lewis, P. M. E., 104
Lewis, T. P., 104
Liang, J. H., 239
Liberles, A., 3, 27, 38, 44, 59, 86, 89, 99, 103, 106, 139, 140, 142, 148, 150, 161, 219
Linder, R. E., 112
Ling, A. C., 112
Liotta, C. L., 97
Lloyd, D., 48
Loening, K. L., 229
Longuet-Higgins, H. C., 65
Louis, J. B., 198
Lowe, J. P., 260
Lowe, J. P., 234
Lum, K. K., 135
Lupton, E. C., Jr., 66, 136

McCrary, T. J., Jr., 260
McDaniel, D. H., 110
McDonald, G. J., 159, 247
McKeever, L. D., 131
McKenna, J., 159
McKervey, M. A., 232
McQuilkin, R. M., 51
Maccoll, A., 26
Maddams, W. F., 243
Malojcic, R., 156
Mango, F. D., 84
Manser, G. E., 135
March, J., 26, 27, 86, 95, 99, 106, 109, 118, 128, 139, 140, 142, 150, 155, 161, 165, 173, 178, 179, 181, 184, 188, 196, 211, 219, 232, 247, 248, 249, 271, 276
Marchand, A. P., 59, 96
Marcus, R. A., 163
Margulis, T. N., 265
Mariani, C., 82
Marshall, D. R., 48
Martin, J. G., 80
Maryanoff, B. E., 264
Masamune, S., 49
Mayo, B. C., 218
Meier, G. F., 248
Melander, L., 150, 158
Memory, J. D., 42
Metcalf, B. W., 51
Mijlhoff, F. C., 266
Miller, J. A., 197
Miller, M. A., 239, 262, 263, 267
Miller, S. I., 59, 130, 131
Minesinger, R. R., 167
Mislow, K., 22, 86, 158, 222, 230, 233, 249, 276
Mitchell, M. J., 57
Mixan, C. E., 260, 264
Moore, W. J., 114, 120, 122
Morandi, C., 264
Moreland, W. T., Jr., 94

Morrison, G. A., 233, 234, 238, 239, 241, 242, 243, 249, 258, 261, 262, 263, 264, 267, 271, 276
Morrison, R. T., 44, 53, 59, 95, 103, 106, 119, 176, 227, 229, 249, 260
Muenter, J. S., 104
Muir, D. M., 260
Mulder, J. J. C., 84
Munson, M. S. B., 104
Murdoch, J. R., 176
Murr, B. L., 156
Murrill, E., 162
Musulin, B., 36
Mylonakis, S. G., 176

Nahlovski, Z., 266
Nahlovsky, B., 266
Nathan, W. S., 105
Needham, T. E., 200
Neese, R. A., 264
Nehring, R., 186
Newman, M. S., 132, 233
Nomura, Y., 97
Nordlander, J. E., 260
Norman, R. O. C., 155
Nowlan, V., 157
Noyce, D. S., 97, 260
Nukada, K., 263

O'Brien, D. H., 140
O'Connor, C. J., 171
O'Ferrall, R. A. M., 153, 159
Ohrt, J., 254
Okamoto, Y., 129
Olah, G. A., 140, 172, 178
Oosterhoff, L. J., 83, 84, 234
Orchin, M., 38, 39, 49, 215, 216
Owens, P. H., 104

Padwa, A., 78
Papaioannou, C. G., 156
Parish, J. H., 260
Parish, R. C., 97
Parker, A. J., 194
Parkin, D. C., 98
Parthasarathy, R., 254
Pasto, D. J., 39, 215
Paul, I. C., 51
Payne, M. A., 162
Pearson, H., 253
Pearson, R. G., 84, 114, 120, 122, 126, 148, 161, 165, 173, 198, 199, 208, 219
Perrin, C. L., 83, 84
Perz, R. J. M., 170
Petersen, R. C., 113
Peterson, M. R., Jr., 106
Peterson, P. E., 192
Petrarca, A. E., 229
Pettit, R., 57
Pihlaja, K., 217
Pinnick, H. R., Jr., 156
Pinschmidt, R. K., Jr., 84, 206
Pinto, J. A., 145
Platt, J. R., 30

Pocker, Y., 153
Pople, J. A., 48, 57, 98, 99, 104, 106, 234
Posposil, J., 135
Prelog, V., 222, 238
Preto, R. J., 157
Price, M. J., 96, 131
Pryor, W. A., 153
Puar, M. S., 194
Putz, G. J., 260

Raban, M., 230
Raber, D. J., 189
Radom, L., 106, 234
Ralph, E. K., 103
Rao, V. M., 263
Rapp, M. W., 156
Rauk, A., 239
Reagen, M. T., 170
Reichardt, C., 188, 208, 213
Reimann, J. E., 181
Richer, J.-C., 259
Riddell, F. G., 242, 261
Riggs, N. V., 28, 34, 36, 42, 43, 49, 54, 86
Rimerman, R. A., 262
Ritchie, C. D., 112, 176
Riveros, J. M., 104
Roberts, J. D., 3, 27, 43, 54, 86, 94, 96
Robinson, J. K., 153
Robinson, M. J. T., 242, 255, 262
Robinson, R., 104
Rochester, C. H., 165, 171, 173
Romers, C., 261
Ronayne, J., 217
Ross, J., 115
Roth, W. R., 82
Ruch, E., 222, 232
Rummens, F. H. A., 217
Rush, J. E., 229
Rutherford, R. J., 214

Sagatys, D. S., 154, 162
Sager, W. F., 112
St. Jacques, M., 262, 263
Saito, H., 263
Salem, L., 59, 82, 84, 206, 265
Salomaa, P., 207
Sanders, J. K. M., 218
Sargent, M. V., 49
Sato, Y., 176
Sauer, J., 80
Saunders, M., 266
Saunders, W. H., Jr., 153
Saville, B., 203
Schaad, L. J., 58
Schachtschneider, J. H., 84
Schadt, F. L., 192
Schaefer, T., 99
Scharpen, L. H., 263
Scheibe, G., 215
Scheinmann, F., 83
Scheppele, S. E., 154, 156
Schleyer, P. von R., 104, 106, 135, 155, 156, 189, 192, 267
Schmeising, H. N., 43, 47

Schmid, H., 83
Schmidt, W., 84, 206
Schneider, W. G., 48
Schofield, K., 146
Schowen, R. L., 181
Schreck, J., 138
Schroeter, S. H., 259
Schubert, W. M., 107–108, 217
Schwartz, M. E., 99
Seeley, D. A., 78
Segre, A. L., 264
Seip, H. M., 266
Seltzer, S., 157
Servis, K. L., 59, 65
Shapiro, S. A., 167, 170
Shen, K., 84
Sheppard, W. A., 97
Sherrod, S. A., 158
Shiner, V. J., Jr., 108, 155, 156, 248
Shone, R. L., 156
Shorter, J., 109, 128, 132, 135
Shoulders, B. A., 263
Sicher, J., 260
Silber, E., 30, 36
Simon, E., 158
Simonetta, M., 82
Smart, B. E., 105
Smith, B. H., 233
Smith, G. G., 135
Smith, W. B., 263
Snatzke, G., 271
Snyder, J. P., 49
Sondheimer, F., 51
Songstad, J., 199
Sonnichsen, G. C., 156
Sorensen, T., 105
Sousa, L. R., 84
Southam, R. M., 260
Spindler, E., 79
Spragg, R. A., 263
Srikrishnan, T., 266
Srinivasan, R., 266
Stang, P. J., 26, 217, 267, 268
Steigman, J., 135
Sternhell, S., 99
Stock, L. M., 94, 97, 105, 108, 118, 178
Storr, R. C., 59
Streitwieser, A., Jr., 3, 13, 27, 34, 38, 42, 43, 44, 47, 49, 54, 56, 86, 104, 140, 188, 248
Sunko, D. E., 156
Sussman, D., 135
Sustmann, R., 79
Swain, C. G., 136, 179, 181
Sweeney, W. A., 107
Szkrybalo, W., 258
Szur, A. J., 39
Szwarc, M., 45

Taagepera, M., 104
Taft, R. W., 99, 104, 131, 132
Takeuchi, Y., 97
Tashiro, M., 178
Taylor, G. R., 96

Taylor, R., 108, 155
Tel, L. M., 239
Thornton, E. R., 179, 181
Tickle, P., 170
Tidwell, T. T., 248
Topsom, R. D., 96, 131
Traylor, T. G., 108
Tribble, M. T., 267
Trindle, C., 84
Tsuda, K., 198
Tyminski, I. J., 262, 263

Ugi, I., 232
Ullman, E. F., 83
Untch, K. G., 57

VanCatledge, F. A., 239, 258
van der Lugt, W. Th. A. M., 83, 84
van Tamelen, E. E., 49
Velluz, L., 271
Verbanic, C. J., 108
Vitullo, V. P., 176
Vogel, E., 51
Vogel, G. C., 200
Vollmer, J. J., 59, 65

Wagner, J. J., 266
Wahl, G. H., Jr., 106, 158
Wallbillich, G., 79
Waller, F. J., 192
Wasserman, A., 80
Webb, H. M., 104
Wells, P. R., 109, 128
Wertz, D. H., 267
Westheimer, F. H., 150, 233
White, A. M., 140

Whitesides, T. H., 84
Whiting, M. C., 260
Wiberg, K. B., 13, 20, 25, 26, 28, 38, 42, 43, 45, 86, 105, 109, 114, 121, 126, 128, 148, 150, 159, 161, 165, 173, 184, 188, 208, 209, 216, 219, 265
Wigfield, D. C., 197
Wilcox, C. F., 96
Wilen, S. H., 223
Wiley, G. R., 130, 131
Willadsen, T., 266
Williams, D. H., 217, 218
Williams, J. E., 267
Wilson, J. D., 49
Wilson, J. M., 170
Winstein, S., 54, 56
Wirkkala, R. A., 108
Wolfe, R. A., 104
Wolfe, S., 239, 254
Wolfsberg, M., 150
Wollrab, J. E., 263
Woodward, R. B., 59, 65, 86, 216
Woodworth, C. W., 104, 135
Wright, J. S., 84, 265
Wysocki, D. C., 57

Yamdagni, R., 104
Yates, K., 173
Yee, K. C., 163
Yokoyama, T., 130
Young, L. B., 104

Zavitsas, A. A., 145
Zimmerman, H. E., 36, 84
Zollinger, H., 153, 155
Zook, H. D., 197

Subject Index

A1, 174
A2, 174–175
Absolute configuration, 224
Absolute reaction rate theory, 120–124
Acetaldehyde, 231, 241
Acetal hydrolysis, 148–149
Acetic acid, 192–193
Acetone, 150, 215, 241
Acetophenones, 142, 175
Acetylation, 179
Acetylene, 26
Acetyl halides, 241
Acidity, 161–162, 163, 165–173, 200–204
 alcohols, 104
 and hybridization, 26
Acidity function, 169–173
 and mechanism, 173–176
Activated complex, 121–123 (*see also* Transition state)
Active site, 146
Activity, 165–175
Activity coefficient, 166–175
Acylation of enolates, 197
Acyl-oxygen fission, 139
2-Adamantyl tosylate, 193
Alcohols:
 acidities, 104
 oxidation, 259
Aldehydes, 241–242
Aliphatic amine, 100
Alkenes, 175
 stabilities, 106–107
N-Alkylacetaldimine, 241
Alkylation of enolates, 197
Alkylbenzenes, 103, 105, 154
Alkyl-group effect, 97, 103–108, 156
Alkyl halide, 91–92, 155, 197, 203
Alkyl hydrogen sulfates, 168
3-Alkyl ketone effect, 262
Alkyl-oxygen fission, 140
2-Alkylpyridines, 135
Allenes, 223, 225
Allyl system, 34–35, 68, 141
 charge densities, 41
 NBMO coefficients, 35–36
Alpha effect, 183, 184, 196, 198–199, 205–206
Alpha isotope effect, 155–156, 157
Alternant hydrocarbon, 33–37, 49
 even, 34–35
 odd, 34–35, 38, 41
Alternation in magnitude of polarity, 98–99
Ambident anion, 196–198, 204, 205
Amine basicities, 103–104
Ammonia, 24–25, 103–104
Anancomeric group, 254–255
Angle strain, 249, 265
Angular momentum, 17

Anilines, 100–101, 130, 169–172, 216–217
Anilinium salts, 141
Annulenes, 50–52, 58
Ansa compounds, 233
Antarafacial:
 component, 68–69
 migration, 67–70
Anthracene, 58
Anthropomorphic rule, 181
Antiaromaticity, 34, 56–58
 criteria, 57
 destabilization, 57
 in pericyclic reactions, 85
Antibonding, 16, 17, 31–33, 43, 198
Anti conformation, 237–239, 244, 269
Aromatic Claisen rearrangement, 83
Aromaticity, 36, 48–55, 57–58
 closed-shell systems, 49
 criteria, 48–49
 in pericyclic reactions, 85
 polycyclic systems, 52–53
 VB theory, 52
Aromatic-solvent-induced shifts, 217–218
Arrhenius theory, 120, 124
Atropisomerism, 233
Attenuation per methylene, 133–134
Axial bond, 250–251
Axial chirality, 233
Azulene, 52–53, 58

Back-donation, 98
Baker-Nathan effect, 103, 105–108
Banana bond, 24
Barrier:
 to inversion, 262
 to ring inversion, 252, 253, 263, 264
 to rotation, 235–236
Basicity, 103–104, 141, 153, 165–173, 184–188, 194, 197, 200–204
Bathochromic shift, 211, 216
Bell-Evans-Polanyi principle, 176–179
Bending vibration, 152–155
 and hybridization, 25–26
Bent bond, 24
Benzaldehyde oxime, 229
Benzaldehydes, 142–145, 232
Benzene, 31–33, 37, 46–50, 58, 105, 118–119, 154, 165–166
 delocalization energy, 31
 MO theory, 31–33
 orbital energies, 31–33
 pi energy, 31
 resonance energy, 46–48
 VB theory, 37
Benzenonium ion, 155, 178–179
Benzoic acids, 88, 102, 109–112, 128, 131, 145
Benzoquinuclidine, 100

283

Benzyl system, 34–36
  bromination, 154
  charge densities, 41
  halides, 140, 195
  NBMO coefficients, 35–36
Beta isotope effect, 108, 155–159
Bicycloaromaticity, 55–56
Bicyclobutanes, 78
Bicyclo[3.1.0]hexenes, 78–79
cis-Bicyclo[3.1.0]hex-3-yl tosylate, 55
Bicyclo[3.2.2]nonatrienyl anion, 55–56
Bicyclooctadienyl anion, 55
Bicyclo[2.2.2]octanes, 131–134
  -1-carboxylic acids, 94–95, 97, 111
Bicyclo[2.2.2]oct-2-ene-1-carboxylic acids, 97
Bimolecular elimination, 260–261 (see also Elimination)
Biphenylene, 53, 58
Biphenyls, 158, 223, 233
Bisected conformation, 241–242
Bishomocyclopentadienide anion, 55
Boat-chair conformation, 266
Boat conformation (see Cyclohexane)
Boltzmann distribution law, 121
Bond anisotropy, 268–269
Bond–bond repulsion, 240
Bond dissociation energy, 45, 145
Bond energy, 44–45, 57–58
  and hybridization, 25–26
Bonding, 15, 17, 31–33, 43
Bond length, 43
Bond order, 43
Bond strength, 45
Born-Oppenheimer principle, 9
Boron hydrides, 19
Boron trifluoride, 135, 201
Branch and Calvin model, 92–93, 134, 162
Bromination, 105, 119, 141, 142, 154, 179
  benzylic, 141, 154
Bromobenzene, 90
1-Bromo-2-chloro-1-iodoethylene, 229
N-Bromosuccinimide, 141, 154
Brønsted α, 161–164
Brønsted β, 199
Brønsted catalysis law, 160–164, 179, 184
B-strain, 247–248
Buffer, 127, 150
Bunnett parameters, 175–176
1,3-Butadiene, 28–30, 33, 48, 57, 60–67, 85
  conformational isomers, 28, 242–243
  coplanarity, 28
  in cycloadditions, 71–77
  delocalization energy, 29
  MO theory, 28–29
  orbital energies, 29
  pi energy, 29
  wave functions, 29–30
n-Butane, 46, 237–239, 244
  conformational equilibrium, 244
1-Butene, 241
cis-2-Butene, 244
trans-2-Butene, 46
tert-Butylamine, 247

tert-Butylbenzene, 105
4-tert-Butyl-1-carboethoxycyclohexanes, 254
tert-Butyl chloride, 189, 192
tert-Butylcyclohexanes, 254
4-tert-Butylcyclohexanols, 259
n-Butyl fluoride, 98

Cahn-Ingold-Prelog system, 223, 225–228, 231, 232
Calciferol, 61, 67
Cannizzaro reaction, 142–143
Carbanion stabilities, 103–104
Carbinolamines, 143–144
trans-2-Carboethoxydecalins, 255
Carbon isotopes, 152
Carbonium ion stabilities, 103, 106
Carbonyl compounds, empirical rules for uv, 39–40
Carbonyl stretching frequency in 2-halocyclohexanones, 262
Carboxylic acids, 162–165
Catalysis, 84, 148–150, 181, 198–199, 203
Cation effect, 197
Center of asymmetry, 223
Charge-controlled reaction, 204–206
Charge delocalization, 156, 190, 198
Charge density, 36, 41, 180, 196, 205
  allyl, 41
  benzyl, 41
  odd alternant hydrocarbon, 41
Charge effect, 162, 171, 196
Charge transfer, 108, 200, 208, 211, 212, 213
Cheletropic reactions, 83–84
Chirality, 222–228
  axis, 223, 225
  center, 223–228
  element, 273
  plane, 223
  rule, 226
Chiroptical methods, 270–275
Chloroacetic acid, 93
Chlorobenzene, 89
p-Chlorobenzoylperoxy radicals, 145
2-Chlorobutane, 226–227
Chlorocyclohexane, 253
Chloromethane, 22
trans-2-Chloro-5-methylcyclohexanone, 275
Chlorophenylpropiolic acids, 96
Circular dichroism, 271–275
Cis, 228–229
s-cis conformation, 242–243
Citric acid, 231–232
Citric acid cycle, 231
Claisen rearrangement, 83
Clinal zone, 239
Coalescence temperature, 252
Cogwheeling, 258
Collision complex, 217
Collision factor, 120
Collision theory, 120, 124
Common rings, 265
Competitive reactions, 117, 118–120
Concerted process, 59–85, 157, 206
Configuration, 224, 275

Subject Index

Configurational equilibria, evaluation of, 245–246
Configuration interaction, 19, 84
Conformation, 234, 275
    bias, 254–255, 267, 268
    n-butane, 237–239, 244
    cyclohexane, 249–258
    effect, 132–133, 156, 158, 198
    entropy, 243–244
    interconversion paths, 266
Conjugation, 108
    effect, 99
Conrotatory, 62–67, 85
Consecutive reactions, 117
Cope rearrangement, 82–83, 156
Correlation diagram, 65, 71–77
Cotton-effect curve, 271–275
Coulombic interaction, 208
Coulomb integral, 10, 13, 18
Coulomb repulsion energy, 14–16
Cram's rule, 232
Cross-conjugated systems, 39–40
Crown conformation, 266
Cryoscopy, 168
Cubanecarboxylic acids, 97
Cumene, 105, 154
Cumulenes, 229–230
Cumyl chlorides (*see* 2-Phenyl-2-propyl chlorides)
C values, 195–196, 200–201
Cycloaddition reactions, 70–79, 206
    secondary interactions, 79–83
    stereochemistry, 71–72
Cyclobutadiene, 34, 37, 56–58
    delocalization energy, 33
    MO theory, 33
    orbital energies, 33
    pi energy, 33
Cyclobutadienyl dianion, 50
Cyclobutane, 71–77, 264–265
Cyclobutenes, 60–67, 85
    fused, 66
Cyclobutenyl dication, 49–50
Cyclodecapentaene, 49–51
Cyclododecane, 265
Cycloheptane, 265
Cycloheptatriene, acidity of, 52
Cycloheptatrienyl cation, 50
Cyclohexadiene, 61–67, 85
Cyclohexane, 249–252, 265, 267, 269–270
    boat, 251
    chair, 249–251
    disubstituted, 255–258
    half-chair, 251
    monosubstituted, 252–254
    ring-flipping, 251–252
    twist, 251
Cyclohexane-1,2-diols, 268
Cyclohexane-1,4-dione, 263
Cyclohexane-1,4-dioxime, 263
Cyclohexanols, 259, 267–268
Cyclohexanones, 197, 213, 218, 261–262, 273–274
Cyclohexatriene, 46–47

Cyclohexenes, 76–77, 263
    3-substituted, 263
    4-substituted, 263
Cyclononane, 266
Cyclononatetraenyl anion, 49–50
Cyclooctane, 266
Cyclooctatetraene, 49, 55–58
Cyclooctatetraenyl dianion, 49–50
trans-Cyclooctene, 223
Cyclopentadiene, 209–210
    acidity, 52
Cyclopentadienyl anion, 50, 52
Cyclopentadienyl cation, 56–57
Cyclopentane, 265
Cyclopentane-1,2-diols, 268
Cyclopentanone, 266
Cyclopropane, 24, 70, 264, 265
Cyclopropenyl anion, 56
Cyclopropenyl cation, 50
Cycloreversion reaction, 70
Cycloundecane, 265

Debye-Hückel theory, 126
trans-2-Decalyl compounds, 254–255
Dehydration, 143–144
Delocalization energy, 29, 37–38, 44, 48 (*see also* Resonance energy)
    allyl, 35
    benzene, 31
    cyclobutadiene, 33
    odd alternant hydrocarbon, 35
Delocalization of electrons, 70, 103–108 (*see also* Electron delocalization)
Deshielding (*see* Nmr spectroscopy)
Desulfonation, 154
Deuterium, 151–152, 154–160, 162
Deuterium oxide, 159–160
Dewar resonance energy, 57–58
Dextrorotatory, 223
Diamagnetic anisotropy, 268–269
Diamagnetic susceptibility, 48–49, 57
Diamond lattice, 266
Diastereomers, 223, 227–230
Diastereotopic faces, 232
Diastereotopic ligands, 230–231
Diazo compounds, 150
Dibenzobicyclo[2.2.2]octa-2,5-diene-1-carboxylic acids, 97
Dibenzoyl peroxides, 144
Dibromobenzenes, 90–91
2,3-Dibromobutane, 245–246
2,3-Dibromosteroids, 260
Dichloroacetic acid, 93
2,3-Dichlorobutane, 228
2,3-Dichloropentane, 227–228
Dielectric constant, 93–94, 172, 204, 205, 208–210, 214
Diels-Alder reaction, 43, 61, 76–77, 79–82, 85, 157, 209–210
Dienes, 80–82
    empirical rules for uv, 39–40
Dienophile, 80–82
Diethylamine, 247
1,3-Dihalocyclobutanes, 265

Dihydropyrenes, 51
Diimide reductions, 83–84
2,2′-Diiodobiphenyl, 233
1,2-Dimethoxyethane, 240
Dimethylamine, 103–104, 247
N,N-Dimethylanilines, 100–101, 140
N,N-Dimethylanilinium salts, 141
2,2′-Dimethylbiphenyl, 233
Dimethylbromobenzenes, 119
N,N-Dimethyl-4-*tert*-butylcyclohexylamines, 259
Dimethylcyclohexanes, 255–258
1,4-Dimethylenecyclohexane, 263
2,6-Dimethylpyridine, 247
1,4-Dioxan, 264
Dipolar aprotic solvents, 194–197, 203
1,3-Dipolar cycloadditions, 79
Dipolar interaction, 93, 94, 196, 215–218, 240, 255, 262
Dipolar intermediate, 79
Dipole moment, 18, 19, 41, 88–91, 96, 101, 103, 106, 208–211, 216, 275
  in conformational analysis, 262
  induced, 89
  MO theory, 42
  nonalternant hydrocarbon, 42
Directional character of hybrid orbitals, 21
Direct resonance interaction, 129–131
Disrotatory, 62–67, 85
Dissymmetric chromophore, 272
Dissymmetry, 222 (*see also* Chirality)
Double bond rigidity, 22 (*see also* Geometrical isomerism)
DRE (*see* Dewar resonance energy)
Drug action, 145–146

Eclipsed conformation, 235, 241–242
Edwards equations, 185–188, 194, 199, 201, 203, 205–206
Effective nuclear charge, 6–7
Eigenfunction, 4
Eigenvalue, 5
Equatorial bond, 250–251
Equivalent ligands, 231
Electrocyclic reaction, 61–67
Electron affinity, 38, 205
Electron correlation, 19
Electron delocalization, 40, 44, 46, 54–58, 103–108, 180, 196 (*see also* Delocalization of electrons)
Electron density, 15, 16, 18, 19, 27, 131, 154, 155, 158, 181, 197, 198
  at atom, 40
  neutral alternant hydrocarbon, 41
  total, 40–41
Electron diffraction, 49, 267
Electronegativity, 22, 89–91, 98–99, 102, 142, 156, 199, 202, 204
  and hybridization, 26
Electronic effect, 180
Electronic transition, 19, 38–40, 210–212, 215–216, 271–272
  empirical rules, 39–40
  selection rules, 39

Electron impact, 38
Electron interaction, 17, 42
Electron repulsion, 38, 198
Electrons, wavelike properties, 3
Electrophilic aliphatic substitution, 209
Electrophilic aromatic substitution, 48, 89, 103, 105–107, 118–120, 129, 144, 154–155, 178–179
Electrophilicity, 178–179, 185, 186, 196, 197
Electrophilic substituent constants, 129–130, 137
Electrostatic effect, 91–102, 103–108, 111–112, 129–137, 146, 156–158, 196, 200, 204, 208
  model, 208
  proximity effect, 135–136
  substituent constants, 111, 136–137
Elimination, 83, 157, 182, 189, 204, 208 (*see also* Bimolecular elimination)
  anti-coplanar, 260
  syn-coplanar, 260
Emulsin, 232
Enantiomer, 223
Enantiotopic faces, 232
Enantiotopic ligands, 230
Endo orientation, 80–82
Enediones, 39
Energy (*see also* Potential energy, Kinetic energy):
  of activation, 120, 151–152, 177–178
  and amplitude, 3
  commutation, 5
  and nodes, 3
  pi, 27–37
  total, 14–16
Enhanced resonance interaction, 129–131
Enol, 142
Enolate ion, 197–198, 204
Enolization, 149–150, 175
Enthalpy:
  of activation, 123–124, 182, 194–196
  of formation, 200
  of solution, 194–195
  of transfer, 182, 195–196
Entropic constraint, 83
Entropy:
  of activation, 123–125
  of mixing, 243–244, 246, 256, 257
  of symmetry number, 243–244, 246, 256, 257
Envelope conformation, 265
Enzyme stereospecificity, 231
Epoxide stereospecificity, 261
2-Epoxysteroids, 261
E_s, 134–135
Esr spectra, 108 (*see also* Hyperfine splitting)
Ester hydrolysis, 132–134, 138–140 (*see also* Hydrolysis of esters, Saponification)
Ethane, 26, 234–236
Ethanol, 100, 103–104, 230–231
Ether hydrolysis, 154
Ethyl acetate, 202
Ethyl acetoacetate, 197
Ethylamine, 247

Ethylbenzene, 105, 154
Ethyl benzoates, 107, 139–140
Ethyl p-tert-butylbenzoate, 107
Ethyl 4-tert-butylcyclohexanecarboxylates, 259
1-Ethyl-4-carbomethoxypyridinium iodide, 211–213
Ethylene, 21–22, 26, 33, 85
 in cycloadditions, 71–77
 dimerization, 71–77
 MO theory, 27
 orbital energies, 27
 pi energy, 27
Ethyl iodide, 236
Ethyl p-methylbenzoate, 107
E_T values, 210, 212–214
E values, 200–201
Exchange integral, 11, 18
Exchange phenomenon, in VB theory, 18
Excited state of hydrogen molecule ion, 13, 16
Exo orientation, 80–82
Extrathermodynamic relationship, 109, 195
Eyring equation, 123–124
E-Z system, 78, 229

Field- and charge-transfer approach, 137
Field effect, 91, 93–95, 96–97, 100, 135 (see also Electrostatic effect)
First-order reaction, 115–116
Fischer projection, 224
Flagpole interaction, 251
^{19}F chemical shift, 131
Fluorobenzenes, 131
Forbidden character in electron excitation, 39
Force constant, 152–154, 180
Formic acid, 192–193
$4n + 2$ rule, 49–54, 56, 85
Franck-Condon principle, 212, 213
Free-electron model, 30
Free energy of activation, 123–124, 161
Free-radical reactions, 144–145
Freezing-point depression, 168
Friedel-Crafts alkylation, 179
Frontier-controlled reaction, 204–206
Frontier orbital, 63–64, 83, 205–206
F-strain, 159, 247
Fumaric acid, 93
Furan, 53–54
Furst-Plattner rule, 261

Gauche-butane interaction, 252, 256, 257
Gauche conformation, 237–239, 244, 269
Gauche effect, 239–242, 264
Gegenion effect, 108
General acid catalysis, 127, 149–150, 159–164
General base catalysis, 150, 161
Geometrical isomerism, 61, 223, 228–229
Glucose, 150
Ground-state compression, 259
Ground state of hydrogen molecule ion, 13–15

Group–group repulsion, 240
Group transfers, 83

H_-, 170–171
H_{2-}, 171
H_0, 168–175
H_R, 171–172, 175
Half-chair conformation, 263, 265
2-Halocyclohexanones, 262
α-Haloether, 198
Halogenation, 150 (see also Bromination)
Hamiltonian operator, 4–8
Hammett acidity function, 169–173 (see also H_-, H_{2-}, H_0, H_R)
Hammett base, 167–170, 174–175
Hammett equation, 108–113, 128–146, 178–179, 184
Hammett substituent constant, 110 (see also σ)
Hammett-Zucker hypothesis, 173–175
Hammond postulate, 176–179, 182, 197
Hard acid, 199–204
Hard base, 199–204
Hard and soft acids and bases, 199–204
Harmonic oscillator, 151
Heat of combustion, 46
Heat of formation, 37, 45, 46, 58
 environmental contributions, 45
Heat of hydrogenation:
 of benzene, 46
 of cyclohexene, 46
 definition, 46
Heisenberg uncertainty principle, 5, 10
Heitler-London approach, 18
Hemiketals, 201
Heteroaromaticity, 53–54
 MO theory, 54
Heteropolar system, 17–18
1,5-Hexadienes, 82–83
Hexatrienes, 61–67, 78–79, 85
Hindered rotation, 233, 234–238
HOMO, 38, 198, 205
 in electrocyclic processes, 63–64
Homoaromaticity, 54–56
tris-Homocyclopropenyl cations, 54–55
Homopolar system, 17–19
Homotopic ligands, 231
HSAB principle, 199–205
Hückel $4n + 2$ rule, 49–52 (see also $4n + 2$ rule)
Hückel MO theory, 13, 28, 34, 37–39, 42, 58 (see also MO theory)
Hückel process, 85
Hund's rule, 8, 9
Hybridization, 20–27, 47, 97, 155
 and acidity, 26
 in ammonia, 24–25
 and bending constants, 25–26
 and bond energy, 25–26
 of carbon, 20–28
 in cyclopropane, 24
 and electronegativity, 26
 in ethylene, 21–23
 index, 22–23

Hybridization—*Continued*
  in methane, 20–21
  nonequivalent orbitals, 22–23
  parameter, 22–23, 26–27
  and stretching constants, 25–26
Hybrid orbitals, nonintegral values, 22–25
Hydration of alkenes, 175
Hydration parameters, 175–176
Hydrazines, 198, 240
Hydrazones, 171
Hydrobromination, 261
Hydrogenation, 46
Hydrogen atom, 5–6
Hydrogen bonding, 93, 125, 150, 172, 181, 190, 194–196, 208, 215, 216, 267–268
Hydrogen chloride, 91
Hydrogen cyanide, 232
Hydrogen exchange, 55
Hydrogenic orbitals, 6
Hydrogen molecule ion, 9–18
  MO solutions, 13–14
  orbital energies, 13
  VB theory, 18
Hydrogen peroxide, 240
Hydrogen rearrangement, 67–68
Hydrolysis:
  of acetals, 148–149
  of esters, 132–134, 138–140, 174–175 (*see also* Saponification)
  of ethers, 154
  of lactones, 174–175
Hydroperoxide anions, 198
Hydrophobic bonding, 146
Hydroxamic acids, 198
Hydroxybenzoic acids, 102
Hydroxylamines, 198
Hyperconjugation, 103–108, 156
  model, 105–106
Hyperfine splitting, 108
  in aromatic radical ions, 42
Hypsochromic shift, 211, 216

I classification, 95, 102
i factor, 159
Imines, 143–144
Indicator, 167–168
Inductive effect, 54, 91–100, 133–135 (*see also* Electrostatic effect)
Inherently dissymmetric chromophore, 272
Interligand interactions, 21
Intermolecular effects, 164
Internal solvation 104
Internuclear angle, 23, 24
Interorbital angle, 23, 24
Interorbital electron repulsion, 21
Inversion of configuration, 70, 225
Inverted molecule, 96
Ion cyclotron resonance, 104
Ionic interaction, 176, 204
Ionic strength, 126–127, 150, 170
Ionization, 163–164
Ionization constant, 109–112
Ionization potential, 9, 38, 108, 205
Ionizing power, 189

Ion-pair formation, 126, 155–156, 172, 196, 212, 260, 261
Isobutane, 46, 105, 244
Isobutylene, 46
Isoenthalpic relationship, 112
Isoentropic relationship, 112
Isoequilibrium relationship, 113
Isokinetic relationship, 113
Isomers, 222
Isopropylation, 179
Isopropylbenzene, 154
Isosteres, 135
Isosterism, 134–135
Isotope effect, 108, 150–160 (*see also* Primary isotope effect, Secondary kinetic isotope effect)
Isovalent hyperconjugation, 106

Karplus equation, 270
Kekulé structure, 53
Ketals, 201
Kinetic control, 118, 209
Kinetic energy, 4, 15, 16
  of hydrogen atom, 6
Kinetic isotope effect, 150–160 (*see also* Isotope effect)
Kinetics, simple types of, 114–117
Kinetic theory of gases, 120
Kirkwood-Westheimer model, 94–97
Klyne and Prelog system, 238–239
Krebs cycle, 231

Lanthanide-induced shifts, 218
Large rings, 265
LCAO method, 9–18
Leaving group, 155–156
Leveling effect, 165
Levorotatory, 223
Lewis acid, 247
Lewis base, 247
Limiting process, 190
Linear combination of atomic orbitals (*see* LCAO method)
Linear free-energy relationship, 108–113, 160–161, 179, 195–196
Liver alcohol dehydrogenase, 231
Lock-and-key hypothesis, 146
Log k_{ion}, 190–191, 208, 210, 213
London force, 204, 208
LUMO, 38, 205
Lyonium ion, 148–150

Magnetic anisotropy, 48, 217
Maleic anhydride, 157
Malic acid, 231
Malonate ion, 204
Malonic ester, 197
McConnell approximation, 268–269
Medium rings, 265
Menshutkin reaction, 195
Merocyanine dyes, 211
Mesityl oxide, 213, 215–216
Meso compound, 228

Metal catalysis, 84
Methanol, 103–104, 236
Methoxymethyl chloride, 204
p-Methoxyneophyl tosylate, 190
Methyl acrylate, 209–210
Methylamine, 103–104, 236, 247
Methyl benzoates, 139–140
Methyl bromide, 185, 192
Methyl chloride, 202
2-Methylcyclohexanones, 218
trans-10-Methyl-2-decalone, 274-275
Methylene chloride, 231
Methylenecyclohexane, 261–262
2-Methylfuran, 157
5-Methyl-2,4-heptadiene, 229
Methyl iodide, 124–125, 135, 140, 204
Methylmercury cation, 200
Methyl tosylate, 193
Methyl vinyl ketone, 243
Microscopic reversibility (*see* Principle of microscopic reversibility)
Microwave spectrometry, 49, 267
Mixing coefficient, 22–23
Mobile pi-bond order, 43
Möbius process, 85
Molar refractivity, 187
Molecularity of reaction, 115
Molecular mechanics calculations, 266, 267
Molecular orbital, 9, 105–106 (*see also* MO theory)
  primary characteristics, 9
  symmetry classification, 71–77
Momentum, 5
Monohomotropylium cation, 54–55
MO theory, 19, 58, 205 (*see also* Hückel MO theory)
  benzene, 31–33
  butadiene, 28–29
  cyclobutadiene, 33
  ethylene, 27
  hydrogen molecule ion, 9–18
  spin density, 41–42
Multicenter additions, 79
Mutarotation, 150
m values, 188–193

NAD^+, 231
NADD, 231
NADH, 231
Naphthalene, 53, 58
2-Naphthol, 197
NBMO, 34–35, 38, 41–42
NBMO coefficients:
  odd alternant hydrocarbons, 35–36
  zero-sum rule, 35–36
Neglect of non-neighbor interactions, 13, 28
Neglect of overlap, 13, 18
Neopentane, 105
Neopentyl iodide, 125
Neopentyl tosylate, 186
Newman projection, 235, 237–239
Nicotinamide adenine dinucleotide, 231
Nitroalkanes, 153, 163–164
Nitrobenzene, 101

p-Nitrophenol, 101
Nitrous acid, 93
Nmr spectroscopy, 99, 268–270
No-bond resonance, 106
Nodes, 3, 30–32, 85
Nonalternant hydrocarbon, 41
Nonbonded interaction, 157–158
Nonbonding character, 31–33
Nonbonding orbital (*see* NBMO)
Nonbonding state, 16
Nonpolar interactions, 145
Nonspecific effect, 160
Nonvertical process, 108
Normal electrostatic order, 105, 107, 108, 134, 136
Normal inductive order (*see* Normal electrostatic order)
Normalization, 6, 22, 35
  constant, 6, 14
  hydrogen molecule ion, 14
Normal mode, 182
Normal substituent constants, 130
Nuclear attraction integral, 10
Nuclear repulsion energy, 15, 16
Nucleophile, 183–206
  definition, 183–184
Nucleophilic aliphatic substitution, 140–141, 155–157, 183–206, 208, 209
Nucleophilic aromatic substitution, 130, 195
Nucleophilic attack, 142, 144
Nucleophilicity, 156, 176, 179, 183–206
Nucleophilic participation, 156
Nucleophilic substituent constants, 130
N values, 191–192

Octanol, 146
Octant rule, 273–275
Odd alternant hydrocarbon, 41
O—H stretching frequency, 267–268
Ω values, 209, 213
One-electron bond, 19
One-electron property, 19
Operator, 4
  lack of commutation, 4
Optical activity, 223
Optical rotatory dispersion, 271–275
Orbital exponent, 6–7
Orbital overlap, 54
Orbital symmetry, 206
  conservation of, 59–85
Order of reaction, 114–115
Orthogonality, 11, 22
Ortho-substituted benzenes, 111, 132–136
Overlap, 9, 54, 102, 136, 180, 186, 198
  integral, 11
  $2p$ orbitals, 11–12
  and *s* character, 25–26
Overlap method, 167–168
Oxibase scale, 186–188
Oxidation half cell, 187
Oxidation of hydrocarbons, 46
Oxidative dimerization, 186–187
Oximate anion, 198
Ozonolysis, 43

Paracyclophanes, 223, 233
Partial rate factor, 118–120, 178–179, 197
Particle in a box, 3, 30
Partition coefficient factor, 146
Partition function, 122–123
Pauli exclusion principle, 8, 9
trans-3-Penten-2-one, 243
Perhomobenzene, 55
Perhomocyclopentadienide anion, 55
Pericyclic reaction, 60
Perimeter free-electron model, 31–32
Periodic table, 8
Periplanar zone, 239
Perturbational MO treatment, 84, 85, 204–206
  frontier orbital, 84, 85
Perturbation method, 11
Phenanthrene, 53, 58
α-Phenethyl methyl ketone, 232
Phenol blue, 211, 213
Phenols, 100, 101, 130, 162, 165, 171
Phenylacetic acids, 109
Phenylpropiolic acids, 96
2-Phenyl-2-propyl chlorides, 129–130, 180
ϕ (phi), 175–176
Photochemical reactions, 60–62, 64, 68, 76–79, 85
  triplet, 83
Photoionization, 38
pH scale, 165–166, 169, 171, 174
π-bonded complex, 155, 179
π-electron approximation, 27, 38
  failure, 42
π energy levels, polygon in a circle, 36–37
π orbitals, 17
π system, 99
Plain curve, 271
Planar chirality, 233
Plane-polarized light, 223
Polar character, 145, 209
Polar effect, 84, 145, 180, 240 (*see also* Electrostatic effect)
Polarity, 19, 88–91, 98–99, 208–212, 214–216
Polarizability, 88–91, 104, 156, 184, 186–188, 194, 196, 198, 199–204, 205, 274
Polarizability constant, 89
Polarizability value, 187
Polarographic oxidation potential, 38
Polarographic reduction potential, 38
Polar substituent constants, 96, 133–135
Polygon-in-a-circle, 36–37
Pople-Gordon classification, 98–99
Pople-Gordon effect, 98–99
Potential energy, 4
  Coulombic, 6
  of hydrogen atom, 6
Potential-energy diagram, 120–121
Potential-energy surface, 150, 164, 179
Precalciferol, 61, 67
Pre-exponential factor, 120, 124
Primary isotope effect, 150–155, 157, 159, 162, 181
Primary salt effect, 126–127

Primary value, 111
Principle of hard and soft acids and bases (*see* HSAB principle)
Principle of microscopic reversibility, 62, 124, 157
Probability factor, 120
Probability of finding electron, 4, 5
Prochiral element, 230
Prochirality, 230–232
n-Propane, 244
1-Propanol, 202
2-Propanol, 202
Propenes, 241
  3-substituted, 242
Propionaldehyde, 241
Proton addition, 154
Proton transfer, 161–164
Pseudoaromaticity, 56
Pseudoaxial bond, 263, 265
Pseudoequatorial bond, 263, 265
Pseudorotation, 265–266
Pyridine, 54, 195
Pyridinium-N-phenol betaines, 211, 213
Pyrolysis of acetates, 261
Pyrolysis of xanthates, 261
Pyrrole, 53–54
Pyrylium cation, 54

Quantum numbers, 6
Quinones, 172
Quinuclidine, 100
Q values, 192–193

Racemate, 223
Racemization, 158, 225, 233
Radiationless transition, 83
Radical, 108
Rate constant, 123
Rate-determining step, 115, 126, 138, 143–144, 148–151, 154–156, 160, 161, 173, 177, 179, 181, 205, 214
Rate expression, 114–117
R classification, 102
Reacting bond, 179–181
Reacting-bond rule, 179–181
Reaction constant (*see* ρ)
Reaction coordinate, 121–123, 152–153, 156, 163, 177–179, 181, 182
Reaction profile, 121–122
Reaction rate, 114–115, 122–123
Reactivity, 154
Reactivity indices, 42
Reactivity–selectivity relationship, 119–120, 162, 176–179
Re-face, 232
Reflection symmetry, 223
Refractive index, 89, 208, 214
Refractivity, 187
Relative configuration, 224
Resolution, 223
Resonance effect, 98–103, 111–112, 129–132, 135–136
Resonance energy, 29, 44, 46–48 (*see also* Delocalization energy)

Resonance energy—*Continued*
 benzene, 46–48
 butadiene, 29
 empirical, 47
 hypothetical localized structures, 46–47
 vertical, 47
Resonance integral, 11, 13, 204–205
Resonance proximity effect, 135–136
Resonance substituent constants, 136–137
Resonance theory, 211
Retention of configuration, 225
Reversible reaction, 116–117
ρ, 109–113 (*see also* Hammett equation)
 temperature dependence, 112–113
Ring current, 48, 57
Ring flipping (*see* Ring inversion)
Ring inversion, 252, 275
 in cyclohexane, 249–252, 263, 270
 in cyclohexene, 263, 270
Rotation (*see also* Conformation):
 about sp^2–sp^3, 241–242
 about sp^2–sp^2, 242–243
Rotation symmetry, 243
R-S system, 223, 225–228, 231

Sacrificial hyperconjugation, 106–107
Salt effect, 126–127, 207, 213–214
Saponification, 107, 132–133, 138–139, 258–259 (*see also* Hydrolysis of esters)
Saturation effect, 93, 131
Sawhorse diagram, 235
s character, by nmr, 26–27
Schiff base, 143–144
Schrödinger equation, 4–8
 hydrogen atom, 6
 hydrogen molecule ion, 9–18
 polyelectronic systems, 8
 polynuclear systems, 8
Schubert-Sweeney model, 107–108
Screening constant, 6
Secondary kinetic isotope effect, 155–160
Secondary salt effect, 126–127
Secondary value, 111
Second-order reaction, 116
Secular equations, 12–13
 benzene, 31
 cyclobutadiene, 33
 ethylene, 27
 general Hückel, 30–31
Selectivity, 119–120, 178–179
Semicarbazide, 144
Sequence rules, 225–226
Shielding, 269
Shielding constant, 6
Si-face, 232
σ (sigma), 110–113, 128, 130, 137, 179
σ^-, 130
σ^+, 129–130, 136, 137, 140, 141, 143–144, 178–180
σ', 111, 131, 133, 136
σ^*, 133–137, 179
σ^0, 130
σ-bonded complex, 155, 178–179
σ orbital, 17

σ–π separation, 38
Sigmatropic rearrangement, 67–70
 stereochemistry, 67–70
Skewed conformation, 235
Slater orbital, 6
Small rings, 264–265
S_N1, 115–117, 129–130, 140–141, 155–157, 177, 180, 182, 189–191, 193, 198, 248, 259–260
S_N2, 105, 116, 117, 124–125, 140–141, 155–156, 177–178, 182, 184–186, 190, 191, 194–195, 197, 203, 204, 259–260
Soft acid, 199–204
Soft base, 199–204
Softness factor, 200
Solubility rule, 201–202
Solvation changes, 124–125
Solvation model, 105, 107–108
Solvation rule, 181
Solvatochromism, 210–214
Solvent activity coefficient, 195–196
Solvent effect, 107–108, 112, 124–125, 130, 132, 135–136, 141, 155, 160–162, 165–167, 170, 172, 175–176, 179–183, 188–197, 201–203, 205, 207–218
 on amine basicities, 248
 on conformational equilibria, 239, 246
 on Cotton-effect curve, 275
Solvent isotope effect, 150, 159–160
Solvolysis, 140–141, 156, 188, 259–260 (*see also* S_N1, S_N2)
Specific acid catalysis, 148–150, 162
Specific base catalysis, 149
Specific rate constant, 114
sp hybrids, 22
sp^3 hybrids, 21
sp^2 hybrids, 21–22
Spin density, 41–42
 benzyl radical, 41–42
 and hyperfine couplings, 42
 MO theory, 41–42
Spin–spin coupling, 26–27
 constants, 270
Spin states, in electronic excitation, 39
Staggered conformation, 235
Standard electrode potential, 186–188
Steady-state approximation, 117
Stereoelectronic effect, 156, 260–261
Stereoheterotopic faces, 232
Stereoheterotopic ligands, 230
Stereoisomers, 222
Stereoselectivity, in pericyclic reactions, 84
Stereospecificity, 206
 in pericyclic reactions, 60
Steric effect, 62, 78–79, 83, 101, 103–105, 119, 120, 132–137, 146, 155–159, 184, 197, 200, 214, 217–218, 258–260
 assistance, 258–260
 congestion, 248
 hindrance, 107, 258–260
 inhibition of resonance, 101
 repulsion, 247
Steric kinetic isotope effect, 157–159, 247
Steric substituent constants, 134–135

Δ² Steroids, 261
Stilbene, 58
Strength factors, 200
Stretching vibration, 151–153, 180, 267, 268
  and hybridization, 25–26
Structure–activity relationships, 145–146
Structure–reactivity relationships, 119–120, 146, 183
Styrene, 58
Substituent constants, 110–113
Substituent effect, 128–146
Substituent-induced structural change, 105
Sucrose, 174
Sulfonation, 154
Sulfurous acid, 93
Supernucleophile, 198–199
Suprafacial component, 68–70
Suprafacial migration, 67–70
Swain-Lupton approach, 136–137
Swain-Scott equation, 184–185, 197
Symbiotic effect, 193, 201
Symbiotic principle, 201
Symmetric chromophore, 273
Symmetry-allowed, 60, 66–80
Symmetry classifications, 65
Symmetry-forbidden, 60, 66, 67, 84, 206
Symmetry number, 243–244

Taft equation, 132–136, 141, 146
Taft substituent constants, 96
Tartaric acid, 224, 228
Tau bond, 24
Tautomeric effect, 99
Temperature dependence of ρ, 112–113
1,1,2,2-Tetrachloroethane, 240
Tetrahedral carbon, 20–21
Tetrahydrofuran, 266
Tetrahydropyran, 264
Tetrahydrothiophene, 266
Tetramethylenehalonium ions, 192
2,3,5,6-Tetramethyl-4-nitroaniline, 101
Tetramethyltin, 209
Thiane-1-oxide, 264
Thiophene, 53–54
Three-electron bond, 19
Through-ring resonance, 101
Toluenes, 91, 105, 118–119, 141, 154
Torsional energy, 234–236
Torsional isomers, 233
Torsional strain, 265
Total bond order, 43
Total energy curve, hydrogen molecule ion, 14–16
Total energy operator, 4
Trans, 228–229
Trans addition, 232
  diaxial, 260–261
s-trans conformation, 242–243
Trans double bonds, 66
Transition state, 67–68, 82–83, 85, 89, 106, 107, 119, 124–125, 132, 138, 140, 146, 147–182, 194–199, 208, 209
  pericyclic, 60

Transition-state theory, 120–124, 129
Transmission coefficient, 123–124
Triarylcarbinols, 168, 171–172
Trichloroacetic acid, 93
Trideuteriomethoxycyclohexane, 253
Triethylamine, 247
Trimethylamine, 103–104, 247
Trimethylboron, 135, 247
3,3,5-Trimethylcyclohexanols, 259
1,2,6-Triphenylhexa-1,3,5-trienes, 78–79
Triphenylmethyl systems, 57, 168, 171–172
Tritium, 152, 154
Tropylium iodide, 213
Tropylium ion, 50, 52
Tub conformation, 57
Tunneling, 154
Twist-boat-chair conformation, 266
Twist conformation, 263 (see also Cyclohexane)

Ultraviolet absorption, 271–272 (see also Electronic transition)
Uncertainty principle, 89
Unsaturated systems (see also Alkenes, Butadiene):
  rotations about sp^2–sp^3, 241–242
  rotations about sp^2–sp^2, 242–243

Valence bond theory (see VB theory)
Valence state, 20–21
Van der Waals radius, 134, 135
Van der Waals strain, 237
Van't Hoff factor, 168
Variation method, 11–12, 18
VB theory, 18, 19, 52, 106–107
  benzene, 37
  hydrogen molecule ion, 18
Vibration frequency, 151–152
Vibrational potential, 182
Vicinal spin–spin coupling constants, 270
Vinyl chloride, 89
Vinyl ethers, 154

w, 175
w*, 175
Wave function, 4, 5
Wheland intermediate, 178 (see also σ-bonded complex)
Winstein-Grunwald equation, 188–190
Woodward-Hoffmann rules, 59–85, 206

X-ray diffraction, 49, 267
X values, 209
m-Xylene, 119

Yeast alcohol dehydrogenase, 231
Y values, 188–191, 193, 208, 209, 213

Zero overlap, 28
Zero-point energy, 5, 151–152, 160
Zero-sum rule, 35–36
Zucker-Hammett hypothesis, 173–175
Z values, 210, 212–214